Computer Numerical
Control for Machining

Computer Numerical Control for Machining

Mike Lynch

Boston, Massachusetts   Burr Ridge, Illinois
Dubuque, Iowa   Madison, Wisconsin   New York, New York
San Francisco, California   St. Louis, Missouri

Library of Congress Cataloging-in-Publication Data

Lynch, Mike.
    Computer numerical control for machining / Mike Lynch.
        p.   cm.
    ISBN 0-07-039223-4
    1. Machine-tools—Numerical control.   I. Title.
TJ1189.L96   1992
621.9'023—dc20   91-31750
                                                      CIP

McGraw-Hill

A Division of The McGraw-Hill Companies

8 9 10 11 12 13 14 BKMBKM 9 9 8

ISBN   0-07-039223-4

The sponsoring editor for this book was Robert W. Hauserman, the editing supervisor was Mitsy Kovacs, and the production supervisor was Donald Schmidt. This book was set in Century Schoolbook by McGraw-Hill's Professional Book Group composition unit.

To Timothy and Jessica—all the inspiration a father could ever need

Contents

Preface

Computer numerical control is among the fastest-growing fields in manufacturing today. Even though this is the case, the number of people currently entering this field is not sufficient to fill the needs of the manufacturing companies that utilize CNC equipment. Manufacturers are finding it more and more difficult to find qualified CNC people. CNC professionals are currently enjoying their best benefits to date. Job security, pay, and responsibility for people working with CNC are at an all-time high.

This makes CNC and its related fields an excellent choice for a person to enter. With the continuing growth of this sophisticated equipment, there appears to be no end to the possible opportunities in sight. While not one of the most glamorous fields, CNC offers its own special rewards and satisfactions.

Normally CNC equipment is among the most sophisticated machinery a company owns. Employees working with this equipment are usually given a high level of respect within the company. The potential for growth is almost limitless.

With a good understanding of CNC, a person can go almost anywhere. Geographically, there is no area of the United States that does not have a high concentration of CNC equipment. The room for personal growth is substantial. Some of the current job titles for people working with CNC include operator, setup person, manual programmer, computer-assisted manufacturing system programmer, CNC coordinator, application engineer, and industrial instructor.

The basic intention of this text is to provide the beginner in CNC with a way to learn about this sophisticated type of equipment with a minimum of effort. Frankly, no single text can take the place of hands-on experience, but this text addresses many of the considerations related to learning CNC.

Prerequisites

We have striven to make this text as simple to understand as possible. When presenting complicated information, we make analogies to everyday life in an attempt to keep things simple. For the same reason, we assume that the reader has little background in CNC as we make presentations throughout this text. While it would be helpful for the reader to possess a knowledge of basic machining practice, it is not mandatory. Even the beginner in CNC will find our presentations to be quite basic and easy to understand.

Organization of This Text

There are three distinct parts to this book. Part 1 addresses manual programming. With the current popularity of computer-assisted manufacturing (CAM) systems, there are people in the manufacturing field who feel that manual programming is a thing of the past. While we would agree that CAM systems make programming much easier and faster, we strongly feel that the beginner in CNC must have a good understanding of manual programming techniques. We equate learning manual programming as it relates to CAM systems to learning basic mathematics as it relates to an electronic calculator. With an electronic calculator, a person can perform arithmetic calculations *much* faster and more accurately than by doing calculations longhand. But most people still feel it is best to learn to make calculations the long way, without the help of a calculator. There will be times when a calculator is not available and calculations have to be done the hard way.

In much the same way, anyone serious about working in the field of computer numerical control must understand the basics of manual programming. While a CAM system may be used to prepare CNC programs, there will be many times when changes must be made to programs at the machine while production is being run. With a good basic knowledge of manual programming, these changes are relatively easy. Without it, they are next to impossible.

Chapter 1 gives the reader an elementary introduction to CNC, and lays the groundwork for manual programming. Chapter 2 begins the key concepts of manual programming, introducing the most common machine tools to which computer numerical controls are applied. Chapter 3 discusses the second key concept, informing the reader of the preparation steps that must be taken before CNC program can be manually prepared. Chapter 4, key concept no. 3, discusses the various motion types related to CNC equipment. Chapter 5, key concept no. 4, describes the compensation types related to CNC machines. Chapter 6, key concept no. 5, shows the reader how to format CNC programs for a variety of CNC machine types. Chapter 7, key concept no. 6, shows the reader the special features of manual programming. Most of the techniques described in this chapter can be thought of as advanced techniques.

Part 2 includes two chapters that discuss a new kind of CNC control, the conversational control. Conversational CNC controls are growing in popularity. They allow a graphic way for the programmer to enter programs at the control. Chapter 8 introduces conversational controls, and shows the reader its pros and cons. Chapter 9 discusses the flow of entering conversational programs at the control. While they vary dramatically from one control builder to the next, we show the most common methods of program input.

Part 3 is devoted to CNC machine tool operation. While it is difficult to explain machine operation techniques through any text, our key concepts approach allows us to give the reader the general techniques related to machine operation. Chapter 10, key concept no. 1 of operation, addresses the various buttons and switches found on the machine's control panel, telling the reader their basic function. Chapter 11, key concept no. 2 of operation, explains the three modes of CNC machine operation and gives the reader examples of their use. Chapter 12, key concept no. 3 of operation, discusses the most important procedures related to operating a CNC machine. Chapter 13, key concept no. 4 of operation, shows the reader how programs must be safely verified.

The Key Concepts Approach

CNC equipment takes many forms. Moreover, each type of CNC machine tool can be equipped with a variety of CNC controls. While control manufacturers do follow some basic standards, there are dramatic differences related to the specific programming and operation for any one type of equipment.

While the examples we give throughout this text are common ways to program and operate CNC equipment, you must be prepared for major differences for specific controls and machines. For this reason, throughout this text we stress that it is more important to know *why* you are doing things, than it is to know the physical techniques required to accomplish the task. If you understand the basic reasons why CNC features are used, you will be able to easily adapt to the specific CNC control and machine you must work with.

For example, the programming commands for the feature called *tool length compensation* vary dramatically from one machining center control to the next. However, the basic reason why tool length compensation is needed *never* changes. If you know why it is used, and if you have a basic idea as to one way it is used, you can easily relate what you know to the kind of machine tool you must work with when the time comes.

The key concepts approach also helps with organization. It lets you limit the number of things you must know in order to become proficient with CNC. We can easily categorize all information throughout this text into logical order. Once you understand the key concepts (six for manual programming, two for conversational programming, and four for machine operation), you should have a good understanding of *all* important features related to CNC.

Specific Examples

Throughout the text, you will find examples that will help in your understanding of the subject matter. We must point out that while our given examples will work for several popular CNC control models, actual techniques vary widely from control to control. For this reason, you must take our examples with "a grain of salt." You must keep in mind that the basic ideas behind each key concept are much more important than any of the given examples.

Because specific techniques vary dramatically from one CNC control to the next, the information and especially the specific examples must be viewed as a guide only. The author welcomes questions and requests for clarifications related to any information presented.

Usage Index

This text has been prepared so that the reader can go from beginning to end and learn a great deal about CNC as it applies to a wide variety of machine tools. Our key concepts approach to learning CNC presents a wide variety of machine types during each key concept. This allows us to show the similarity between CNC controls and makes it easy for the reader to learn features that are applied to a variety of machines. If you are reading to gain a broad knowledge of many types of CNC equipment, by all means read the book straight through from cover to cover.

However, there will be those readers who have a specific and immediate need to learn about only one of the CNC machine tool types discussed in this book. For those readers, we have listed the four most common CNC machine types along with the page numbers where related information is given. You can use this as a road map to guide you through only the information in which you are most interested and skip information that is not of immediate importance.

Vertical machining centers: 1–48, 51–57, 89–103, 105–126, 127–165, 191–231, 247–287, 291–307, 309–322, 325–345, 347–357, 359–378, 407–415.

Horizontal machining centers: 1–45, 48–57, 89–103, 105–126, 127–165, 191–231, 247–287, 291–307, 309–322, 325–345, 347–357, 359–378, 407–415.

Turning centers: 1–45, 57–72, 89–103, 105–120, 123–126, 127–131, 165–176, 191–196, 231–238, 247–250, 261–270, 287–288, 291–307, 309–322, 325–345, 347–357, 359–361, 378–394, 407–415.

Wire EDM equipment: 1–45, 72–86, 89–103, 105–120, 123–126, 127–131, 160–165, 176–190, 191–196, 238–246, 247–260, 262–270, 291–307, 309–322, 325–345, 347–357, 359–361, 394–405, 407–415.

Mike Lynch

Acknowledgments

Special thanks to these machine tool and control manufacturers for the photographs throughout this text.

Cincinnati Milacron, Inc.

Motch Corporation

Okuma Machinery, Inc.

Sodick, Inc.

Trumpf, Inc.

Fanuc U.S.A. Corporation

Thanks also goes to Beth Lewin. Her help with grammatical correctness made this book much more readable.

Computer Numerical
Control for Machining

Manual Programming

1

Basic Principles
of CNC

Motion control—from within your body, you take it for granted. You're thirsty. You pick up a glass, hold it under the faucet with one hand and turn on the cold water with the other. When the glass reaches the level of water you desire, you turn off the faucet and bring the glass to your mouth to quench your thirst.

This seemingly simple series of motions requires many thousands (if not millions) of motion commands from your mind to your various muscles. But you take these internally controlled motions for granted. It is not necessary for you to consciously think through every motion your body makes.

For those things that humans are incapable of doing by themselves, they have developed tools. Thousands of years ago, a human took hammer and chisel in hand and carved images into rock. These tools helped, but the level of motion control they allowed was very poor. An extremely high level of skill was necessary to construct anything useful.

This has always presented humans with a severe problem. The tools they develop to help with tasks never seem to match the dexterity they are capable of within their own bodies. Most tools still require a degree of manual labor to be supplied by our bodies. But as the motion control offered by the tool improves, the level of manual labor required is lowered.

With the passing of time, motion control has improved. The industrial revolution brought the most concentrated improvement in motion control in history. The hand loom, the sewing machine, and the forge are examples of inventions from this time that have improved motion control. In fact, almost every innovation from the industrial revolution is directly related to an improvement of motion control.

Early machine tools incorporated a series of slides, lead screws, cranks, and handwheels to control the motion of a cutting tool as a workpiece was being machined. By turning a crank a certain number of revolutions, an operator moves the slide of a machine a given linear distance. Over the years, the accuracy of these manual linkages has improved dramatically, but the actual methods of causing motion have not changed much—not until the advent of numerical control (NC).

And motion control is still the name of the game. But today, with computer numerical control (CNC), humans finally have the ability to command that motion take place in an accurate and programmable manner. That is, CNC allows a level of accuracy and repeatability that was previously next to impossible to achieve with any other form of motion control system. And CNC machines can be programmed to do different tasks, thus the ability to machine a variety of workpieces with the same machine tool also improves.

The common factor *all* types of CNC machines share is their ability to control motion accurately and in a way that can be repeated over and over again. These motions are also, for the most part, automatic, meaning the CNC machine tool needs no operator during the actual cutting cycle.

In manufacturing, this means the manual labor required to produce parts is now replaced by the ingenuity of the person preparing the program for the CNC machine. Instead of turning cranks and using manual labor to machine the workpiece, the CNC control is told by the program to drive the machine through a series of motions in much the same way your mind drives the muscles of your body.

However, the sophistication of the CNC machine cannot hold a candle to your body. With your mind and muscles, your range of movement (dexterity) is almost limitless. Given your mind's ability to control the motion of your muscles in any required fashion, there is almost nothing outside your range of motion.

By comparison, the CNC machine will have a *very* limited range of motion. Typically, between only two and five linear and rotational directions of motion are possible with today's CNC machines. And machines that boast five directions of motion are considered very sophisticated machine tools indeed.

On the other hand, the muscles in your body cannot begin to match the power available from current CNC equipment. Though their range of motion is limited, all but the very smallest of CNC machines are much more powerful than the human body.

The actual motions being made by the machine tool will vary from machine type to type, but the method by which the motion is controlled remains essentially the same. The method by which CNC machines control motion will be our first area of study.

The Directions of Motion

In the discussion of the first key concept for each specific machine type, one of the first things to be addressed will be the directions of motion. In terms of CNC, each direction of motion is called an axis. Depending on the machine's application, CNC machines commonly have from two to five axes of motion.

An *axis* of motion is simply a motion direction under the influence of the CNC control. A CNC axis can be either linear or rotary. The equipment for each axis consists of a mechanical device that performs the motion, a drive motor that supplies the power for the axis, and a ball screw that transfers the power from the drive motor to the mechanical device.

Rotation of the ball screw makes motion of the mechanical device possible. The drive motor is the link between the ball screw and the CNC control. Also called a servo motor, it allows extremely precise rotation. Figure 1.1 shows the relationship of drive motor, ball screw, and mechanical device for a linear axis.

As an example, suppose the CNC control is told (by a program) to move an axis a given linear distance. The control sends a signal to the drive motor, specifying a certain number of pulses, or rotation increments, for the drive motor, equivalent to the desired motion amount. The drive motor rotates that specified amount. This, in turn, rotates the ball screw. The ball screw drives the mechanical slide of the axis the precise amount commanded by the CNC program.

The quality of the drive motors, ball screws, and mechanical devices

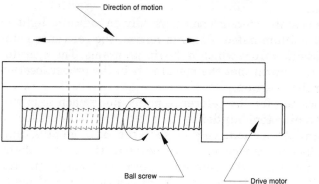
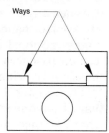

Figure 1.1   Drawing showing components of a linear axis.

determines the positioning accuracy of the CNC machine. Generally speaking, the application of the CNC machine tool itself will determine the accuracy required for these components. For example, a CNC jig boring machine is expected to hold tighter tolerances than a CNC machining center. Therefore, the CNC machining center need not be as accurate as the CNC jig boring machine. This means the quality of the components that make up each axis on the CNC machining center need not be as high as for the CNC jig boring machine.

To give you an idea of the accuracies we are speaking of here, today's typical CNC machining centers have positioning accuracies of ±0.0004 in throughout each axis' range of travel. Today's CNC jig boring machines, in contrast, boast a positioning accuracy of ±0.00012 in throughout the range of travel.

The reference point for each axis

In most types of CNC equipment, each axis of motion (linear and rotary) will have a *reference point*. This reference point is a very accurate position somewhere along the travel of the axis. The purpose of the reference point is to provide a good starting point for each axis. The name for the reference point varies among control builders. It is known as *zero return* on some controls; on others, it is called *grid zero*; on yet others, it is called the *home position*.

Throughout this text we will call this position simply the machine's *reference point*. The reference point is a very important location to the programmer. For example, the CNC machine must be sent to its reference point in all axes as part of the machine's power-up procedure. Also, some controls require that the reference point be the location from which every program must begin. As each key concept is presented, you will be exposed to more reasons why the reference point is so very important.

When any axis is at its reference point, usually an indicator light on the control panel is illuminated. For this reason, the reference point makes an ideal position from which to begin programs. The operator can easily check to confirm that the machine is in the proper location before activating the program.

The specific location for the reference point along each axis will vary from one machine tool builder to the next. Most builders locate the reference point of each axis very close to the extreme plus limit for the axis. Others place the reference point in the center of the axis travel. In any case, the reference point is usually (in the machine-tool builder's opinion) a convenient position from which to work.

Linear axes

For a linear axis, motion occurs along a straight line. The mechanical device performing the motion is usually very similar to the ways of a manual milling machine. As the ball screw is rotated by the drive motor, it causes a linear motion along the ways of the axis.

A very crude analogy can be found in a common table vise. When you want to clamp a part in the vise, you rotate the handle of the vise clockwise or counterclockwise to close or open the vise. The handle is linked to a lead screw that is coupled with the movable jaw of the vise. As the handle is rotated, the lead screw also rotates, causing linear motion of the movable jaw. Essentially, this is the same action that is taking place in the linear axis of a CNC machine tool. However, a table vise does not require precise motion, and this is where the similarity ends. On a linear axis of a CNC machine, the motion must be controlled very precisely in four ways.

First, the motion will take place along a perfectly straight line (hence the name linear axis). The precision of the axis (how straight the motion will be) depends on the application of the CNC machine tool.

Second, each axis must be sufficiently stable to withstand machining operations that are to take place. That is, the axis must be rigid enough to allow straight motion even when machining is taking place. The application of the CNC machine tool determines how rigid each axis must be.

For example, the CNC machining center is intended to perform quite powerful machining operations. During these operations, the axis will be driving the workpiece into a cutting tool. The tendency will be for the axis to stray from a straight line motion, so the rigidity must be very high. On the other hand, a CNC wire electrical-discharge machining (EDM) machine requires very little rigidity in the axes, since the force exerted during machining is quite low. The design of the mechanical device driving each axis reflects the usage of the machine tool.

Third, all CNC machines require a high degree of positioning accuracy in their linear axes. The distance to be traveled (or ending point of each motion) is of extreme importance. For example, if a program commands that an axis move precisely one inch, the axis must respond correctly. The quality of the ball screw and drive motor determine how precise the axis will be in this regard.

Fourth, almost all CNC machine tools allow the rate of motion to be controlled accurately through programmed command. For machining operations, this allows the programmer to specify the feed rate to be used during machining. The method by which the feed rate is controlled will vary according to the kind of CNC machine tool. More information will be given on this in the discussions of the first key concepts.

Linear axes are commanded in linear increments of length measurement. Depending on preference, the increments can be in millimeters or inches (most CNC equipment currently manufactured allows the user to input commands in either measurement system). Since most beginners to CNC in the United States are most familiar with the inch system, this text will use this system for all examples.

Each linear axis will have a maximum travel distance that cannot be exceeded. This is one of the main factors contributing to the size and capacity of the CNC machine tool itself. This travel distance determines the maximum size of workpiece that can be machined. On each end of the travel, there is some form of protection device that will keep the axis from exceeding its limit. Normally some form of limit switch is mounted close to the limit of a linear axis. If the machine is commanded to exceed its limit manually or by programmed control, an alarm will sound. This keeps the linear axis from exceeding its boundaries.

Rotary axes

The same basic principles apply to rotation axes. The biggest difference is in the mechanical device causing the motion. Instead of being linear, the motion is along a circular path.

A rotary axis is employed in one of two ways. First, some CNC machine applications require that the workpiece rotate during machining. This kind of rotary axis is employed simply to rotate the workpiece to an attitude that allows machining or to turn it while it is being machined.

Second, some CNC machine applications require the cutting tool to be swiveled during the machining operation. For example, in the aircraft industry, when an airfoil is on a machining center during a milling operation, it is desirable (if not mandatory) to keep the milling cutter as close to perpendicular as possible to the surface being machined. The complex shape of an airfoil requires that the milling cutter be swiveled to maintain this desired perpendicularity.

During the discussion of linear axes, we mentioned the limits of each axis. For rotary axes, if a full 360° rotation is allowed (as would be the case with a rotary table used to rotate the workpiece), no limit is necessary. In this case, the rotary axis could continue to rotate past 360° without causing any limit to be exceeded. However, if the rotary axis had not been designed to rotate a full 360° (as would be the case for an axis designed to swivel the tool), the same form of limits used for linear axes would be used for the rotary axis.

This has been a cursory look at what makes up an axis. Remember that it is not of major importance that you understand the mechanics of how a CNC machine is put together to work with CNC. This can be equated to driving an automobile. You do not have to understand how an internal combustion engine works to be able to drive a car. But this elementary look at what makes up an axis should help you with your overall introduction to CNC. Now that you know the physical method by which each axis moves, let's look at how these motions are commanded in the program.

The Rectangular Coordinate System

You now know the overall makeup of an axis. You know that as a drive motor for an axis is rotated a specified number of rotations, it rotates a ball screw. As the ball screw rotates, it drives the axis in the desired manner. But how do you determine the number of rotations a drive motor must turn to cause the desired linear or rotary motion? How many rotations of the drive motor equal 1 in of linear travel? And how would you command the speed of the drive motor's rotation to attain the proper motion rate (feed rate) of the axis for machining? Thanks to the rectangular coordinate system, you do not have to address these questions.

It would be extremely difficult to program a CNC machine if the programmer had to command the drive motors to rotate a specified amount to cause the desired motion distance to be traveled (although this was how the first NC machines had to be programmed). For this reason, all CNC control manufacturers conform to the same easy method of axis control. They utilize a tool of mathematics called the rectangular coordinate system.

This coordinate system is also used for making graphs. Even though you may not have realized it at the time, if you have ever had to make or read a graph, you have had to work with the rectangular coordinate system. Since the way the rectangular coordinate system is used to make a graph is so similar to its use with CNC, let's first review how graphs are made and used. Along the way, we'll show the similarities as they relate to CNC.

Graphs allow the demonstration of a great amount of data in a way that can be absorbed by the reader quickly and easily. This is why the term *graphic* is used to describe anything that is easy to visualize. The making of a line graph involves plotting points along horizontal and vertical lines. Figure 1.2 shows an example of a graph.

Suppose you were assigned to develop a graph that plots a company's productivity for last year. You could label the horizontal line of the graph as time (in months) starting in January and ending in December. You could then label the vertical column as productivity (in

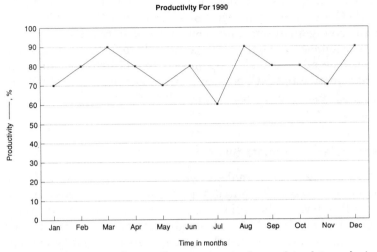

Productivity For 1990

Figure 1.2   The rectangular coordinate system is also used to plot graphs in the same way that CNC coordinates are plotted.

percentage of productivity) starting with 0 percent at the bottom and ending with 100 percent at the top.

With your company's productivity data from last year in front of you, you find the company was 70 percent productive in January. To plot the point for this piece of data, you would look to the position that is 70 percent on the productivity line and January on the time line. There you would plot a point. Next, you find your company was 80 percent productive in February. So you look to the position that is at 80 percent on the productivity line and February on the time line. Plot the point. With this technique, you can continue plotting the productivity for the entire year. When you are finished, you would connect the dots with straight lines or develop curves that come tangent to the plotted points.

When you plot or view a graph, you are using the rectangular coordinate system. Yet there are a few points that are almost taken for granted when you study a graph that require further explanation to help with your understanding of CNC. First is the *point of origin* for each axis of the graph. The vertical and horizontal base lines of the graph have to start somewhere. With a graph, the place where the two lines intersect is the starting point. More specifically for our example, January was the starting point for the horizontal line, and 0 percent productivity was the starting point for the vertical line.

Second, each base line of a graph must represent something. The person developing the graph determines what each base line will represent. Also, depending on how the data are to be shown, each base line is broken down into increments of measurement for what the base line represents. Our simple example labeled each base line with a designation and determined an increment for each. In our case, the vertical line was labeled with productivity, and the increment was in 10 percent values. The horizontal line was labeled with time, and the increment was in 1-month periods.

The coordinate system for *all* linear axes of CNC machines is surprisingly similar to our simple graph example. But instead of an imaginary or intangible value representing conceptual ideas, each point being plotted in a CNC program is actually a physical location the machine will be told to move through. As with a graph, the points that are being plotted in the CNC program are also called coordinates.

With a CNC machine, the time and productivity graph base lines become the directions of motion (axes) on the CNC machine tool itself. It is along these axes that points will be "plotted." The increments will be units of measurement. In the inch system, the increments will be related to inch. In the metric system, the increments will be related to metric.

The actual value of the increments themselves will vary, depending on the kind of CNC machine tool. For linear axes, in the inch system, the smallest increment is typically 0.0001 in (one ten-thousandth of an inch). In the metric system, the smallest increment is usually 0.001 mm (one thousandth of a millimeter). For rotary axes, the smallest increment will be in degrees. Commonly, 0.001° is the smallest increment for a full rotary axis. With these tiny units of measurement, you can see that each axis is broken up into tiny increments. When two or more axes of the CNC machine are considered together, these small increments constitute an extremely fine grid.

A letter (*X, Y, Z*, etc.) is used to designate each axis on the CNC machine. This letter is called the letter *address* for the axis. The actual letters to be used will vary from machine type to machine type. As we present key concept no. 1, we will introduce the letter addresses for the axes of common CNC machines.

The *origin* point is the location from which the coordinates are taken. With the graph example, we said the origin for the graph is where the vertical and horizontal lines crossed. With CNC, the same is true. In CNC, we call this origin position *program zero* (some people call it *work zero* or *part zero* or *part origin*).

Figure 1.3 shows two axes (*X* and *Y* in this case) superimposed over a workpiece drawing. Note how similar this is to the graph example given earlier. Now the base lines of the graph relate to the axes of a CNC machine. The origin of this coordinate system corresponds to actual surfaces on the workpiece itself. In this example, coordinates needed in the program will come right from the print. Unfortunately,

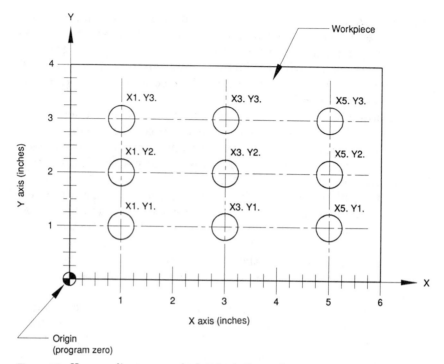

Figure 1.3 How coordinates are calculated relative to the program zero point.

this will not always be the case. There will be times when calculations must be made to come up with coordinates required in the program.

Understanding program zero

As stated earlier, the program zero point can be related to the origin position on the graph. It is a logical reference point from which to work. The usage of program zero allows the programmer to devise a convenient location from which coordinates going into the program will be taken.

This convenient location is left completely to the programmer's discretion. Careful selection of the program zero point makes the calculating of coordinates needed in the program much easier for the programmer. In many cases, the coordinates going into the program can be taken directly from the print. The most logical position to make program zero is the location on the part from which dimensions are taken.

If the print for the workpiece incorporates *datum surface dimensioning* techniques, the program zero point for your program will simply be the datum surface for each axis.

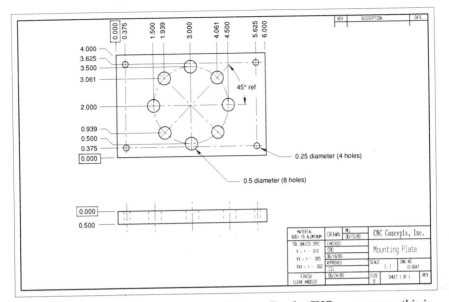

Figure 1.4 Datum surface dimensioning techniques. For the CNC programmer, this is the easiest style of dimensioning to work from.

Figure 1.4 shows an example of datum surface dimensioning. Note that each surface marked with zero would make an excellent program zero point for the corresponding axis. In this case, the lower left-hand corner of the part would be selected as program zero. Then, all coordinates needed for the program could be taken right from the print.

If the print does not incorporate datum surface dimensioning, you look for the position in each axis from which most of the dimensions are taken. For those positions not dimensioned from the program zero point, you will have to calculate the coordinates accordingly.

Figure 1.5 demonstrates an instance when calculations must be made to come up with some of the coordinates needed in the program regardless of where the program zero point is placed. Note that most of the dimensions on this drawing are coming from the lower left-hand corner of the part. However, some of the dimensions (the length and width of the pocket) are not dimensioned from the lower left-hand corner of the part. While the lower left-hand corner would still be the best choice for program zero, some of the coordinates going into the program would require arithmetic calculations to be made.

Unfortunately, some prints have dimensions taken from many surfaces, making it difficult to determine any one position that could be considered the best program zero point. In this case, you must determine the location surfaces from which the part will be held in the

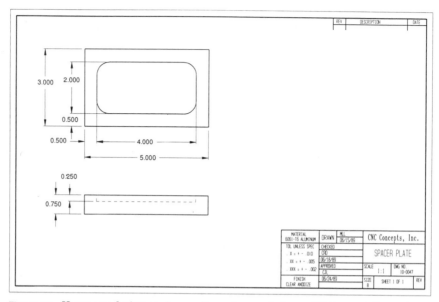

Figure 1.5 How most design engineers dimension workpieces. While most dimensions come from the same surface, there are some that come from other surfaces, making the calculations required for the CNC program a little more difficult.

work-holding setup. This would make the best choice of where your program zero point should be.

Figure 1.6 shows an example of a drawing where dimensions are scattered, taken from many surfaces of the part. This print would require almost every coordinate going into the program to be calculated. While this print would be considered a poor drawing, and today's design engineers try not to scatter dimensions like this, unfortunately the beginning programmer must be prepared to deal with this kind of drawing.

Once the program zero point is decided, *all* coordinates in the absolute mode will be specified from this point. At worst, calculations may have to be made to come up with program coordinates, but at least there will be consistency with regard to the method of calculating coordinates.

When working from the program zero point, the programmer is working in the *absolute* mode. (More on the absolute mode a little later.)

Figure 1.7 shows a series of simple parts that should drive home how you decide on the program zero point. The part surface from which the most dimensions are taken is chosen as the program zero point in that axis.

Understanding plus and minus

Think about the productivity graph example again for a moment. We said the origin for the graph was the point where the vertical and hor-

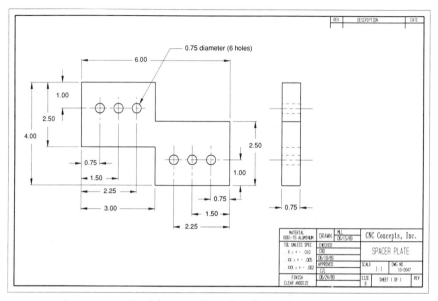

Figure 1.6 A programmer's nightmare. Note that dimensions are coming from everywhere. In this case, there is no logical program zero point. Almost every coordinate going into the program would require a calculation.

izontal base lines crossed. In our case, it was *January* along the horizontal base line and *zero percent productivity* along the vertical line.

Every point we plotted happened to be *after* January and *above* zero percent productivity. It may have seemed a coincidence that all plotted points happened to fall up and to the right of the origin for the graph. In reality, the person developing a graph will usually locate the origin for the graph so that all points on the graph can be shown in this manner.

The area up and to the right of the origin point is called a *quadrant*. But this area is only one of four quadrants in the rectangular coordinate system. There are also quadrants up and to the left, down and to the left, and down and to the right of the origin point. For the sake of organization, each of these quadrants is labeled with a quadrant number. The quadrant up and to the right is quadrant number 1, up and to the left is quadrant number 2, down and to the left is quadrant number 3, and down and to the right is quadrant number 4. Figure 1.8 shows the four quadrants of the rectangular coordinate system.

Though a typical graph will use only quadrant number 1 of the rectangular coordinate system, you will be using all four quadrants when working with CNC. To specify coordinates in any quadrant except quadrant number 1, a minus sign must be used for at least one of the coordinate positions.

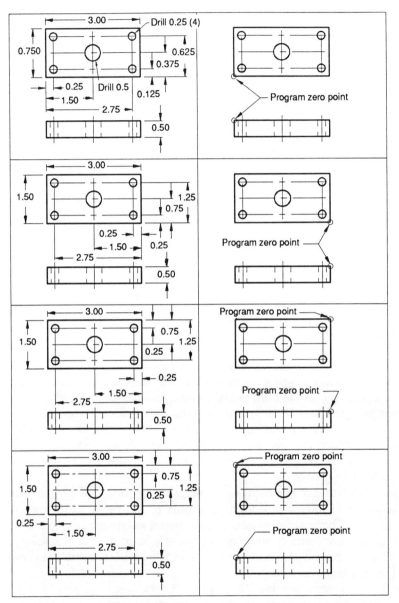

Figure 1.7   How to decide on the program zero point. Note that surface on the drawing from which dimensions are taken makes an excellent program zero point.

Note that a plus sign is always assumed by the CNC control. That is, if a coordinate is plus in one axis or another, you *never* program the plus sign with the coordinate value. Only if the coordinate value is minus do you need to include the sign for the value. Figure 1.9 shows the relationship of the four quadrants with respect to the plus and minus signs.

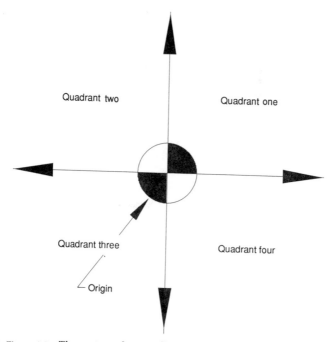

Figure 1.8 The rectangular coordinate system is broken into four quadrants for each plane.

To specify a negative coordinate, you must first specify the letter address for the axis being commanded, followed by the minus sign (–), then the value of the coordinate.

Figure 1.10 shows an example of when the programmer must be concerned with the plus and minus signs for coordinates. In this case, a ring is to be machined. For simplicity's sake, program zero has been assigned as the center of the ring. As you study the coordinate sheet related to the point numbers (1, 2, 3, etc.), notice that any X value to the right of program zero is plus, to the left is minus. Y values above program zero are plus, below are minus.

It is very important that the beginning programmer understand this idea. One of the biggest sources for mistakes in a CNC program is the incorrect use of plus and minus signs for coordinate values. In the absolute mode, the decision as to whether a coordinate is to be plus or minus is *always* made relative to the program zero position. In the incremental mode, the sign simply points in the direction in which the motion is to occur.

Absolute mode vs. incremental mode

There are two methods of commanding how motion is to take place. One is relative to program zero and the other is relative to the ma-

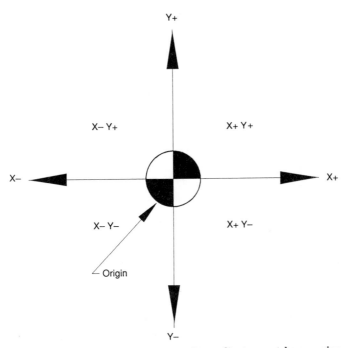

Figure 1.9 In the absolute mode, each coordinate must have a sign (plus or minus) to show the position of the coordinate relative to the program zero point.

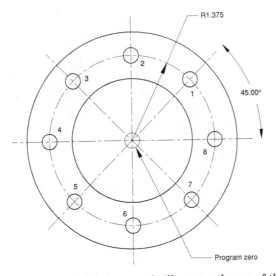

#	X	Y
1	0.9723	0.9723
2	0	1.375
3	−0.9723	0.9723
4	−1.375	0
5	−0.9723	−0.9723
6	0	−1.375
7	0.9723	−0.9723
8	1.375	0

Figure 1.10 Bolt hole example illustrates the use of the plus and minus sign for each coordinate.

chine's current position. The *absolute* mode requires that the programmer input values relative to program zero. The *incremental* mode requires that the programmer input values from the machine's current position.

It is *very* important that you understand the difference between the absolute mode and the incremental mode. In the incremental mode, you would ask yourself, *"How far* do I want the axis to move?"* The motion is commanded from the machine's current position. If moving in the minus direction of the axis, a minus sign (–) must be included in the axis motion. For example, if you wanted to command that the X axis move 1 inch in the minus direction from its current point, the word $X-1.0$ would be used.

In the absolute mode, you would ask yourself, *"To what position do I want the axis to move?"* In the absolute mode, this position (or end point) is *always* relative to the program zero point. Programs written in the absolute mode are *much* easier to work with for several reasons.

First, the person viewing a program written in the absolute mode can easily locate the current position of the tool at any point of the program. By comparison, incrementally written programs are nothing more than a whole series of motions from the tool's previous position. It can be very difficult to tell where the tool truly is relative to the part at any one point in the incremental program.

Second, if a motion mistake is made in a program written in the absolute mode, only one command of the program will make an improper positioning movement. If the next commanded motion is correct, the program will be back on track. By comparison, if a motion mistake is made in an incrementally written program, *every* motion command from that point on will cause motion to an incorrect location.

Third, it is easier to develop the coordinates for the program when programming in the absolute mode. If the program zero point is chosen carefully, many times the coordinates for the program will match print dimensions. By comparison, when working in the incremental mode, the programmer must make at least one calculation for every motion the program is to make. CNC programming can be tedious and error-prone enough in the absolute mode. Any way that can minimize the number of calculations necessary to write a program will help assure the success of the program.

Fourth, preparation is more easily done. When working in the absolute mode, the programmer can come up with the coordinates going into the program easily before actually writing the program, thus facilitating better preparation. By comparison, the incremental mode almost forces the programmer to come up with coordinates as the program is being written. While experienced programmers may be able to do this, the beginner should do as much preparation as possible *before* the program is written.

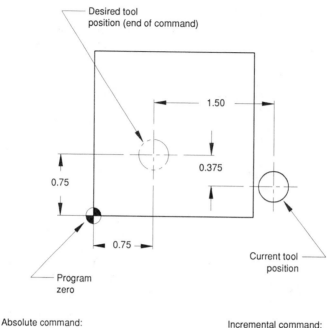

Absolute command:
    X0.75 Y0.75

Incremental command:
    X-1.5 Y0.375

Figure 1.11   In the absolute mode, coordinates can be easily specified relative to the program zero point regardless of the tool's current position. In the incremental mode, the programmer must specify each movement relative to the tool's current position.

Figure 1.11 is a drawing that illustrates the difference between incremental and absolute commands. Notice that, prior to the movement command, the tool is resting to the right of the workpiece. The phantom circle shows the desired end point of the motion. There are two ways to command the motion.

In the incremental mode, the programmer would have to calculate the distance from the tool's current position. In this case, the tool must be moved 1.5 in along the X axis and 0.375 along the Y axis. Notice also that the direction of motion (plus or minus) is also important. That is, the X axis will be moving in the minus direction. In the incremental mode, the command

    X-1.5 Y.375

would cause the tool to move to the desired end point.

By comparison, in the absolute mode, the programmer simply gives the coordinates relative to program zero. It is much more likely that the coordinates will match the dimensions on the print. Now, no concern whatsoever is given to the current position of the tool. The programmer need not even know where the tool is prior to the movement command. Notice the program zero point is located at the lower left-hand corner of the workpiece. In the absolute mode, the command

X.75 Y.75

would cause the desired motion.

Figure 1.12 stresses the difference between the incremental and absolute modes. It also stresses the possibility of making a series of motions in either mode. Again, notice how the absolute mode allows you to work from one location while the incremental mode forces you to consider the motion from the current location.

We must warn you that beginners have the tendency to think in the incremental mode. That is, as a programmer, you may find yourself asking the wrong question related to how to determine the end point of a motion command. As you begin to develop your own programs, you may find yourself asking, "How far does the tool have to move?"

#	Absolute mode	Incremental mode
1	X1.0 Y1.0	X0 Y0
2	X2.0 Y1.0	X1.0 Y0
3	X2.5 Y0.5	X0.5 Y−0.5
4	X3.5 Y0.5	X1.0 Y0
5	X3.5 Y2.5	X0 Y2.0
6	X5.0 Y2.5	X1.5 Y0
7	X5.0 Y2.0	X0 Y−0.5
8	X6.0 Y2.0	X1.0 Y0
9	X7.0 Y3.5	X1.0 Y1.5

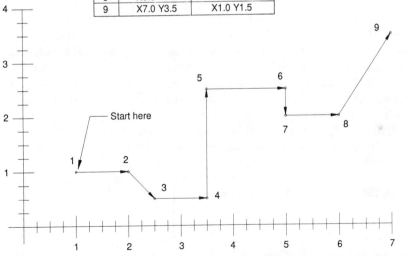

Figure 1.12 This chart helps to understand the difference between the absolute mode and the incremental mode.

instead of "To what position must the tool move?" If you prepare to write the program properly (as we teach in key concept no. 2), you should not find yourself falling into this trap.

While incremental programming is still possible even on current CNC controls, it is a method of programming reserved for special occasions. That is, most programmers today prefer not to program in the incremental mode. And we strongly recommend that beginners work exclusively in the absolute mode for the reasons given above.

Most CNC machines use two *G codes* (more on this later) to specify the incremental or absolute modes. For most types of CNC controls, G90 specifies the absolute mode and G91 specifies the incremental mode. At this point it is important that you understand the difference between incremental and absolute modes. The examples we give in this text will always be using the absolute mode. And we recommend that beginners get in the habit of using this mode exclusively.

The Language of Manual Programming

The method by which motion is made on a CNC machine is very important to your understanding of CNC. However, the motions that a CNC machine makes are only part of the whole picture of what a CNC machine is intended to do. All forms of CNC machines have other special considerations that must be understood for the particular machine. Understanding the machine's motion alone is not enough.

Each CNC machine will have programmable functions specific to the machine tool. Here are some examples. The CNC machining center will have an automatic tool changer (ATC) to allow automatic loading of tools. The spindle of the CNC turning center must allow precise speed to be programmed. The CNC wire EDM machine must allow the electronic cutting conditions to be manipulated through program commands. The CNC turret punch press must have a way of commanding that a hole be pierced. These are but a few examples of the many things which the programmer must control on various CNC machine tool types.

The CNC programmer must be able to tell the control what must be done. CNC programs are made up of a series of commands that inform the control what must be done step by step. That is, a CNC program is executed *sequentially,* one command at a time. When a program is executed, the control will encounter the first command in the program, execute it, then go on to the second command. The control executes each command in the same order encountered.

Another CNC term for a command is *block.* A command (or block) is made up of *words.* Each word in the CNC program is made up of a letter address (X, Y, Z, R, etc.) and a numerical value. The letter ad-

Block or command

Figure 1.13   Excerpt from a CNC program that shows the various elements of a command.

dress tells the control the kind of word that is being referenced. Figure 1.13 shows an example of a CNC command.

Many of the letter addresses are easy to recognize and remember. For example, letter addresses like F for feed rate, T for tool station, X for X-axis motion, and S for spindle speed are easy to associate. Unfortunately, other letter addresses will not make as much sense and are not as easy to remember. They will require memorization.

Most CNC controls allow the words included in a command to be in any order. That is, the programmer can give the command in any way that feels "comfortable." The control will sort out the command and execute it regardless of the order in which the commands are given. For example, on most CNC controls, the command

N005 G00 X10. Y10. M03

would be executed *exactly* the same as the command

N005 M03 Y10. G00 X10.

Fortunately, there are relatively few words used with the typical CNC control on a regular basis. Generally speaking, a CNC program uses only about 40 to 50 different words. So if you can think of letter addresses as a foreign language that contains only 40 to 50 words, learning them shouldn't seem too difficult. Also, during key concept no. 5 of programming, we will be showing you how CNC programs are formatted. In this concept, you will find that very few CNC words need to be memorized. Many times remembering the words to be used in the program will be as simple as recognizing them from an example format.

We compare writing a CNC program to writing a set of instructions. A CNC command is like a sentence in the set of instructions. The words of each CNC command are like the words that make up each sentence. The CNC command must end with a special character [called an *end-of-block* (EOB) character] just as each sentence must end with a period. Just as the set of instructions will convey information about the construction of a toy, a piece of furniture, or some other object, the CNC program gives the CNC control information about how the machine must respond. The sequential way by which a person reads a set of instructions (one instruction at time), is identical to the way the CNC control reads the CNC program.

However, when a person reads a set of instructions, certain things may be implied or taken for granted. For example, the reader may be told to assemble two components with a screwdriver. It is assumed that the reader *has* a screwdriver and knows how to use it. Also, many times the set of instructions will be quite vague, forcing the reader to make an interpretation of what is intended. Depending on the reader's previous experiences, this interpretation may or may not be correct.

By comparison, the CNC control requires a *very explicit* set of instructions. Every detail of what the programmer wants to make the machine do must be programmed. Though the control will interpret the program, each command given will have only one possible result.

Figure 1.14 shows a segment of a CNC program. While the words and commands of this program make little sense at this point, remember that we are illustrating the sequential order by which a CNC con-

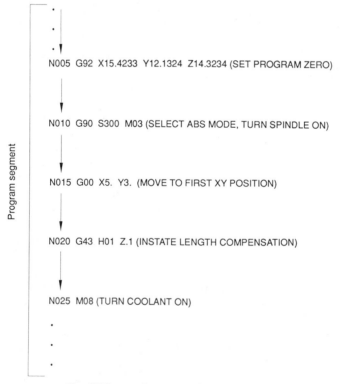

Program segment

N005 G92 X15.4233 Y12.1324 Z14.3234 (SET PROGRAM ZERO)

N010 G90 S300 M03 (SELECT ABS MODE, TURN SPINDLE ON)

N015 G00 X5. Y3. (MOVE TO FIRST XY POSITION)

N020 G43 H01 Z.1 (INSTATE LENGTH COMPENSATION)

N025 M08 (TURN COOLANT ON)

Figure 1.14 The CNC control executes the commands in sequential order.

trol executes the program. It reads the first command it comes across, executes it, then goes on to the next.

CNC programs and the control's memory

The series of commands to be executed by the control is (as stated) called the *program*. In the days of NC (before CNC), one of the major limitations of this kind of equipment was the fact that the program had to be run from a tape. Holes punched in this tape allowed the program to be coded in a way the tape reader of the NC machine could understand. As the tape ran along the reader, the control would follow the commands coded on the tape. The program on tape had to be perfect. If something was wrong with the program, the tape had to be changed before the program could be used.

While today's CNC machines may still have tape readers (some no longer do), the purpose of the tape reader is to read the tape into the *memory* of the control. That is, no longer is the program actually executed from the tape reader. It is executed from within the control's memory, in much the same way software is executed from within a personal computer. This is one of the main differences between NC and CNC. If a problem exists with the program in a CNC control and the program must be changed, it is a simple matter of changing the program within the control's memory right at the control. The program stored in the CNC control can be easily modified.

Most CNC controls allow multiple programs to be stored. These programs will be kept even when the power is turned off. Some controls even have floppy-disk drives or hard drives for program storage. The operator can easily call up the desired program and execute it at any time.

What must go on in the manual programmer's mind

The person preparing a CNC manual program must be able to visualize the motions a CNC machine will make as it machines a workpiece. To be able to do this, the manual programmer must have a good understanding of the machining process to be performed. This usually requires some previous experience with basic machining practice as it relates to the machining operation to be performed.

Most CNC machines have been designed to enhance or improve what was previously possible on some other form of manual equipment. To truly understand how the CNC machine must be programmed, it is helpful (if not mandatory) to have experience with the kind of machine the CNC machine tool is replacing. This is evidenced by the fact that machinists make the best CNC programmers. Machinists already know what they want the CNC machine tool to do. It is a relatively simple task to learn how to communicate this to the CNC machine tool.

For example, CNC machining centers closely resemble manual milling machines. Operations performed on the typical manual milling machine must still be performed on the CNC machining center. By the same token, operations that would be performed on an engine lathe or turret lathe are very similar to operations to be performed on the CNC turning center.

With prior machining experience, the CNC programmer can simply draw on previous knowledge to help during programming. When the programmer can visualize the operations as they would be performed manually, it will be a simple matter to code the desired series of commands into a language that the CNC control can understand. Once this visualizing can be done, preparing the CNC program becomes much easier. Unfortunately, if a programmer cannot visualize what the machine should do, preparing the program is next to impossible.

The flow of CNC program execution

To help with visualizing the desired commands required for a CNC program, let's look at a very simple example. Suppose you had to drill one 0.500-in diameter hole to a depth of 0.75 in in a piece of steel. Say the hole must be located 1 in from the lower left-hand corner of the part in both directions. Say this operation is to be performed on a drill press. Figure 1.15 shows a drawing of this simple part.

For some readers, this may seem very simple—so simple that you may begin to take some basic things for granted. But let's itemize those things that you would have to do to perform this operation. We will begin by assuming that the drill is already in the spindle of the drill press and that the part is already held in a vise with a center-punched hole where the hole is to be drilled. Here is a step-by-step list of what you would have to do:

1. Select the desired spindle revolutions per minute (RPM) and turn on the spindle.
2. Move the part directly under the drill.
3. Move the quill of the drill press down until the drill is just above the center-punched hole.
4. Applying the proper pressure on the quill, plunge the drill into the hole to the desired depth.
5. Retract the drill from the hole.

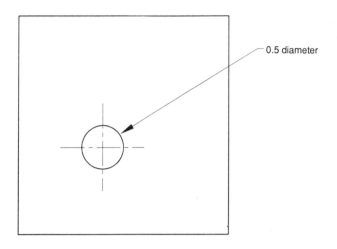

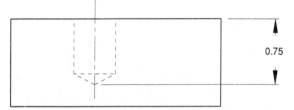

Figure 1.15 Even the simple operation of machining one hole is a good analogy for a simple CNC program.

6. Move the vise out from under the drill.

7. Turn off the spindle.

For an experienced machinist, this may seem simple almost to the point of being insulting. But it shows how the CNC programmer *must* think through every operation to be performed in the CNC program.

Now let's take a look at how the CNC program would look to perform the same operation on a CNC machining center. Note that we are stressing the sequential order (step by step) of how the CNC program would be executed, *not* the actual commands themselves. We are also stressing the itemized way by which the program is prepared. As you view the program, note the similarity to the above set of step-by-step instructions. To help make clear what is going on, we will place a message in parentheses defining what is happening in each command.

Again, at this point we are *not* stressing the meaning of each command in the program. That will come throughout the key concepts. So please do not be intimidated if you do not recognize the actual commands. Read the information in parentheses to find out what is taking place in each command.

This happens to be a program written in the format of a very popular control. However, the format will vary somewhat from control builder to builder.

```
O0001 (program number)
N005 G92 X _____ Y _____ Z _____ (set program zero)
N010 G90 S600 M03 (select absolute mode, turn spindle on clockwise at 600 rpm)
N015 G00 X1. Y1. (move quickly to center of hole in X and Y)
N020 G43 H01 Z.1 (move quickly down just above part)
N025 G01 Z-.75 F5. (feed into hole at 5 in per minute)
N030 G00 Z.1 (move quickly back out of hole)
N035 G91 G28 X0 Y0 Z0 (send machine back to starting point)
N040 M30 (end the program; this turns off spindle)
```

Notice how similar the messages to the right of each command are to the previously given procedure. Notice how the programmer must itemize each movement the machine makes. You can think of the movements a CNC machine makes as being like a connect-the-dots game for children. You will be driving the machine through a series of locations throughout the program in a point-to-point manner.

Though the control executes a program command by command, you must also know that the control is constantly looking ahead in the program to find out what is coming up next. This is done by the *look-ahead buffer* of the CNC control. The look-ahead buffer is required for two reasons. First, when the control is flowing through a series of motion commands, there are many times it would be detrimental for the motion to actually stop between commands. For example, if milling a contour on a machining center, you would not want the machine to come to a complete stop between commands. If the machine did stop between commands, there would be noticeable "witness marks" on the part caused by the milling cutter each time the machine stopped.

Second, there are CNC features that require the control to be able to tell what is coming up in the next command. The feature cutter radius compensation, for example (discussed during key concept no. 4), requires that the control know what is to happen in the next command before the current command can be correctly finished. Although this is the case, it is still helpful to visualize the sequential method of executing a CNC program with little concern for the look-ahead buffer.

Introduction to programming words

Here we will introduce many of the CNC words involved with programming the most common CNC machines. However, if you are a beginner looking at these special words for the first time, do not be intimidated by the number of different words. Also, do not you try to memorize every word. Our intention here is simply to acquaint you with each word, giving you a way to become familiar with those words the typical CNC machine will understand. When we get into the *program formatting* concept (key concept no. 5), we will give you a way to simply look at the word and try to remember its function. You will not have to come up with the programming words on your own.

There is a wide variety of CNC machine tools to which these words can be applied. This means that there can be conflicting usages for some of these word types. For those words that have more than one possible application, we show only the most common usage. Again, remember that we are attempting only to introduce each letter address. Only a brief description is given for each word.

You will find that several of the words are seldom used, meaning that you will have little or no need for them. Other words are constantly used, and you will have them memorized after writing four or five programs. You will find that most of these words are appropriately named, having the letter represent something that is easy to remember. For example, S for spindle speed, F for feed rate, and T for tool station. This system for remembering the function of most words should not be too difficult.

O   This letter address is usually used to represent a program number. Most current CNC controls allow the user to store multiple programs in the memory of the control. For these controls, the programmer will be assigning the program a number from 0001 through 9999. The O word will be the very first word in the program. No decimal point is allowed with the O word. Do not confuse this letter address with the numerical value of zero. One very common mistake, especially for beginners, is to confuse the letter O with the number zero. For controls that use the O word, the program number is the *only* time the letter O will be used in a program command. The number zero is used for *all* numeric words to represent zero.

N   This letter address is used to specify a sequence number. A sequence number is used for program command identification. It allows the programmer to organize each line in a program by number. This permits easy editing of a program at the machine since the command that needs to be modified can be easily searched. Sequence numbers are not required to be in any particular order and can even repeat in the program. In fact, they do not have to be included in the program at all. But for organizational purposes, we recommend that the beginner include them in the program and place them in an understandable order. For all example programs, we will be going by fives in this text to allow for extra lines to be inserted if needed (N005, N010, N015, etc.). No decimal point is allowed with the N word.

G   This letter address specifies what is called a preparatory function. Preparatory functions allow various modes to be set from within the program. They typically inform the control as to what is coming up in the program. Things like incremental vs. absolute mode, inch vs. metric mode, and rapid vs. cutting mode are among the kinds of things that G words will specify. There are many G words, but only a few that are used on a consistent basis. For a list of commonly used G codes, see the list at the end of these programming words. No decimal point is allowed with the G word. Also, many CNC controls permit more than one G code to be programmed in one command. However, they usually have some limitation as to the maximum number. Most CNC controls allow up to three G codes per command. If combining two or more G codes into one command, the programmer must be sure that the G words are compatible. For example, only one motion G word is allowed in a command. That is, only one of G00, G01, G02, or G03 is allowed in a command. More on motion types in key concept no. 3.

X This letter address is used to designate a movement along the X axis. The movement commanded is dependent on whether the control is currently in the absolute or incremental mode. In the absolute mode the end point specified is relative to the program zero point. In the incremental mode, the motion commanded is from the current position of the tool. For current controls, the X word can be specified with a decimal point or without one. With the decimal point, an X position of 10 in would be specified as

X10.

Without the decimal point the X word reverts to an older method of programming called *fixed format*. Without the decimal point an X position of 10 in would be specified

X100000

and the decimal point would be *assumed* to be four places to the left of the rightmost digit (this assumes four-place fixed format). For beginning programmers, it is much easier to program with the decimal point, but you must be sure to include it in every X word, or the machine will revert back to the fixed format.

Y The letter address Y specifies a positioning movement along the Y axis. All decimal-point-related functions are the same for the Y as for the X word.

Z The letter address Z is used to specify a position along the Z axis. All decimal-point-related functions are the same for the Z as for the X word.

U For wire EDM machines, the letter address U is used to specify motion in the U axis. The U axis is the motion of the upper guide left and right (parallels X). With the V axis, the U axis allows cutting tapers and machining complex shapes on the wire EDM machine. All decimal-point-related functions for the U word are the same as for the X word.

V For wire EDM machines, the letter address V is used to specify motion in the V axis. The V axis is the motion of the upper guide in and out (parallels Y). With the U axis, the V axis allows cutting tapers and machining complex shapes on the wire EDM machine. All decimal point related for the V are the same as for the X word.

A If the CNC machine has a rotary table of some kind (*not* just an indexing device), the rotation axis of the device is considered to be a true axis of motion, just like *X*, *Y*, and *Z*. Depending on the style of the CNC machine, the rotary axis can be designated with the letter address A, B, or C. If the machine you will be working with has a rotary table, you must consult the builder's manual to find out what they have named the rotary axis.

B See the description for A.

C See the description for A.

R Most CNC machines allow a motion type called circular motion. Newer CNC controls allow the circular motion command to include the radius of the arc being formed. The letter address R is used to specify the arc radius in a circular command.

I, J, K These letter addresses are the old way to specify the arc in a circular move. While they are still effective, we strongly recommend that the beginner concentrate on using the R word to specify the arc in a circular move. (It is much easier!) The I, J, and K words specify the distance and direction in X, Y, and Z (I = X, J = Y, K = Z) from the starting point of the arc to the center of the arc. Key concept no. 3 gives an example of this technique. While I, J, and K do allow a full circle to be generated in one CNC command, they are quite difficult for a beginner to understand. The R word is much more understandable. The I, J, and K addresses follow the same decimal point rules as the X word.

P The P letter address is used to specify the length of time in seconds for a dwell command. Dwell commands are used to make the axis motion (for all axes) pause for a specified length of time. A time of three seconds would be specified as P3000 (with *no* decimal point). Note that most controls require fixed format for the P word. The P word has *three places* to the right of where the decimal point would be. Other examples: P2500 = 2.5 second, P500 = 0.5 second, and P10000 = 10 seconds. No decimal point is allowed with the P word.

L The letter address L is most often used with special programming features such as subprogramming techniques and canned cycles, to specify a number of repetitions for the operation being performed.

F The letter address F specifies the desired feed rate in a cutting motion command. This allows the programmer to specify the rate at which the axes will traverse during machining. The type of CNC machine determines the designation for the F word. For example, most machining centers allow the feed rate to be specified only in feed per minute, while most turning centers also allow feed rate to be specified in feed per revolution. The F word usually allows a decimal point, so a feed rate of 3½ inches per minute would be programmed

F3.5

A feed rate of 0.005 inches per revolution would be programmed as

F.005

However, without the decimal point, the fixed format is slightly different from the X word and the various axis-related letter addresses. Most CNC controls utilize a two-place fixed format for the F word. So, in the inches per minute mode, F350 would be the fixed format for a feed rate of 3.5 inches per minute.

E The letter address E is used on many turning center controls to specify the pitch of a thread during a threading command. While the F word can also be used, it allows only four places of accuracy. With the E word, up to six digits of accuracy are allowed. For example, with 12 threads per inch, the pitch is 0.0833333333 (1/12). With the F word this pitch can be specified only to four places (as F.0833). With the E word six places of accuracy can be specified (as E.083333).

S The letter address S specifies a spindle speed. As with the F word, the form of CNC machine will determine the designation for the S word. For example, most machining centers allow spindle speed to be designated only in revolutions per minute. Most turning centers also allow spindle speed to be designated in surface feet per minute (SFM) or meters per minute (MPM). The S word does not allow a decimal point. In the RPM mode, S300 would represent 300 RPM. In the SFM mode, S300 would represent 300 SFM.

T The T letter address specifies a tool station. For those CNC machine tools that allow tooling to be automatically changed, the T word designates which tool is desired. The actual format for the T word will vary dramatically on the various types of CNC equipment. In almost all cases, no decimal point is allowed with the T word.

M The letter address M specifies a series of miscellaneous functions. You can think of M words as programmable on/off switches that turn on and off things such as coolant and the spindle. For a list of all M words, see the list under "Typical M codes." One last point to make regarding M words is that each machine tool manufacturer will select its own set of M words. While there are many standardized M words, you must be aware that you will have to consult your machine tool manufacturer's manual to find the exact list for the machine tool with which you will be working. Also note that no decimal point is allowed with the M word and most controls allow only one M word per command.

D, H These letter addresses are usually used for offsetting purposes. Almost all forms of CNC controls allow some form of compensation. We will discuss compensation in key concept no. 4. The D and H words will be used to specify the offset number in which the offset value is stored. Typically, no decimal point is allowed with these words.

As you have seen, there are many different word types involved with CNC programming. And, for special types of CNC machine tools, there may even be more. However, at this point, we are simply trying to acquaint you with the types of words with which you will be working. Again, we are *not* asking you to memorize each word. Also, if one or more of these words seems a little foggy at this point, don't worry.

As we go further, we will explain, in greater detail, the function of each word.

G and M words. Here we continue the description of the various words used in programming. We will state again that it is *not* necessary to try to memorize these commands. This section of the manual can be used as a reference guide to help you remember what each G and M code does. While this section simply defines the function for each G and M word, we will give you more specific information on their usage as we get into the key concepts of programming.

G words. As mentioned briefly above, G words specify what are called preparatory functions. They prepare the machine for what is to come. They set modes. We must point out that what follows is a list of only the most commonly used G words. This list is not complete for any one type of machine. Also, each control builder is constantly striving to make its control better or more useful than their competition's. The builders add special features that make their controls more attractive.

The programmer must check in the programming manual for the specific machine to find the list of G codes and how each one is handled.

G00 Rapid motion

G01 Straight-line cutting motion

G02 Circular cutting motion (CW)

G03 Circular cutting motion (CCW)

G04 Dwell command

G05 *X* mirror image

G06 *Y* mirror image

G07 *X-Y* exchange

G09 Cancel mirror image and *X-Y* exchange

G17 *X-Y* plane selection

G18 *X-Z* plane selection

G19 *Y-Z* plane selection

G20 Inch mode

G21 Metric mode

G28 Reference point return command

G32 Threading command

G40   Cutter radius compensation cancel

G41   Cutter radius compensation left *CLIMB MILLING*

G42   Cutter radius compensation right *CONVENTIONAL MILLING*

G43   Tool length compensation command

G45   Tool offset expansion command

G49   Cancel tool length compensation

G50   Program zero designator

G54–G59   Fixture offsets

G80–G89   Hole machining canned cycles

G90   Absolute mode

G91   Incremental

G92   Program zero designator

G94   Inches per minute feed rate mode

G95   Inches per revolution feed rate mode

G96   Constant surface speed spindle mode *(LATHE)*

G97   RPM spindle mode *(MILLING)* usually

Note that the above list is incomplete. Most CNC controls will have many special functions that are handled by G codes. However, the specific G codes for these special features will vary dramatically from control to control. In key concept no. 6 we will introduce many of these special features. At that time we will also give the most common G codes involved.

Typical M codes.   Note that the M codes in the list below are the most standardized M words for various types of CNC controls. This is not a complete list. Before working with any one kind of CNC machine tool, the programmer must check the programming manual for the machine to find the M codes for that machine. Note how many of these M codes are similar to on/off switches.

M00   Program stop

M01   Optional stop

M02   End of program (does not rewind program)

M03   Spindle on in a clockwise direction *C*

M04   Spindle on in a counterclockwise direction *Ↄ*

M05   Spindle stop

M06   Tool change command

M07   Mist coolant on

M08   Flood coolant on

M09   Coolant off

M30   End of program (rewinds memory)

Decimal point programming vs. fixed format. Current CNC controls allow the programmer to include a decimal point for those word types with which it is feasible. For example, if a position along the X axis of 3.3750 must be specified, the programmer would include the word X3.375 in the command for the motion. Many CNC words require the placement of a decimal point.

However, there is an older method of assigning the decimal point location that has been used for years. The "dinosaurs" of NC controls did not allow a decimal point to be programmed. They required that a fixed format be used for all words that required the decimal point assignment. Usually this format required that the value be carried out to four places to the right of where the decimal point was to be placed. For example, the X position of 3.3750 would be programmed as X33750 in fixed format. The CNC control would assume the decimal point to be four places to the left of the rightmost digit.

The requirements of fixed format have lead to many costly mistakes on the programmer's part. It is very common to miss a required digit in a command, causing the value of the command to change dramatically. For this reason, current CNC controls now allow a decimal point to be included with CNC words with which the decimal point is feasible.

In order to keep current controls compatible with older controls, most CNC control manufacturers continue to allow the older fixed format to be used. If the control sees a word that allows a decimal point with no decimal point, it simply reverts to the fixed format. This means a programmer who intends to use a decimal point *must* remember to include a decimal point in *every* word. For example, to specify an X position of 3 in, the programmer must program

X3.

in the command. If the decimal point is left out (X3), the control will revert to the fixed format and the decimal point will automatically be placed four places to the left of the rightmost digit. In this example, the intended value of 3 in will be taken as 0.0003 in. This should show you how important the correct placement of the decimal point is within a programming word.

Additionally, the fixed format itself may change from word to word for a particular control. This means the number of digits required to the right of the decimal point may change with the type of word. For example, the X word may require that the programmer include four digits to the right of the decimal point. But the F word may require only two places to the right of the decimal point.

Another possible variation of fixed-format programming has to do with whether the *leading* zeros or *trailing* zeros be programmed. All of this can become very confusing to the manual programmer (especially if more than one machine must be programmed) and is the reason we urge beginners to use the decimal point if it is allowed.

Remember that certain CNC letter addresses have no use for the decimal point. They are programmed as whole numbers (integers). For example, the N word for sequence number is programmed without a decimal point even on controls that allow decimal point programming. Other examples of this kind of word include S for spindle speed, T for tool station, and D for tool offset number.

If the control you will be working with allows decimal point programming, we strongly recommend that you use it. However, you *must* remember to include it in *every* word that allows it. If you forget the decimal point for even one word, the control will revert to the fixed format for the word and take your commanded value incorrectly.

General Flow of the Programming Process

You have now been exposed to many of the basic principles of CNC. Before we send you on to the first key concept of programming, we want to show how the typical company using CNC equipment would process a job to be done on a CNC machine.

It is helpful for the beginner to understand the general method by which CNC programs are processed in order to grasp the specific task of programming. It helps to see the big picture before diving into a specific task.

Here we give a generic example of how a typical shop would handle the CNC programming task. By no means will this flow be the same for all shops. Smaller job shops tend to have one or two people handling all steps in this flow. Larger manufacturing companies will usually break these steps up to be handled by several departments.

Deciding which CNC machine to use

If the company has more than one CNC machine tool, there may be some question as to which CNC machine to use. On the basis of the required number of workpieces, the workpiece accuracy, the material to be machined, the required surface finish, and the shop loading on any one CNC machine (among other things), the decision is made as to which CNC machine to use.

The machining process is developed

If more than one operation must be performed by the CNC machine, someone with machining background (perhaps the programmer) must come up with a sequence of operations to be used to machine the part. The program will follow the same sequence.

Tooling is ordered and checked

For the previously developed process, the required tooling is obtained. It is helpful to determine early on in the process what tooling, if any, must be ordered.

The program is developed

In this step, the programmer codes the program into a language that the CNC machine can understand. We will present six key concepts during part one of this text to help the beginner understand the programming process.

Setup documentation is prepared

Part of the programmer's responsibility is to make it clear how the machine is to be set up. Drawings can be made to describe the work-holding setup. If multiple tools are to be used in the program, a tool list, including the machine station numbers to be used for each tool, is prepared.

Program is loaded into the control's memory

Once the program is prepared, there are two common methods used to load it into the control's memory. One way is for someone to physically type it through the keyboard of the control panel. Generally speaking, this is a cumbersome way to get the program into the control's memory. For one thing, the keyboard of the control panel is quite difficult to work with. The keys are not usually positioned in a logical way (at least most are not like those on a typewriter), and usually the panel itself is mounted vertically. Thus, typing the program at the machine can be tiring. Also, most CNC controls require that the machine sit idle while the program is being typed. A medium-length program could take 30 minutes or more to enter. A CNC machine makes a *very* expensive typewriter!

A more popular way to enter programs into the control's memory is also much more efficient. The program is not typed at the machine. Instead, a separate device is used for the typing of the program. For example, a personal computer could be used to type programs. The software used for this purpose resembles a common word processor. In fact, some word processors can be used. Once the program is entered through the computer's keyboard, it can be saved on floppy disk or on a hard drive. When the time comes to use the program at the machine, the program is transmitted to the machine from the computer. This transmission takes place almost instantaneously, even for lengthy programs. Use of this method makes it much easier for the person typing the program, who can sit in a much more comfortable environment. Most importantly, almost no machine time is wasted while the program is entered.

The setup is made

Before the program can be run, the work-holding setup must be made. Using some form of setup instructions prepared by the programmer, the setup person makes the setup. Tooling is assembled and loaded into the proper locations. Sometimes measurements must also be made and values must be entered into offsets (more on offsets in key concept no. 4). For example, a machining center setup would require at least measurements related to program zero and tool lengths.

The program is cautiously verified

It is very rare that a program requires no modification at the machine. Even if the programmer does a good job programming the motion required, there will almost always be some optimizing necessary to reduce the cycle time to the minimum. In key concept no. 4 of "Operation" (in Part 3 of this book) we will discuss program verification.

Production is run

At this point the machine is turned loose to run the production. While the programmer's job could be considered finished at this point, there may be some long-term problems that do not present themselves until several workpieces are run. For example, in metal cutting operations, tool wear may be excessive. That is, tools may have to be replaced more often than the company would like. In this case, the speeds and feeds in the program may have to be adjusted.

Corrected version of the program is stored for future use

Most workpieces run on CNC machines are run on a regular basis, especially in manufacturing companies. At some future date it will be necessary to run a workpiece again. If changes were necessary during verification, it will be necessary to transmit the corrected CNC program back to the computer for storage on floppy disk or on a hard drive. If this step is not done, the program will have to be verified a second time when the part is run again.

Why Learn Manual Programming?

As you have seen, the "conventional" CNC control requires a rather cryptic form of CNC program. Without previous exposure to CNC programming, the beginner will find even the simplest CNC program almost impossible to understand. For this reason, there are those in the industry who truly feel that manual programming is a thing of the past. They contend that there are alternatives to manual programming that eliminate the need to learn it.

Admittedly, there are alternatives to manual programming. Though this is the case, the beginning programmer must grasp a basic understanding of manual programming in order to continue on to the alternatives. There are three important points we want to make regarding the alternatives to manual programming.

First, there is a form of CNC control called *conversational* control. This newer form of CNC control allows the operator to input programs in a more understandable and graphic way than with manual programming. Quite a bit later (during Part 2), we will discuss the most popular forms of conversational controls in detail. However, there is quite a controversy in the industry about whether conversational controls should be used (more on why in Part 2). This, coupled with the fact that conventional CNC controls requiring manual programming currently outnumber conversational controls by at least 10 to 1, means the well-versed programmer must have a good understanding of how manual CNC programs are prepared.

Second, even conventional CNC controls need not be programmed manually. There are many excellent computer-aided manufacturing (CAM) systems available to help the manual programmer develop programs. These CAM systems eliminate much of the drudgery of manual programming. However, the experienced programmer must be well-acquainted with manual programming techniques even if he or she is using a CAM system to develop programs. There are many times when a good working knowledge of manual programming will help, especially when CAM-generated CNC programs must be modified at the machine when the program is to be run. The better programmers understand manual programming techniques, the faster and better they will be at correcting mistakes and optimizing programs during program verification. Without a good knowledge of manual programming, the programmer must send a CAM program back to the computer to fix even simple mistakes. This wastes precious machine time.

Third, CNC as we know it today has evolved over 30 years of development and constant change. The machine tool and control builders have constantly strived to offer new features that make CNC equipment easier to work with and more cost-effective. For example, many techniques available today were not even possible as little as 2 years ago. Many of these improvements have been related to how the manual program is developed. Older NC and CNC controls were quite inflexible. The program's format had to be "just so." Newer CNC controls are much more "forgiving" with regard to how the program must be prepared. Also, many of the advanced techniques currently allowed in manual programming rival even a good CAM system or conversational control. For this reason, and because this evolution is still occurring, it is wise for the beginner to understand manual programming. With these programming enhancements, and especially with simpler applications, it is getting to the point that an experienced manual programmer can outperform the CAM system programmer.

Understanding how this evolution took place truly gives a person an insight into why a particular feature has been developed. Manual programming is the ground level of CNC programming. Even with the best CAM system, a programmer requires a good understanding of manual programming in order to take full advantage of all the CAM system can do.

Key Concept No. 1: Know Your Machine

The first key concept in programming is that you must understand the basic makeup of the CNC machine tool with which you will be working. When considering any CNC machine's makeup, there are two points of view, depending on your concern. One way to consider the machine is from the programmer's standpoint. It is from this standpoint that our first key concept to programming will be viewed.

The other way to view the CNC machine is from the operator's point of view. In general terms, the operator must be much more intimate with the machine, knowing the functions of all buttons and switches, being familiar with the basic sequences to run the machine, and being able to handle any problem that occurs. Much later in this text (in Part 3) we will look at the CNC machine from the operator's standpoint.

You will find, by comparison, that the programmer doesn't have to know that much about the actual running of the CNC machine tool. The programmer's main concern will be quite simple in comparison to the operator's. Here we will present a variety of popular CNC machine tools and explain those features of the machine with which the beginning programmer must be familiar.

The basic design and components of the CNC machine are of vital importance to the beginning programmer. Even more importantly, the CNC programmer must possess a very good understanding of what the machine was designed to do. Without this very basic understanding, programming a CNC machine would be similar to attempting to give a speech on a subject of which you have no knowledge.

For example, a speaker who is to make an effective and credible presentation about airplane safety must already possess an extensive background on the subject, or must research the subject well enough to be able to speak authoritatively.

Similarly, the CNC programmer must possess a knowledge of what the particular CNC machine tool was designed to do, or do the necessary research to learn what the machine tool was designed to do. For instance, a programmer who will be working with a CNC machining center must understand the kinds of operations of which this machine is capable. Milling, drilling, tapping, boring, and reaming are among the machining operations that the programmer must thoroughly understand. A beginning programmer without machining knowledge must research the machining operations, talking to people with experience and getting advice on how to proceed with each specific machining operation.

In like manner, a beginning programmer who will be working with a CNC turning center should possess a knowledge of turning, boring, threading, grooving, and other lathe operations.

Wire EDM programmers should have a good understanding of the EDM process as it relates to the kind of workpieces to be produced. Truly, the same can be said for *any* kind of CNC machine. The more you know about what the CNC machine was designed to do, the easier it will be to learn to program and work with it.

Fortunately, most CNC machines have been designed to enhance or replace another type of manually operated machine. Since many readers will have had experience with the type of manually operated machine that the CNC machine is replacing, we will begin each discussion in this key concept with how the CNC machine resembles the manually operated machine being enhanced or replaced.

If you possess the knowledge of what a particular type of CNC machine tool is designed to do, learning to program that particular type of machine will be much easier. In this case, you already know *what* you will want the machine to do. It is a relatively simple matter to learn how to translate what you want the machine to do into language that the CNC machine tool can understand.

On the other hand, if you do not understand what the CNC machine tool is designed to do, you may be in for real problems. You could be compared to the speaker who doesn't know the subject matter. While there is room for some trial-and-error techniques with CNC equipment, there is also considerable danger of injury to the operator and damage to the machine if the programmer or operator is not well-acquainted with basic machining practice as it relates to the CNC machine tool in question.

Note that there are many excellent texts that instruct the reader about basic machining practice and the various machining operations performed in the machine shop. Our text will only scratch the surface of basic machining practice. If you need help with basic machining practice, we refer you elsewhere.

In this key concept, you will become acquainted with the most popular forms of CNC equipment. You will be able to see many similarities from one type of CNC machine to the next. Once you understand the makeup of one CNC machine tool, it will help you in your understanding of other CNC machine tools.

Our discussions will follow a consistent order for each CNC machine tool to be discussed. There are several basic categories for CNC machine tools. We will first introduce each category, drawing comparisons with the manually operated machines to CNC. Then, in each category, we will describe the most common types of CNC machine tools.

At this point we will discuss the CNC machine's construction, axis directions, special programming functions, and how program zero is described. Here is a little of what will be discussed about each type of machine tool:

Construction

The basic design and construction will vary dramatically from one type of CNC machine to the next, depending on the machine tool's application. For example, a large vertical machining center is designed and constructed to provide rigid and powerful machining operations. The construction of the way system, drive motors, and ball screws must be able to withstand a great deal of pressure during machining.

On the other hand, a wire EDM machine undergoes very little stress during machining. Its way system, drive motors, and ball screws need not be designed in as rigid a manner. However, the wire EDM machine must allow very accurate movements through a programmed path. In this regard, its design is adjusted accordingly.

In this section, for each machine, we will attempt to give you a very general understanding of how the machine is constructed, though we will not make you a machine designer. We will simply scratch the surface of how the machine is constructed. While not critical to your understanding of programming, this basic knowledge of machine construction will help you to understand the basic function of the machine.

Directions of Motion

One of the most basic methods of classifying a CNC machine tool is by the number of programmable axes. This usually helps to clarify the level of difficulty related to the machine's usage. For example, a two-axis CNC machine is considered relatively easy to work with. By comparison, a five-axis machine is considered to be quite difficult to work with. This classification of difficulty is understandable, since the level of difficulty to work with a CNC machine is directly related to the number of axes the programmer must be concerned with.

We will show the various directions of motions available for each CNC machine tool to be discussed. You will find a dramatic consistency from one CNC machine type to the next with regard to the directions of motion. When learning about any form of CNC machine tool, one of the first things a beginner should try to memorize are these directions of motion and the corresponding letter address for each axis (X, Y, Z, etc.).

Special Programming Functions

There are many functions about each particular CNC machine tool a programmer must be aware of in order to prepare CNC programs. Generally speaking, these special programming functions will differ with the kind of machine tool. For example, with a vertical machining center, the programmer must be aware of how the automatic tool changer is activated, how the spindle is controlled, and how coolant is turned on and off. On a CNC turning center, the special functions will be different.

Here we will introduce and describe the special functions of concern to the programmer. We will also introduce common programming techniques to control the function. It must be stressed at this point that we are not expecting beginners to memorize and comprehend every detail of how these features are activated. We are simply introducing these features. During key concept no. 5 ("Program Formatting"), we will discuss more thoroughly the programming of these special features.

How Program Zero Is Assigned

The use of program zero allows the programmer to work from a logical location on the workpiece. If program zero is chosen wisely, dimensions going into the program can usually be taken from the workpiece print. This basic reason for using program zero remains the same for *all* forms of CNC machines.

However, the method by which program zero is assigned will vary from one machine type (and control type) to the next. Some CNC machines make the assignment of program zero almost transparent to the user. Others force the programmer to assign program zero in every program (or even for every tool used in the program). Some CNC machines allow program zero to be assigned in several ways, giving the user a choice based on their particular application. In this section for each type of machine, we will discuss the method (or methods) by which program zero is assigned.

Machining Centers

The first category of CNC machine to be discussed is machining centers. We begin with machining centers because they are among the most popular types of CNC machine tools. Most machine shops that have any kind of CNC equipment will have at least one CNC machining center.

The name *machining center* itself implies that this form of equipment has a great deal of machining versatility. Many forms of machining operations can be performed by a typical machining center. These include many forms of hole machining operations as well as most forms of milling operations.

The most basic thing that can be said for any machining center is that the cutting tool is rotated and the workpiece to be machined remains stationary. While this may seem to be an elementary statement, it helps one to understand and limit those machining operations that are possible on a machining center.

Vertical machining centers

The word *vertical* in vertical machining center refers to how the spindle is oriented on the machine tool. With a vertical machining center, the spindle is utilized in a vertical attitude.

The type of manual machine the vertical machine tool most resembles is the very common "knee-style" milling machine. That is, the vertical machining center is capable of performing the same kinds of operations as the knee-style milling machine. Drilling, tapping, end milling, face milling, and boring are among the kinds of operations the vertical machining center can perform. Figure 2.1 shows a picture of a typical CNC vertical machining center.

However, most vertical machining centers are much more rigid than the typical knee-style milling machine, so that they can perform operations that require more power. This means the vertical machining center can usually machine at a much faster rate than the knee-style milling machine.

Vertical machining centers vary with regard to the amount of control they allow the programmer. Some require that tools be loaded manually and that spindle speeds be selected manually, demanding a great deal of operator intervention. This kind of vertical machining center may also be referred to as a CNC milling machine. Other variations of machines that resemble CNC vertical machining centers in-

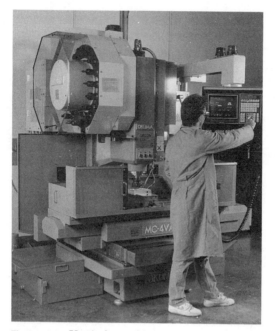

Figure 2.1 Vertical machining center (*Courtesy Okuma Machinery, Inc.*)

clude CNC drilling centers, CNC drilling and tapping machines, CNC jig boring machines, and CNC boring mills.

Full-blown vertical machining centers allow the programmer to control virtually everything that needs to be controlled through the program. They incorporate an automatic tool changer device that automatically loads tools into the spindle. They allow selection of spindle speeds, feed rates, coolant, and all other required functions through programmed command. This lowers the amount of operator intervention as the program is executed. In most cases, once a cycle has been started, the operator will have nothing to do until the cycle is completed. With vertical machining centers, once the setup has been made and the program is verified, the operator will be required only to maintain the tooling and load parts. Everything else is automatic.

Most experienced programmers will agree that the vertical machining center is the easiest form of machining center to work with. There are two reasons for this. First, the typical vertical machining center will allow only one side of the workpiece to be machined per program (unless some form of rotary device is added to the machine). This means the typical CNC program will be relatively short.

Second, most machinists have had to work with a knee-style milling machine at one time or another, and the basic layout of the machine will be quite familiar.

Directions of motion. Vertical machining centers allow at least three directions of motion (axes). The table can move left to right, in and out, and the headstock (or quill) can move up and down. The table motion left and right (as viewed from the front of the machine) is on the X axis. The table motion in and out (toward and away from the operator) is on the Y axis. The headstock (or quill) motion up and down is on the Z axis. Figure 2.2 depicts this series of motion directions.

Depending on the application for the machine, the vertical machining center can also be equipped with a rotary device. This rotary device can be used to rotate the workpiece during the machining cycle. There are two basic kinds of rotary devices. One, called an *indexer,* simply allows the part to be rotated to expose the surface to be machined. Generally speaking, a simple M code is used to activate this kind of rotary device.

The second kind of rotary device is a full rotary axis under the influence of the CNC control. Vertical machining centers equipped with this kind of rotary device will have four axes of motion (three linear axes and one rotary axis). This kind of rotary device allows machining to occur at the same time the workpiece is being rotated, meaning precise control of the feed rate of rotation is possible.

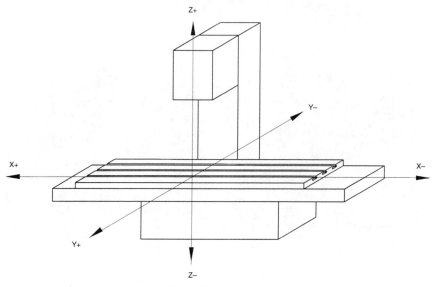

Figure 2.2   Axis motions of vertical machining center.

If a full-blown rotary axis is equipped on the vertical machining center, the orientation of the axis determines the name for the axis. If the centerline of rotation is parallel to the X axis, as most are, the axis should be named the A axis. If the centerline of rotation is parallel to the Z axis, facing up toward the spindle, the axis should be called the C axis.

Unfortunately, not all machine tool builders conform to the standards set by the American National Standards Institute (ANSI) for naming axes on CNC equipment. This means that you may come across vertical machining centers that have a rotary axis, but are not named as we have stated.

Horizontal machining centers

As with the vertical machining center, the word *horizontal* in horizontal machining center has to do with the orientation of the spindle on the machine tool. With a horizontal machining center, the spindle is utilized in a horizontal attitude. Figure 2.3 shows a picture of a horizontal machining center.

While there are manual horizontal milling machines, the horizontal machining center has little in common with them. Generally speaking, the manual horizontal milling machine is reserved for performing

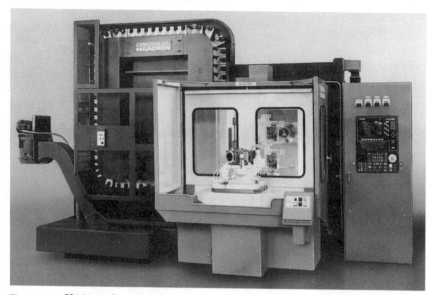

Figure 2.3 Horizontal machining center with pallet changer. (*Courtesy Cincinnati Milacron, Inc.*)

very powerful operations that require a high degree of rigidity. Examples of this kind of operation include slot milling and "duplex" milling.

By comparison to the manual horizontal milling machine, the horizontal machining center is much more flexible. The same kinds of operations that can be performed on the vertical machining center can still be performed on the horizontal machining center. Also, most horizontal machining centers incorporate some form of rotary device as part of the machine tool's table. This means the workpiece can be rotated to expose several surfaces to be machined during the machining cycle.

Horizontal machining centers tend to be more difficult to work with than vertical machining centers for two reasons. First, since the horizontal machining center has the capability to rotate the part to expose more than one surface for machining, the typical program for a horizontal machining center tends to be much longer than for the vertical counterpart. While the actual difficulty per operation is no worse than for a vertical, the sheer bulk length of the horizontal machining center program can be intimidating.

Second, horizontal machining centers tend to require more elaborate setups than their vertical counterparts. Depending on the nature of the workpiece to be machined, the horizontal machining center usually requires special work-holding fixtures in order to hold the part. This means that the horizontal machining center requires more thought and preparation when it comes to tooling.

Because of the basic construction of the horizontal machining center, this kind of machine tends to be weaker than vertical machining centers of matching capacity. That is, horizontal machining centers are less rigid. This sometimes requires compromises with regard to the cutting conditions allowed for a particular machining operation.

As with vertical machining centers, horizontal machining centers allow the programmer to control everything that needs to be controlled through the program. They incorporate an automatic tool changer device that automatically loads tools into the spindle. They allow selection of spindle speeds, feed rates, coolant, and all other required functions through programmed commands, lowering the amount of operator intervention during program execution. With this kind of machining center, once the setup has been made and the program verified, the operator will be required only to maintain the tooling and load parts. Everything else is automatic.

Most current horizontal machining centers also come with a pallet changing device, incorporating at least two pallets. This device allows the operator to load a workpiece onto one pallet while the machine is machining a workpiece on another pallet. This keeps the part loading time to a minimum.

Directions of motion. Horizontal machining centers allow at least three directions of motion (axes). The table can move left to right and in and out, and the headstock (or quill) can move up and down. The table motion left and right, as viewed from the front of the machine, is on the X axis. The headstock motion up and down is on the Y axis. The headstock (or quill) motion in and out (or the table motion in and out) is on the Z axis. Figure 2.4 shows a drawing of this layout.

Most horizontal machining centers are also equipped with a rotary device. This rotary device can be used to rotate the workpiece during the machining cycle. There are two basic kinds of rotary devices. One, called an *indexer,* simply allows the part to be rotated to expose a surface to be machined. Generally speaking, a simple M code is used to activate this kind of rotary device.

The second kind of rotary device is a full rotary axis under the influence of the CNC control. Horizontal machining centers equipped with this kind of rotary device have four axes of motion (three linear axes and one rotary axis). This kind of rotary device allows machining to occur at the same time the workpiece is being rotated, meaning precise control of the feed rate of rotation is possible. If a full-blown rotary axis is equipped on the horizontal machining center, it will be called the B axis.

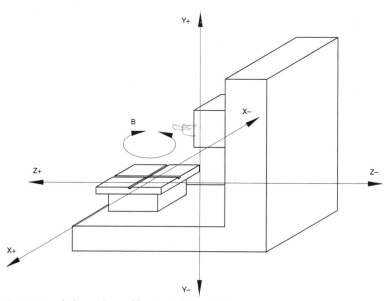

Figure 2.4 Axis motions of horizontal machining center.

Special programming functions of machining centers

Full-featured CNC machining centers allow almost every function to be programmed. This means there will be very little operator intervention once a program is activated.

While it may seem a little out of context to be discussing programming words at this point, we want to introduce those specific things on each machine type that are under the influence of program command. The beginner need not concentrate on memorizing every word presented. It is more important to simply understand the reasons for each programmable feature.

Coolant. Coolant is the flow of fluid onto the cutting tool during machining. This flow of cutting fluid aids in the machining process and keeps the part cool to avoid dimensional changes in the workpiece caused by thermal instability after machining. There are two commonly used kinds of coolants, flood coolant and mist coolant.

Flood coolant is a constant flow of liquid to the tool and workpiece, and most machinists would agree that it is the most positive form of coolant. Mist coolant is a mixture of air and fluid that is blown at the tool and workpiece. While its cooling attributes are considered better than flood coolant, mist coolant is more difficult to direct and not as popular as flood coolant.

Almost all CNC machining centers allow flood coolant. A series of coolant nozzles are manually adjusted during setup. When turned on, the coolant flows out and around the tool and workpiece. The command to turn flood coolant on is

M08

on almost all machining centers.

M09

is the command to turn off the coolant.

Some, but not all, CNC machining centers also allow mist coolant. A single spray nozzle can be adjusted to blow the mist toward the cutting tool during machining. If the machine has mist coolant,

M07

is the command to turn it on.

M09

is still the command used to turn it off.

Spindle speed control. The spindles of all CNC machining centers are programmed in RPM. Current CNC machines allow the programmer to specify the actual RPM desired. An S word is used to command the desired speed in RPM. For example, S350 would command 350 RPM.

Older machining centers used a *code number* to select spindle RPM. This kind of designation was used prior to about 1980. With this designation, the programmer was forced to look up the desired code number on a spindle chart. With this system, the code number usually did not make much sense. Also, the programmer would always have to compromise the desired speed, rounding it to the closest available code number. For example, S01 might be used for 100 RPM, S02 for 140 RPM, S03 for 180 RPM, and so on. If the calculated RPM for an operation came out to 153 RPM, the programmer was forced to decide between S02 and S03 (140 and 180 RPM). This meant the actual RPM used for any one operation was not perfect for machining.

Of course the current method for commanding spindle speed is much better (direct RPM programming). Now there is no need to use the code number system and the programmer can specify the desired RPM down to 1-RPM increments.

To turn the spindle on and off, a series of M words is used. An M03 word turns the spindle on in the clockwise, or "forward," direction. This is the proper direction for right-hand tools (most tools used on machining centers are right-hand). An M04 word turns the spindle on in the counterclockwise, or "reverse," direction, which is correct for left-hand tools. Note that both M03 and M04 must be interpreted by looking *from* the spindle. An M05 turns the spindle off. Also note that the M05 is seldom required because an M06 tool-change command or M19 spindle-orient command will also stop the spindle.

Spindle power ranges. While the fact may be "transparent" to the user, most current CNC machining centers, especially larger ones, have more than one spindle power range. Like a manual transmission in an automobile, the transmission in a CNC machining center allows extra power in the low range and higher speed in the high range. This allows heavy machining to be done at low RPM without overloading the spindle.

Most current CNC machining centers key on the spindle speed in RPM to automatically determine the proper spindle range. Usually there is a cutoff RPM at which the spindle range will be changed. The machine-tool builder's programming or operation manual will tell at what point the spindle range will be automatically changed. For example, assume the low range runs from 30 to 1500 RPM and the high range runs from 1501 to 3000 RPM. Whenever a speed of under 1500 RPM is specified, the machining center will change to the low range before the spindle comes on, if it is not already in the low range. In a like manner, if a speed of over 1500 RPM is specified, the machine will change to the high range.

Since this range changing is automatic, the programmer need not be overly concerned about which range is being selected by the machine. There is only one time when it is very important to know about this automatic range changing. If the programmer is going to be machining with a tool that requires a great deal of horsepower, it is wise to select an RPM that will place the machine in the low spindle range. For the machine in the previous example, if the programmer calculated a speed of 1520 RPM for a tool performing a powerful machining operation, it would be very wise to compromise on the RPM and make it under 1500. This would force the machine to select the low range. If this is not done, it is possible that the powerful machining operation will cause the spindle to stall.

Feed rate. During cutting operations, the programmer will have complete control of the traverse rate at which the tool will machine. This motion rate is called the *feed rate* and is commanded by an F word.

For machining center applications, feed rate is almost always commanded in feed per minute. In the inch mode, this is inches per minute and in the metric mode, this is millimeters per minute. Very few machining centers allow feed rate to be specified in feed per revolution [inches per revolution or millimeters per revolution (MPR)].

If the programmer desires a feed rate of 5.5 inches per minute, it is programmed as

F5.5

in the motion command. Note that current CNC machining center controls allow a decimal point to be included with the feed rate word.

Automatic tool changer. As mentioned earlier, almost all machining centers allow tools to be loaded from a magazine or turret automatically. This, of course, keeps the operator from having to do so manually. Programming for the automatic tool changer on current machining centers is quite easy. Usually, only two programming words are involved.

The T word tells the control the tool station to be placed in the *waiting* or *ready* position. This is the tool that will next be placed in the spindle. The T word by itself does not make the tool change, it simply rotates the tool magazine or turret to bring the desired tool to the waiting position.

For most machining centers, an M06 word is used to actually make the tool change. That is, the tool in the waiting position will be placed in the spindle when an M06 is commanded.

Most machining centers allow the programmer to command the T word in the same command as the M06. For example, the command T03 M06 would first rotate tool station number three to the waiting position. Once there, tool no. 3 would be placed into the spindle.

Most machining centers require that the spindle tool be rotated to a special position *prior to* the tool change. This is to allow the keyways of the tool in the spindle to line up with the keys of the tool changer arm. This special position is called the *orient position,* where the spindle is properly oriented with the tool changer arm.

If the machining center utilizes this key and keyway alignment system, the M06 will automatically rotate the spindle to its orient position. However, this orient rotation does take time. If no concern is given during programming, this spindle orient time will be wasted time. It could take from 1 to 3 seconds for the spindle orientation to take place, depending on the machine-tool builder.

For this reason, machining centers that incorporate the spindle orientation system also allow an M19 word. The M19 is a command that orients the spindle. If an M19 is included in the motion command that returns the machine to its tool changing position, the spindle will start rotating to its orient position as it is moving to the tool change position. This means that when the machine arrives at its tool change position, the spindle will probably be properly oriented, saving from 1 to 3 seconds per tool change. While this may not sound like much time for one tool change, it will add up quickly when you consider the number of tool changes the machine will make over its years of use.

Pallet changers. Though they are more common on horizontal machining centers, some vertical machining center manufacturers are now beginning to incorporate pallet changers. This device allows the operator to be loading one part while the machining center is machining another.

The design and programming of pallet changers vary dramatically from builder to builder; therefore, we cannot be very specific about their programming. One common type uses an M60 word to make the pallet change. When an M60 is executed, the pallet changing device is activated. As a safety precaution, the operator is required to press a *ready* button telling the control a part is loaded and ready to be machined. This keeps the pallet changer from activating while the operator is still loading a workpiece.

How program zero is assigned

Machining center controls vary with regard to how program zero is assigned. However, one thing is consistent in all controls. The user will need to tell the control where the program zero point is for every program to be run. This usually involves taking a measurement on the machine during setup, measuring the distance from program zero to the centerline of the spindle in X and Y, and to the spindle nose in Z while the machine is at its reference return position. More information will be given about how to actually take this measurement in the "Operation" part of the book.

For the programmer, there are two common methods used to inform the control as to where the program zero point is. The oldest way, and the most common, is to use a G92 command right in the program. This command includes the measured values from the program zero point to the position of the machine at which the program begins.

In essence, the G92 command simply forces the control to reset the absolute position displays to the numbers coming from the G92 command. No movement is caused by the G92 command, the control is simply being told the distances from program zero to the machine's current position. For example, suppose the operator measured the distances for each axis and found them to be as follows:

12.4543 from program zero to the starting position in X

10.4322 from program zero to the starting position in Y

15.3025 from program zero to the starting position in Z

With these numbers as the measured distances for each axis, here is the corresponding G92 command that would be programmed and executed while the machine is resting at its reference return position:

N005 G92 X12.4543 Y10.4322 Z15.3025

Notice that the X, Y, and Z values of the G92 command simply match the measured values from program zero to the machine's reference return position. The control is simply being told the distance from program zero to the starting position by the G92 command.

The G92 command is usually close to the beginning of the program. That is, before any axis motion is commanded, the control is usually given the G92 command. At the time the G92 command is executed, the machine must be resting at its starting point. Some programmers even include a G92 command at the beginning of every tool. While this may not be necessary, it makes rerunning tools a little easier.

There is one major problem with using the G92 command to assign the program zero point. The G92 command assumes the machine to be where it is supposed to be when the G92 command is executed. If the machine is supposed to be at the machine's reference point when the G92 command is read, the control assumes this to be the case when the G92 command is executed. If, for some reason, the machine is out of position (maybe the operator moved one or more axes for one reason or another) when the G92 command is read, the control will assume the values in the G92 command to be the current distances to the program zero point. If the axes were out of position in a direction bringing the workpiece closer to the spindle, real problems will occur. This is the greatest cause of crashes on machining centers. When using the G92 to assign program zero, the operator must be very careful to assure that the machine is at its proper starting position before activating the machining cycle.

The second way to assign program zero on machining centers is by *fixture offsets*. Some control builders also refer to this feature as *work coordinate system setting*. This newer way to assign program zero provides several advantages over the older G92 command.

First, fixture offsets allow the user to assign the program zero point separately from the program. With fixture offsets, the values are stored in the control in a series of *offsets* (more on offsets in key concept no. 4). With fixture offsets, the operator will not have to know how to edit the program in order to enter the measured values. Also, each entry for the fixture offset (X, Y, and Z) will be entered only one time.

Second, fixture offsets make it easy to incorporate more than one program zero point in a setup. Maybe the programmer wishes to machine two or more parts during a machining cycle. With G92, it can be quite difficult to keep reassigning the program zero point for the various parts on the table. Fixture offsets make it immensely easier. While each control manufacturer may differ slightly in usage of this feature, most manufacturers use a series of G codes to specify which fixture offset is being used. One popular method uses the series of G codes from G54 through G59. With this method, up to six fixture offsets are allowed. G54 would specify that fixture offset 1 be used. G55 would specify fixture offset 2 be used, and so on. Any X, Y, or Z motion commanded in the absolute mode from this point on would be taken from the program zero point specified by G54 through G59.

For example, suppose there are two workpieces on the table to be machined by the same program, one to the left side and one to the right. Say the programmer uses G54 to assign program zero for the part on the left and G55 to assign program zero for the part on the right. If G54 is instated, any motion commands will be relative to the part on the left. If G55 is instated, any motion commands will be relative to the part on the right.

A third advantage of fixture offsets has to do with the machine's starting position. Most controls that incorporate fixture offsets do not require the machine to be in any one specific location before the cycle is activated. The control will constantly and automatically keep track of the machine's position relative to the reference position, making the commanded motions accordingly. This makes using fixture offsets to assign program zero safer than the G92 command.

Turning Centers

Another very common type of CNC equipment is the turning center. Most machine shops that use CNC equipment have at least one turning center of one kind or another. There are four kinds of CNC turning centers that we will address in this section:

1. Two-axis turning centers
2. Four-axis turning centers
3. Twin-spindle turning centers
4. Vertical turning centers

Two-axis turning centers

As related to manual machine tools, the two-axis turning center (Fig. 2.5) most resembles a standard engine lathe. As with the engine lathe, the workpiece is held in a work-holding device and is rotated. The cutting tool is driven into the part so that machining occurs.

The work-holding device.   Three components of the turning center are of primary concern to the programmer. First is the headstock and

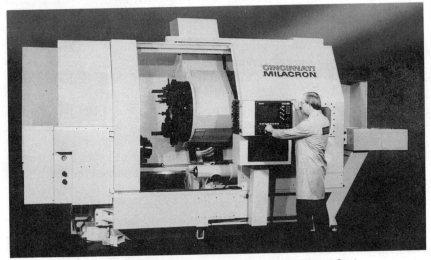

Figure 2.5   Two-axis turning center. (*Courtesy Cincinnati Milacron, Inc.*)

work-holding device. The headstock encloses the spindle of the turning center. Internal to the headstock is the spindle drive motor, which supplies the power to rotate the spindle. On the side opposite the drive motor, the spindle is attached to the work-holding device.

The type of work-holding device varies according to the application of the turning center. The most common work-holding device for a turning center is a hydraulic three-jaw chuck. The hydraulic three-jaw chuck is usually activated by a foot pedal, meaning the operator can use both hands to load the part into the chuck and activate the foot pedal when ready. This kind of device supplies an extremely high degree of clamping power to hold the workpiece. The primary advantage of the hydraulic three-jaw chuck is the high level of clamping power it can supply. Large, heavy workpieces can be securely held without fear of their coming loose during machining. However, the primary advantage of the three-jaw chuck can also be its biggest disadvantage, depending on the application. If the part being machined is flimsy, like a piece of thin-walled tubing, the power of the hydraulic chuck will crush or warp the workpiece as soon as it is clamped. While the clamping force for the hydraulic chuck can be adjusted, its range of adjustment is relatively limited.

Another popular kind of work-holding device is the pneumatic three-jaw chuck. This chuck is similar to the hydraulic chuck, but it is powered by air, not oil, and the amount of clamping power is lower. The pneumatic chuck's clamping force can be more finely adjusted than that of a hydraulic chuck, meaning weaker workpieces can be held without fear of being crushed or warped.

The last form of work-holding device we will mention is the collet chuck. Collet chucks can also be pneumatic or hydraulic. The same advantages and disadvantages apply. The collet chuck permits holding of the workpiece completely around the periphery of the part (as compared to a three-jaw chuck that contacts the part at only three points). The most common application for the collet chuck is with bar-feeding work. With a bar feeder, the workpiece is in the form of a long bar, usually 6 or 12 ft long, when it enters the bar feeder. The CNC turning center machines a part and cuts it to length in the machine. The bar feeder then pushes the rough stock into the work area for the next part to be machined. This allows unsupervised operation for the number of parts that can be machined for the length of the bar.

No matter what kind of work-holding device is used, the basic purpose is the same. The work-holding device will hold the part securely enough to allow the necessary machining to take place. It will hold the part during rotation, actually rotating with the part, allowing the necessary machining speed to be achieved.

The tool-holding device. The second basic component the programmer must be familiar with on a turning center is the tool-holding device. Most turning centers allow several cutting tools to be used during the machining of the part. There are two popular methods used for the tool-holding device.

The most common tool-holding device is a *turret*. The turret allows several tools (usually 8, 10, or 12 tools) to be held. During the activation of a CNC program, the turret can quickly rotate the desired cutting tool into the machining position. Most turrets can rotate in either direction, following the shortest distance to the desired tool.

Another method used by some machine tool builders to hold cutting tools is the gang-style tool-holding device. This form of tool-holding incorporates a table that looks very similar to the table on a knee-style milling machine. The tool holders themselves are mounted to this table in any way that the operator or setup person sees fit. This means that the tools can be mounted in very close proximity to one another, and the time for changing tools is kept to the bare minimum. While this form of tool-holding device is more difficult to work with, it is the most efficient method of holding tools for turning center applications.

The work support device. The third turning center component with which the programmer must be familiar is the work support device. Many, but not all, CNC turning centers incorporate a tailstock that can be used to support the workpiece during machining. This tailstock resembles the tailstock that is found on the typical engine lathe, and its function is the same. The tailstock includes a center that is actually engaged with the part.

Use of the tailstock is required only when the workpiece to be machined is rather long and skinny, as with shaft work. Usually the length-to-diameter ratio of the part determines whether the tailstock may be necessary for support. If the length of the part is over approximately 3 times the diameter of the part, the setup usually requires the use of the tailstock.

Most machine-tool builders that provide a tailstock on their CNC turning centers make the tailstock *programmable*. This means the motion of the tailstock body and the motion of the tailstock quill can be activated by program command. This allows certain machining operations that do not require the tailstock to be performed, and then the tailstock and center can be engaged automatically during the machining cycle and with no operator intervention. Usually a series of M codes are used to activate the tailstock and quill.

We must point out that not all CNC turning centers are equipped with a work support device. If the application for the CNC turning center does not include shaft work, the tailstock is unnecessary, and the cost of the turning center can be kept down if it is not included.

Another type of work support device that is sometimes found on CNC turning centers is the *steady rest*. However, this is not nearly as common as the tailstock. As you know, the tailstock provides support at the very end of the workpiece. If the workpiece is extremely long, the middle of the part may have the tendency to bow during machining, even if the tailstock is used. For this reason, a steady rest can be engaged to support the workpiece in the middle of the part, just as the tailstock provides support for the end of the part. The steady rest incorporates two or three rollers that are brought into contact with the outside diameter of the workpiece. These rollers rotate with the part and provide resistance to keep the part from deflecting as the machining takes place.

On CNC turning centers that have a steady rest, the steady rest is usually programmable, meaning that it can be engaged through program command. Like the tailstock, the steady rest is usually commanded with a series of M codes.

These three components of the CNC turning center are of primary importance to the CNC programmer.

Directions of motion. The CNC turning center will have at least two directions of motion. One of the axes will be the diameter-controlling axis and the other axis will be the length-controlling axis. The diameter-controlling axis is the X axis. The length-controlling axis is the Z axis (Fig. 2.6).

Depending on the machine-tool builder, the X axis may represent one of two things. Most machine-tool builders have the X axis representing the diameter of the workpiece being machined. If the control reads an X value of

X3.

it would send the tool to a *diameter* of 3 in. This is the most common method of representing the X axis.

However, some machine-tool builders make the X axis represent the *radius* of the part being machined. If the control reads an X value of

X3.

it would send the tool to a diameter of 6 in (radius of 3 in). While this is the older way of handling the X axis, some builders still use this method.

The Z axis allows machining along the lengthwise surfaces of the workpiece. This is the direction of motion toward and away from the work-holding device.

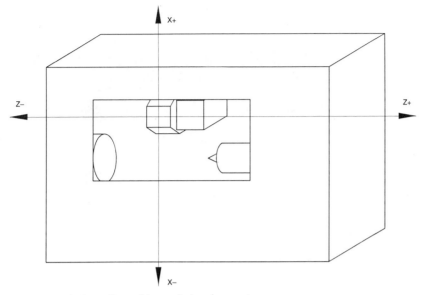

Figure 2.6 Axis motions of two-axis turning center.

Some, but not many, CNC turning centers have an additional axis. They allow the spindle to act as a rotation axis. The rotation axis of the chuck on a three-axis turning center is called the *C* axis. This type of turning center incorporates rotating tools in the turret to allow operations that are not normally possible on a turning center. This kind of turning center is usually referred to as a three-axis turning center with "live" tooling.

Turning centers with a *C* axis can double as milling machines, allowing secondary operations to be performed. For example, after the turning operations are performed, the three-axis turning center can continue working, milling slots, drilling and tapping holes, and milling flats on the part. This keeps the number of secondary operations required after the turning center operation to a minimum and assures that accuracy can be maintained between turning and milling operations, since the number of setups is kept to a minimum. While it is not yet commonly found in machine shops, the three-axis turning center with live tools is becoming more popular.

Four-axis turning centers

The same basic components that make up a two-axis turning center are included in a four-axis turning center (Fig. 2.7). So we ask you to read the two-axis turning center information first.

The major difference in a four-axis turning center is the addition of

Figure 2.7   Four-axis turning center. (*Courtesy Okuma Machinery, Inc.*)

a second turret. This allows two tools to be machining on the same workpiece simultaneously. The two turrets are programmed independently, and usually two programs are required, one for the upper turret and one for the lower turret.

The level of programming difficulty is *much* greater with a four-axis turning center. The biggest problem for the programmer will be developing the most efficient process to machine the part with two tools. The programmer must be able to visualize machining operations that can be performed together.

Directions of motion.   As with the two-axis turning center, the diameter-controlling axis is the X axis and the length-controlling axis is the Z axis. The situation is simply duplicated for the second turret. Notice from Fig. 2.8, however, that the X axis for the lower turret is reversed. While at first you may think this is backwards, remember that X is a diameter-controlling axis. The bigger the diameter, the larger (more positive) the X value. Since the lower turret is on the opposite side of the spindle centerline, the axis must be reversed.

Twin-spindle turning centers

You can think of twin-spindle turning centers (Fig. 2.9) as being two single-spindle turning centers in one. That is, the basic advantage of a twin-spindle turning center is that it can double the productivity of the two-axis turning center for the right application.

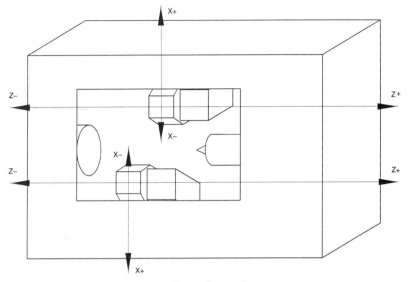

Figure 2.8 Axis motions of four-axis turning center.

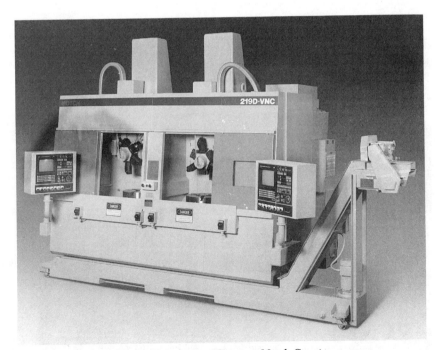

Figure 2.9 Twin-spindle turning center. (*Courtesy Motch Corp.*)

The actual configuration of each twin-spindle turning center will vary with the machine-tool builder. Some have the two spindles side by side. Others have the spindles opposed (facing away from each other), or have the two spindles facing toward each other.

One limitation of twin-spindle turning centers as compared to single-spindle two-axis turning centers is that most twin-spindle turning centers cannot incorporate a tailstock. This limits their application to bar and chucker work. Other than that, just about everything you will learn about single-spindle turning centers applies to the twin-spindle turning center.

Vertical turning centers

The vertical turning center (Fig. 2.10) has the spindle oriented in a vertical condition. The workpiece is placed into the work-holding device from above, meaning that the weight of the part will actually help to hold it, as compared to a spindle oriented in the horizontal position.

Vertical turning centers are generally used for very large workpieces, although there are small versions of vertical turning centers. Vertical turning centers can also be found in single-spindle and twin-spindle versions.

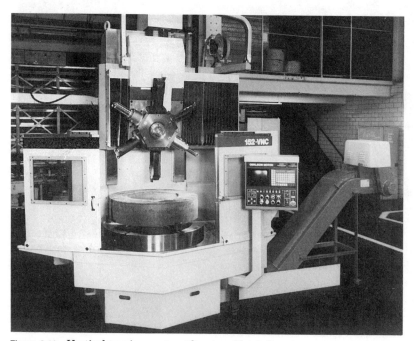

Figure 2.10   Vertical turning center. (*Courtesy Motch Corp.*)

Special programming functions of turning centers

While it is very important to understand the basic construction of the turning center, it is also important to realize that there are many things under the control of program commands. While it is not mandatory that you memorize these commands at this time, it is important that you be able to recognize the various functions that a programmer can control in a turning center program.

Coolant. Coolant is the fluid flowing around the cutting tool edge. The purpose of coolant is to aid in the machining process and to keep the part cool during machining. On a turning center, flood coolant is the only type used. The command to turn on the flood coolant is

M08

To turn it off, M09 is used (same commands as for a machining center).

Note that many programmers never use an M09 to turn the coolant off. During a turret index, the coolant will automatically turn off temporarily, so no M09 is necessary. Additionally, the end-of-program command (M02 or M30) will also turn the coolant off. With most turning centers, the work area is totally enclosed. This means coolant can always be left on without fear of drenching the operator of the work area.

Spindle speeds. The commands to control the movement of the spindle are M03, M04, and M05. An M03 turns the spindle on in a clockwise, or forward, direction. Generally speaking, this is the direction that would be needed for right-hand tools. An M04 turns the spindle on in a counter clockwise, or reverse, direction. This is the direction for left-hand tools. Both M03 and M04 are determined from the spindle side of the machine.

An M05 command is used to turn the spindle off. With a turning center, the M05 command is seldom used. During tool changes, it is wise to leave the spindle running to minimize cycle time, save electricity, and minimize wear and tear on the machine. Additionally, the end-of-program command (M02 or M30) will automatically turn the spindle off for most turning centers.

Constant surface speed vs. direct RPM programming. A CNC turning center will allow the user to specify the desired spindle speed in RPM with an S word. For some types of operations, specifying the desired RPM will be the best method. But, if you have machining experience, you know that a formula must be applied when you are developing the desired RPM. Here is the formula to calculate RPM:

$$RPM = \frac{3.82 \times SFM}{\text{cutting diameter}}$$

where   RPM = desired revolutions per minute
                3.82 = constant
                SFM = surface feet per minute speed
Cutting diameter = tool diameter (for drills, taps, reamers, etc.) or diameter being machined (for turning or boring operations)

All reference books for determining speeds and feeds in the inch system will designate the speed in surface feet per minute. In the metric system, the speed designation will be in meters per minute. Here is an example: A machinist trying to determine the proper RPM to turn a 3-in-diameter in mild steel with a carbide insert tool would look in the reference book under mild steel. For a carbide insert tool, the book may recommend a speed of 450 SFM.

Note that most reference books will not mention the diameter to be machined. The speeds and feeds reference book will assume the machinist has the ability to apply the formula shown previously. For the previous example, you would multiply 3.82 times 450 and come up with the value 1719. Then you would divide 1719 by the diameter to be machined (3 in). 1719 divided by 3 is 573. The proper speed at which to run the spindle would be 573 RPM.

On a machining center, since the cutting diameter is constant (the tool diameter), applying the above formula is not very difficult. Usually only one speed calculation per tool is necessary for each machining center operation. However, on a turning center one tool may be used to machine a wide range of diameters. For instance, a rough turning tool may be used to rough-turn the part from the stock diameter down to a very small diameter in small increments. Say, for example, the stock diameter is 4 in, and the part must be turned down to a 2-in diameter. If the programmer wanted to take a depth of cut of 0.125 in per pass, the first cutting diameter would be 3.75 in. The programmer would have to calculate an RPM based on this diameter. After this pass, the next pass would be at 3.5 in. This means another RPM calculation. Then a pass at 3.25-in diameter, yet another RPM calculation, and so on.

To make matters worse, many turning center machining operations require that the tool stay in contact with the workpiece while the cutting diameter changes. Consider how a turning center would machine a workpiece in the facing mode. During facing, the tool is usually sweeping the end of the part from its outside diameter to its center.

This means that the diameter being machined is constantly changing even while machining is taking place.

To help with this problem, control manufacturers for turning centers allow the programmer to specify speed directly in SFM (meters per minute in the metric system). This mode of spindle control is called *constant surface speed* (CSS). In this mode, the programmer still specifies the speed with an S word. A preparatory function (G word) informs the control as to which spindle mode (RPM or CSS) is desired. Most turning center control manufacturers use G96 to designate CSS and G97 to designate RPM. For example, if the control read this command:

G97 S500 M03

it would turn the spindle on clockwise at 500 RPM. But if it read this command:

G96 S500 M03

it would turn the spindle on clockwise at 500 SFM in the inch system or 500 meters per minute (MPM) in the metric system. From the time a constant surface speed command is given, the control will constantly and automatically scan to determine the cutting diameter and apply the formula given above. This is done so quickly that the spindle will appear to be speeding up and slowing down at a smooth rate.

As you can see, the constant surface speed mode saves the programmer a great deal of work. No longer does the RPM calculation have to be done. There are also two other advantages of constant surface speed about which you should know. First, the surface finish on the workpiece itself will be better if constant surface speed is used. As the tool changes diameter while still in contact with the workpiece, the spindle RPM will change smoothly, making the witness marks on the part consistent throughout.

Second, tool life will improve in the constant surface speed mode. Since the tool will always be machining at the perfect RPM, no compromises in spindle speed are necessary. The cutting edge of the tool will last longer.

Though there are many advantages to using constant surface speed, there are two operations for which the programmer must not use constant surface speed. The first instance is when machining with a center cutting tool, like a drill, tap, or reamer. Since this kind of tool is programmed to go to the very center of the part (zero diameter), if constant surface speed is used the spindle will run up to the machine's maximum. For this kind of tool, the programmer must calculate an RPM.

The second instance of when the programmer must program in RPM is while chasing a thread. This operation requires the threading tool to be able to retrace its motion over and over again at slightly smaller diameters with each pass. Because of the accuracy required of the machine in this operation, most CNC turning center controls do not allow threading to be done in the constant surface speed mode.

Feed rates. Feed rate is the tool's traverse rate during a machining operation. The programmer must have the ability to accurately control the tool's rate of motion during machining. CNC turning centers allow feed rate to be controlled in two ways. In the inch system, the two ways to control feed rate are by inches per revolution and inches per minute. In the metric system, the two ways to control feed rate are by millimeters per revolution and millimeters per minute.

As with spindle speed, two preparatory functions (G words) control which mode of feed rate is to be used. Most CNC turning centers use G98 to specify inches per minute mode and G99 to specify inches per revolution mode. An F word is used to specify the actual feed rate desired. For example, if the control reads the command

G98 F10.

the taken feed rate would be 10 IPM. But if the control reads the command

G99 F.005

the taken feed rate would be 0.005 IPR. We must point out that the G98 or G99 word will rarely be in the command that actually specifies the feed rate. Usually, the programmer will include the G98 or G99 in the program close to the beginning of each tool. Since these commands are modal, they will stay in effect until changed. Also, the feed rate (F word) by itself does not actually cause motion. It simply tells the control the desired feed rate when a motion command is given.

For turning center application, inches per revolution mode is much more useful for one main reason. Since constant surface speed (CSS) is often used, the spindle speed in RPM will be constantly changing. For example, if a constant feed rate of 10 IPM is programmed, the tool will move along the programmed path at a constant rate relative to time. However, since the RPM of the spindle is constantly changing, witness marks left on the part would not be consistent. In the inches per revolution mode, as the RPM changes, so does the rate at which the tool traverses. This leaves consistent witness marks, and a more consistent finish on the surface being machined.

Generally speaking, there is only one time when the programmer needs to program in the inches per minute mode on a turning center. If the programmer wishes to make a feed rate movement while the spindle is stopped, inches per minute feed rate mode must be used. With the spindle stopped, any feed rate in the inches per revolution feed rate mode would cause *no* movement.

One time when an inches per minute feed rate is necessary is with a bar application. While bar feeding, most CNC turning centers require that the spindle be stopped. With the spindle stopped, the chuck jaws are opened by program command and the part is pulled along with the turret. If this motion is to be done at a given feed rate, the feed rate *must* be in inches per minute. However, if this technique is used, the programmer *must* remember to reinstate the IPR mode before the next cutting command. Otherwise, the control will assume that subsequent feed rate commands are in IPM.

Other than this, we recommend that the CNC turning center always be programmed in the inches per revolution mode. Note that most CNC turning centers are initialized in the inches per revolution mode. *Initialized* means that this mode is instated when the power is first turned on to the machine. If the machine is initialized in the inches per revolution mode, many programmers will not include any command in the program for IPR mode (G99). Instead, they will let the machine assume the IPR mode at power-up.

Spindle power transmission. Many CNC turning centers, especially larger ones, have more than one power range. These power ranges serve a purpose similar to those in an automobile transmission. At lower speeds, more thrust is required. At higher speeds, the thrust is not as important.

For CNC turning centers that have more than one spindle power range, one M code is used to specify each range. Usually the spindle ranges overlap to a great extent with regard to the available RPM in each range. This allows the machining of very small to very large diameters without having to change ranges.

For example, a turning center may have two spindle ranges. The low range may be commanded by an M41 code, and the RPM available in this range may be from 30 to 1500 RPM. The high range may be commanded by an M42, but the RPM may be from 30 to 3500 RPM. In this case, the high range completely overlaps the low range. The low range would be especially helpful during heavy roughing operations at lower RPM, when a great deal of torque and power is required. The high range would be needed during finishing, when speed is of primary concern. However, with the previous example, remember there would be very little horsepower available in the high range at lower speeds. This means the high range could be used only for light-duty machining, such as finishing operations, at slower speeds.

The actual M codes to specify spindle ranges will vary dramatically from builder to builder. The programmer must refer to the operation or programming manual to find the corresponding M codes for spindle power ranges.

Turret index. Most CNC turning centers incorporate a turret as the tool-holding device. This turret can accommodate several tool stations. As many as 10 or 12 tool stations may be available, depending on the machine tool builder. The turret can be commanded to rotate to the desired tool station at any point in the program. While the turret on most machines will automatically select the shortest direction to rotate to the selected tool, some machines even allow the rotation direction to be commanded in the program by M codes.

A T word is used to rotate the turret. The most common method of using the T word is a four-digit format. The first two digits of the T word specify the station number to be commanded. The second two digits specify something called a *tool offset* to be used with the tool. (More on tool offsets in key concept no. 4). As an example, the command T0101 would tell the control to rotate the turret to station no 1. At the same time, offset no. 1 would be instated.

If the turret is bidirectional, the above command will cause the turret to rotate in the shortest direction to station no 1. If the rotation direction is also programmable, two M codes will control the rotation direction. One of the M codes will command that rotation take place in a clockwise direction, and the other will specify counterclockwise. If this is possible, the machine-tool builder's manual will list the specific M code numbers.

Chuck jaws. Almost all current CNC turning centers allow the chuck jaws to be opened and closed by program command. The most popular use for this feature is with a bar application.

With this kind of application, the rough stock material is in the form of a long bar (usually 6 or 12 ft long). The turning center will machine one part from the end of the bar. When finished, a *cutoff tool* cuts the machined part from the end of the bar. The bar is then fed out to allow the next part to be machined. This allows semiautomated operation, since the only real operator intervention required is the loading of the long bar.

For this application, the chuck jaws must open and close automatically during the bar-feeding cycle. For turning centers that allow this function, two M codes are used, one to open and one to close the chuck jaws. Most turning centers require that the spindle be stopped before these M codes will work. The machine tool builder's manual will specify the actual M code numbers for jaw open and close.

Tailstock motion. Though not all CNC turning centers have a tailstock, most that do allow it to be activated through program command. The actual configuration of the tailstock and quill construction will vary widely with the machine-tool builder. One very popular style allows the body of the tailstock to be activated forward or backward by two M codes. Two limit switches are set by the operator during setup to the correct front and back positions. When the control reads either of these M codes, the body of the tailstock moves until the corresponding limit switch is hit.

In like manner, two more M codes are used to activate the quill. One M code moves the quill forward and the other moves the quill back. Unlike the body movement, when the quill is activated in the forward direction, it continues its motion until the center contacts the workpiece. The pressure applied by the quill is maintained during machining.

For machines with this kind of tailstock system, the four involved M codes can be found in the machine-tool builder's manual.

How program zero is assigned

There are two ways used to assign program zero on turning centers. In essence, both methods require that the programmer tell the control how far it is from program zero to the tip of each tool being used in the program. This involves taking a measurement of some sort on the machine to determine these distances for the X and Z axes. Since turning-center tooling varies dramatically from tool to tool, a different distance will be measured for each tool.

Assigning program zero in the program. One method that can be used to assign program zero is to include G50 (G92 on some controls) commands in the program that tell the control the previously measured distances from program zero to the tip of the tool. When this technique is used, a different G50 command is usually required for each tool in the program. In the program, the G50 command will be close to the beginning of each tool. The actual techniques to measure these G50 values will be given in the "Operation" part of this book.

When the turning-center control reads a G50 (G92 on some controls) command, it simply resets the absolute position displays for the X and Z axes. In effect, this recalibrates the axis displays to the actual position of the tool to be used.

For example, the operator measures the distance from program zero to the tip of a turning tool. If the values come out to 10.4365 in the X axis and 12.2334 in the Z axis, the correct G50 command for this tool is

N005 G50 X10.4365 Z12.2334

This command tells the control that it is 10.4365 in the X axis and 12.2334 in the Z axis from program zero to the tip of this particular tool (the X value is usually a diameter dimension).

While the G50 technique to assign program zero is still quite popular among turning-center users, its use does have its limitations. One dramatic limitation of using this technique to assign program zero has to do with the machine's starting point. If the machine is not at the planned starting point when the G50 command is read, the tool will not go to its correct ending point. This can cause real problems. If the axes have been moved closer to the workpiece, the tool will run right into the workpiece.

Another limitation of assigning program zero with G50 in the program concerns the operator's level of knowledge. When using this method, the operator must be well-versed with program editing techniques, since the operator must be able to manipulate the various G50 (or G92) commands in the program. Depending on the format used for programming, there can actually be four program edits required for each tool (X and Z in two locations of the program).

Assigning program zero with offsets.   The second method of assigning program zero on turning centers is quite a bit newer and becoming more popular. Its use overcomes the previously mentioned limitations of assigning program zero in the program with G50. With this technique, the distances from the tip of each tool to program zero are stored in an *offset* and are separate from the program.

The basic method by which the measurement is taken from the tip of the tool to program zero remains essentially the same. Instead of having to modify the program, the measured values are stored in offsets.

CNC Wire EDM Machines

The CNC wire EDM machine (Fig. 2.11) is one of the most fascinating CNC machine tools currently available. It allows the machining of extremely hard materials that were previously impossible. For example, the CNC wire EDM machine can machine hardened tool steel and even carbide. Prior to EDM, these materials had to be ground by expensive and labor-intensive techniques that required form dressing of the grinding wheel.

There is really no manual piece of equipment that you can become familiar with that will help in your understanding of the wire EDM machine. The closest analogy is a band saw. Think of a band saw with an extremely narrow blade. As you know, the band saw can machine a free-form contour. That is, the operator of the band saw can manipulate the workpiece in a way that the cutting edge of the saw blade forms the desired contour. The narrower the blade, the more elaborate the contour that can be machined. You can think of the wire EDM machine as having the ability to machine with a band saw blade made of

Figure 2.11   Wire EDM machine. (*Courtesy Sodick, Inc.*)

a piece of wire. There is almost no limitation with regard to the contour that can be machined. However, the cutting action is completely different. With the band saw, metal is sheared during machining. With wire EDM, the material is actually melted away during machining.

Because the wire EDM process is unique, we will give a comprehensive description of the process to acquaint you with the capability of a CNC wire EDM machine.

Description of the EDM process

The beginning EDM programmer or operator usually does not have to be an expert on developing conditions for EDM machining. The machine-tool builder will usually supply a set of machining conditions for the various materials to be machined.

However, the beginner does need to know the basics of the EDM process to be able to correctly fine-tune the machining of a workpiece. Here we give you a very basic description of the factors that contribute to a good set of machining conditions.

A good analogy of what occurs in the EDM process is what happens within your automobile engine. The spark plug that ignites the fuel in the combustion chamber is constantly receiving a great amount of voltage and current to form a spark. As time goes on, and the spark plug wears, you will notice a certain amount of pitting on the electrodes of a worn spark plug. This pitting is caused by exactly the same elements that work in the EDM process. The heat generated by the spark causes a small amount of the electrode to melt, and the melted material is blown away from the electrode of the spark plug. Of course, the spark plug is designed to fight against this pitting, but in the EDM process, the melting of material and the washing away of the melted particles is desirable, and the machine is designed accordingly.

In the EDM process, wire is passed through the workpiece. This wire is made of a highly conductive material such as brass, copper, or zinc. Both zinc and copper are quite ductile, meaning they have the tendency to stretch under pressure. For this reason, brass and zinc-coated brass are currently the most commonly used materials for the wire. The wire is called the *electrode*.

The wire (electrode) passing through the workpiece is constantly moving through the part. The wire has an intermittent voltage applied. As the wire comes close to the workpiece, not actually touching it, a spark will occur. This spark is similar to the spark you experience from static electricity when the air is dry and you touch a metallic object. The spark in the EDM process generates enough heat to actually melt a small portion of the workpiece. The melted material will instantaneously solidify and form small particles. After the spark, the electricity to the wire is removed and flushing with water or oil will remove the particles from the machining area. Then the process is repeated, over and over again. It is this constant and persistent process that eventually machines the workpiece to desired specifications.

As with other forms of machining (milling, turning, etc.), the surface finish and accuracy obtained from only one machining pass may not be acceptable. Multiple passes may be required to machine the part to acceptable surface finish and accuracy tolerances.

With almost any form of metal-cutting machining operation, there are typically only five cutting-condition factors that affect the way machining will take place. With operations such as milling and drilling, these five factors include feed rate, spindle speed, tool material, part material, and coolant. Typically, the only real factors that the operator or programmer has control of are the spindle speed and feed rate. This makes it quite easy to adjust the machining conditions correctly.

Unfortunately, this is not the case with wire EDM. Many more factors affect the EDM process. The most important factors are explained below. For the beginner, remember that we are just introducing each factor, and there are a lot of them. Try not to become bogged down with memorizing each one. The machine-tool builder will always include a set of conditions that will machine a variety of materials. As time goes on, the beginner will learn more about these factors and be able to make adjustments to the EDM process.

We must point out that only the most common factors are discussed. There may be more factors related to any one specific wire EDM machine. Also, as machine tool builders continue to enhance their equipment, the need to manipulate some of these factors may be reduced. You may find that the machine with which you must work does not have some of the factors mentioned here. Their use may be transparent to the user.

Each of these factors directly contributes to machining speed, part finish, and accuracy. Generally speaking, as you modify conditions to improve one of these objectives, another will usually suffer.

General factors

First we will look at some of the general things that contribute to the cutting condition. Most of these factors can be controlled manually and are quite general as to the actual effect they have on the cutting condition.

Wire diameter. The first consideration to which the machinist should give thought is the diameter of the wire to be used to machine the part. Generally speaking, the larger the diameter, the faster the machining can occur. However, many times the configuration of the workpiece will dictate the maximum wire diameter possible to machine the part. This usually has to do with the minimum inside radius allowed on the part. If the inside radii must be very small, the wire radius must be even smaller to generate them.

Most machine tool builders offer a variety of wire diameters from which to choose. Common wire diameters include 0.004, 0.006, 0.008, 0.010, 0.012, and 0.014 in. Note how small these wire diameters are. This is why extremely fine work can be done on the wire EDM machine.

Generally speaking, the programmer will choose the largest wire size available that allows the workpiece to be machined within the minimum inside radius restriction. For example, if you see a specification on the print of 0.005 inch for an inside radius, you may think that you could use a 0.010-in diameter wire, since it has a 0.005-in radius. But you must also consider the fact that, as the wire machines the workpiece, a certain amount of *overburn* will occur since the wire will never actually touch the part. This overburn will be equal to the arc distance of the spark itself. In this case, if the 0.005-in inside radius is critical, you would have to use 0.008-in diameter wire.

The wire itself comes in the form of rolls. A roll is mounted on the machine in a way that allows the wire to be passed through some form of tension-control roller. The tension control allows the operator to determine how tight the wire is during machining. To support the wire and keep it at the desired angle, two guides are incorporated. An upper guide is mounted just above the workpiece and is usually adjustable to allow for different workpiece thicknesses. A lower guide is located just below the part. After passing through the lower guide, the wire is pulled to a disposal bin. Note that the wire is good only once. After it has been used in the EDM process, it will be pitted and not acceptable for use again. In fact, if the wire has been used for roughing (the first pass), most experienced EDM users would agree that the wire would be on the verge of breaking as it enters the disposal bin. It would certainly not be useful for further EDM.

Flushing. During the wire EDM process, particles of material will be formed as each spark burns away material from the workpiece. If these particles are allowed to remain close to the wire, they will deter machining. The particles would cause a premature spark to occur. For this reason, the particles *must* be removed from the machining area. This is accomplished with flushing. In the wire EDM process, a column of water is formed around the wire. This column of water is highly pressurized, so the particles will be washed away with the water. Two flush nozzles are mounted (one around the upper guide and the other around the lower guide) in a way that allows the flushing to be very close to the top and bottom of the workpiece. Usually you want the upper flush nozzle about 0.005 in above the top of the part. The lower flush nozzle usually slides up to contact the bottom of the part automatically when the flushing is turned on.

Flushing is considered one of the most important factors in the EDM process because, if the small particles melted by the EDM process are not removed from the work area, *double burning* will occur; the particles will be remelted when the next spark occurs. Generally speaking, the better the flushing, the faster the EDM process will occur.

There are several things you must know about flushing. You usually want high flushing (high flush pressure) when you are roughing to quickly machine the part. During subsequent *trim passes* (for finishing), you will want to lower the flushing pressure to avoid having the pressure of the flush deflect the wire. Also, during finishing, the flushing is not as critical, since the amount of material is not as great as that removed in the roughing pass.

Also, when entering your cut from the solid wall (not from a start hole) or when machining an internal shape (die) after the die slug has been removed, the flushing must be reduced to keep it from being dispersed by the wall of the part itself. So any time the wire is driven around an area of the part that is not solid on both side of the wire, the flushing must be reduced.

For most wire EDM machines, flushing can be programmed as part of the condition itself (high or low) or the operator can take control of the flushing by adjusting a valve on the machine.

Wire speed. The wire speed is the rate at which the wire is passed through the workpiece. Generally speaking, a faster wire speed is used for roughing and a slower wire speed is used for finishing. During roughing, the wire will usually be under the influence of a great deal of power (electricity) as material is blown away as quickly as possible. This results in a great deal of pitting on the wire itself and, if taken to extreme, will cause wire breakage. You can relieve much of this pressure by increasing the speed of the wire as it passes through the part.

For finishing, the above-mentioned pressure is nonexistent. For finishing passes, the wire speed can be reduced to conserve wire.

Wire tension. The wire tension is the amount of pulling force applied to the wire as it is driven through its cutting motions. You can think of the wire tension as how much stretching force is applied from the upper to the lower guide. Actually, the tension is caused by the amount of restriction that the wire is given by the tension roller at the top of the wire threading area. The pinch rollers just above the wire bin pull the wire from the wire roll at the top of the machine. The tension roller restricts the wire in a way that permits the tension to be adjusted accurately.

Generally speaking, you will want a lower wire tension during roughing and a higher tension for finishing. During roughing, the wire will be under a great deal of stress just because of the powerful cutting condition required for roughing. If tension is too great during roughing, it could cause wire breakage.

For finishing passes, it is wise to increase the wire tension to ensure that the wire is as straight as possible during machining. This ensures that the wall of the workpiece will be as straight as possible.

Water conductivity. Generally speaking, the water conductivity level contributes to the cutting condition. The water is intended to be *dielectric,* meaning that it should have no electric conductivity. However, in real life, this is next to impossible to achieve. In the filtering system of the EDM machine, there is a deionization chamber for removing the charged ions from the water, making the water less conductive. While this system works quite well, there will be some conductivity remaining in almost all water.

Most companies that have wire EDM equipment will use simple tap water as the dielectric. They depend on the deionization system to remove any conductivity from the water. However, if well water is used as the supply source for water, and if there is a great deal of iron in the water, this is not acceptable. For this reason, many companies are forced to buy distilled water for their EDM equipment.

The water conductivity will directly affect the cutting conditions. Generally speaking, the lower the water conductivity, the better the cutting condition will be. However, the negative side to a low water conductivity is that the workpiece will be more susceptible to rust. There are water additives available that reduce the tendency for rust.

Workpiece material and hardness.   Another obvious factor that contributes to the EDM process is the material to be machined. Generally speaking, any material that will conduct electricity can be machined on an EDM machine. However, some materials will machine more easily than others. Usually, the harder the material is, the easier it is to machine on an EDM machine. A hardened piece of tool steel will machine much more easily than the same material in its soft state. This has to do with the way that the melted particles form. With soft material, the particles formed will usually be larger and harder to flush. With the same material in its hardened state, the particles will be smaller and almost disintegrated during the EDM process, making them much easier to flush.

There are some exceptions to this general statement. Carbide, for example, is a very hard material, yet it is considered one of the more difficult materials to machine on an EDM machine. This is because carbide is very brittle, and has the tendency to crack easily when exposed to the intense thermal change encountered in the EDM process.

Some of the materials most commonly machined by the EDM process include aluminum, titanium, tool steels (O1, D2, etc.), graphite, and carbide.

Specific parameters of the EDM process

Now let's look at EDM parameters. As mentioned previously, this list is quite generic. Some wire EDM machines may have additional parameters related to cutting conditions, while others may make it unnecessary for the user to change some of those we discuss.

On time.   The on time parameter controls the length of time that electricity is applied to the wire per spark. Depending on the machine-tool builder, there are various techniques to control on time. Usually, there is a series of numbers that specify the various on times available. Generally speaking, the lower the number, the shorter the on time.

As the name implies, on time is used to control the amount of time per spark power is applied to the wire. As the on time is increased, the tendency is toward faster machining. Usually, the on time for roughing is higher than the on time for finishing.

The negative side to increasing the on time is that the finish generated by the cutting is worse. Additionally, accuracy will suffer. Also, if the on time is too great, the wire will be prone to breakage.

Off time.   The off time parameter controls the length of time per spark that the electricity is turned off to the wire. You can think of off time as the opposite of on time. Off time is very important because it is during the off time that the particles are flushed away from the machining area. Without off time, the particles will not be flushed and double burning of the particles will occur. Increasing the off time will generally mean slower cutting, but part finish and accuracy are improved.

Adaptive control parameter.   Almost all forms of CNC wire EDM controls have a cutting condition parameter related to what will happen if something goes wrong during the EDM process. Some machine-tool builders may keep the adaptive control system completely out of the hands of the user, but most allow the user to select the sensitivity of this system. Also, most controls allow the user to specify what will happen if a problem occurs.

If, for example, the flushing is poor in one area of the part or another, as the wire enters that area, the EDM process will suffer. If the control simply tries to plow through the poor flushing area, several negative things may happen. The part finish may be poor in this area, double burning of particles may occur and cause a belly effect in the part, and, taken to extreme, the wire may break.

If the EDM control has some form of adaptive control, when a problem is encountered as mentioned above the cutting conditions for the EDM process will be modified (lowered) until the problem has passed.

Current (amperage).   The current is the amount of amperage applied to the wire. As with on time, the higher the current, the faster the machining will take place, but accuracy and finish will suffer.

Voltage.   As with current, voltage works to supply the electrical power that is induced into the wire. You can think of voltage as being a crude setting for the power range and current as a way of fine tuning the actual power to the wire. An analogy can be drawn from a multiband radio. You can think of the AM/FM selector switch as being voltage, and the fine-tuning knob on the radio as the current. With a radio, you select the crude band you want to be in and fine tune to the desired station with the tuning knob. With EDM, you can make a crude selection of the power band you want to be in with the voltage setting and fine-tune with the current setting.

Voltage also has a great deal to do with the workpiece thickness. The thicker the workpiece, the greater the allowable voltage without fear of wire breakage. This is because a thick part allows the power in the wire to be dispersed over a greater area. Picture the wire pulled tight. If you took a thin, sharp knife, it would be easy to cut the wire. But now picture taking a thick round bar. It would not be as easy to cut the wire. This analogy shows the relationship of voltage to the cut. Thin workpieces require a lower voltage, while thick workpieces allow a greater voltage.

Gap width control.   Most wire EDM machines allow the user to specify, in one form or another, the gap width used during machining. This helps control the rate of cutting. A small gap width has a tendency to push the cut along while a large gap width has a tendency to slow down machining.

Capacitance.   Capacitance is a method for adding more power to the wire while limiting the possibility of wire breakage. Since capacitance generally has an ill effect on part finish, many wire EDM machines no longer utilize capacitance as one of the cutting condition parameters. The higher the value of the capacitance setting, the greater the capacitance that is added to the cutting condition.

What it all means

As previously stated, we are not asking that you memorize every one of these condition parameters. Most wire EDM machine-tool builders supply a standard set of conditions to machine a wide variety of materials, therefore the wire EDM machine programmers do not have to develop the conditions by themselves. However, no matter how many sets of conditions are supplied by the machine-tool builder, the user will eventually have to modify condition parameters for the best results. While it may not be of primary concern, the beginner should eventually strive to understand the meanings of the specific cutting condition parameters for the wire EDM machine to be used.

Axis directions

The most basic forms of CNC wire EDM equipment have two axes of motion. The table of the machine can move left and right and in and out. The left-right axis, as viewed from the front of the machine, is the X axis. The in-out direction is the Y axis. Figure 2.12 shows the relationship of all axes on the wire EDM machine.

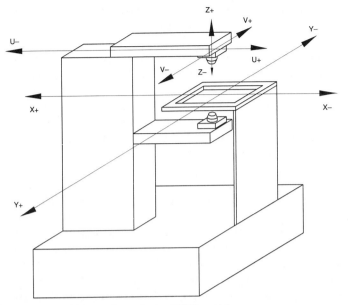

Figure 2.12   Axis motions of wire EDM machine.

Wire EDM machines with only two axes of motion are capable only of machining square-walled workpieces. However, depending on the machine's application, many workpieces require that a taper be formed on the walls of the part. To do this, the wire EDM machine must have two more axes. The upper guide must also be able to move left-right and in-out to allow taper to be formed on the walls of the workpiece. The left-right motion of the upper guide is the U axis, and the in-out motion of the upper guide is the V axis.

Although most CNC machines require manual intervention in this regard, some wire EDM machines also allow the up and down motion of the upper guide to be programmed. If the up and down motion of the upper guide is programmable, it is called the Z axis. This would result in five axes of motion.

Special programming functions of wire EDM machines

There are also several special programmable functions of which the wire EDM programmer must be aware. It is not mandatory that you memorize each of these commands at this time. It is more important for you to understand the basic features than to know the programmed coding.

Many of these features are options on current CNC wire EDM machines. This means you may not find one or more of these features on every wire EDM machine. However, each feature is currently available for purchase from wire EDM machine manufacturers.

Flushing.   As mentioned earlier, flushing is done by the column of water that surrounds the wire during machining. This flushing can be activated on most wire EDM machines by programmed command. A series of miscellaneous functions is used to turn on and off the flushing. Some machines use M codes to accomplish this while others use T words. The T word on a wire EDM machine is virtually identical to the M word in function. That is, it is a programmable on-off switch.

It is important that the pressure of flushing be adjustable during machining. For example, high pressure is required during roughing to flush particles from the burn area. Low-pressure flushing is required during finishing to minimize the possibility of wire deflection. Most wire EDM machines allow the pressure of the flushing to be adjusted by programmed command. Some have a simple high-low flush control. One M (or T) code turns on the low flush and another turns on the high flush. Yet another turns off the flush. For this kind of flushing system, the actual pressure at the high and low settings is controlled manually, with a series of valves.

Other wire EDM machines have quite an elaborate series of commands to more precisely control flushing pressure by programmed command. With this system, the operator need not manually adjust valves. For example, T1 might represent the lowest flushing pressure. T2 may step up the pressure a small amount. T3 would be yet higher pressure, and so on. This system would allow the programmer more control of flushing by programmed command.

Each wire EDM machine builder will use what it considers to be the best flushing system for its particular machine. The programmer will have to adapt to this system and check the programming or operation manual to find the commands related to flushing.

Wire tension.   The wire tension affects how tight the wire is. During roughing operations, it is wise to have a relatively low wire tension to avoid placing undue stress on the wire. For finishing, the wire tension should be quite high to allow a straight wall on the workpiece.

While some lower-cost wire EDM machines allow only manual adjustment of the wire tension, many allow it to be controlled through program command. The method of programming will vary dramatically from one machine tool builder to the next.

Some machines will use a variety of M or T codes to specify the desired wire tension. Others will make the wire tension part of the cutting condition, along with on time, off time, current, voltage, etc.

Wire run speed. Wire run speed is the speed at which the wire is passed through the workpiece during machining. While roughing, when there is a great deal of stress on the wire, it is desirable to have a faster wire run speed to allow the wire to pass through the part without breaking. During finishing, it is desirable to have a slower wire run speed to conserve wire.

Like wire tension, the method of controlling wire run speed will vary from builder to builder. On lower-cost machines, the wire run speed may be controllable only manually, by the operator. Other machines will allow the wire run speed to be controlled through programmed command. There can be a special series of M or T codes designated to specify wire run speed. This feature may also be part of the cutting condition, along with on time, off time, current, voltage, etc.

Wire straightness alignment device. During normal machining, it is important that the wire be vertical or perpendicular to the workpiece being machined. If not, there will be undesirable taper in the workpiece being machined. While there are times when controlled taper is desired, it is mandatory that the user have a way of assuring the straightness of the wire.

If the machine has U and V axes, these axes are used to adjust the wire straightness. Some machines require the operator to adjust the wire straightness manually. In this case, some form of alignment gauge is used. The operator selects a light cutting condition and actually brings the wire into contact with the gauge. The spark will be visible to the operator who tries to make the spark as uniform as possible from top to bottom by manually moving the U and V axes.

Other wire EDM machines allow the vertical wire alignment to be done automatically by programmed command. A wire alignment gauge is permanently mounted to the machine's table in a location that will not interfere with normal machining. To vertically align the wire, the operator makes the corresponding CNC command. On some machines, this may be a simple M or T code. On others, the operator may have to activate a special program that performs the wire alignment.

Automatic wire threader. Given the very delicate condition of the wire and the severe stress induced during machining, especially during roughing, the wire will be prone to breakage. If this happens during the machining of the part, the wire must be rethreaded and the cutting cycle restarted.

Lower-cost wire EDM machines require that the operator thread the wire. This means an operator must always be available during machining in case the wire breaks. But many companies wish to run their wire EDM equipment unattended. They do not wish to tie up an operator to simply wait for wire breaks. Or maybe they wish to run the machine through the night when there is no one in the shop. In this case, if the wire were to break, the machine would sit idle until an operator came to rethread it.

For this reason wire EDM machines can be equipped with automatic wire threaders designed to sense when a wire break occurs, go back to the starting position, rethread the wire, move quickly back through the programmed path to where the wire broke, and continue machining the part.

The design of these wire threaders varies dramatically from builder to builder, as does the method by which they are commanded. Some machines use one or two simple M or T codes to cut and thread the wire. Others require the execution of a special program. In any case, all will sense the wire breakage automatically, so that no operator intervention is required when the wire breaks.

Automatic restart.   Wire EDM is a very slow process. Even the latest "fast" machines will machine only at about 20 cubic inches per hour. For a workpiece that is 1 in thick, a 4- by 4-in square would take 48 minutes. It is conceivable that an elaborate workpiece may take as long as 8 or more hours to machine. Because of the extremely long cutting cycles, power outages become a concern, especially if the machine is running in an unattended condition during the night.

The automatic restart feature is available on many wire EDM machines to allow for power outages. If this feature is activated, when a power outage occurs, the machine will actually power-up by itself when the power eventually comes back on and continue machining the workpiece.

Hole-machining device.   Generally speaking, the workpiece being machined on a wire EDM machine is in a hardened state. EDM is one of the few machining operations that can be performed on hardened materials.

Many wire EDM operations require a *start hole* from which the wire will begin. Die openings require this kind of start hole. This means that the start hole must be machined in the workpiece before the wire EDM operation.

If the wire EDM machine has a hole-machining device, there is no need to machine the start hole when the material is in its soft state. The start hole can be machined into the material in its hardened state, right on the wire EDM machine.

The programming and usage of this device varies from builder to builder. The programmer must check the builder's programming and operation manual to find more about its use.

How program zero is assigned

Most CNC wire EDM machines require that program zero point be assigned in the program by using a G92 command. The G92 will include the distance from program zero to the centerline of the wire at which the program is to begin.

Depending on what kind of part is being machined, the assignment of program zero ranges from crude to very accurate. With the machining of outside shapes, such as a punch used in a die set, there is nothing very critical about the assignment of program zero. The operator only needs to be concerned that the shape to be machined fits in the rough stock used to form the part.

However, the location of a die opening is usually *very* critical indeed. In this case, the operator must ensure that the die opening maintain a close tolerance for dimensions from the edges of the die to the opening.

Most wire EDM machines have a series of pickup routines designed to make assigning the program zero point much easier. For example, a corner pickup routine would automatically have the wire find the edges of the part and, when finished, display the distances in X and Y from the corner of the part being picked up to the center of the wire. Another commonly used pickup routine is designed to find the center of a hole.

These automatic routines use the continuity of the wire (like a continuity tester used in electronics) to determine when the wire comes into contact with the workpiece. Basically, the wire is told by the automatic routine to move quickly toward the edge of the part. When the wire touches the surface, it forms a short circuit. When the control senses the short circuit, it makes the wire back up and reapproach the surface at a much slower rate. This time when the wire touches, the control can stop the motion in such a way that the surface of the wire will be perfectly flush with the edge of the part.

Once one of these pickup routines has been used, the absolute position displays on the control screen will be constantly showing the distance from the surface picked up to the center of the wire. Now it is a relatively simple matter to jog the machine to the location where machining is to begin.

Many wire EDM machines also allow multiple program zero points to be assigned. The method by which this is accomplished varies from one machine tool builder to the next; therefore, we cannot be very specific. For now, it is much more important that you understand the function of program zero. More on how program zero is assigned will come in the operation section of this book.

Turret Punch Presses

The CNC turret punch press (Fig. 2.13) is used to pierce holes of various shapes into plate. The workpiece material and thickness, as well as the size of the hole to be pierced, determine the power required of the machine tool. This kind of machine has a turret that can hold several punch-and-die combinations. Common shapes to be punched include circles, squares, rectangles, and other standard geometric shapes. The turret punch press can also incorporate special shapes as long as the punch-and-die combination is properly made.

The part to be machined is held securely by two finger clamps. While clamped, the part is moved between the punch and die being held by the machine's turret. When a desired location has been reached, the ram of the turret punch press can be told by programmed command to punch a hole.

The turret can be quickly and automatically rotated to the desired station. Once the desired axis position is attained, the hole is punched. If more than one hole is required in the same shape, the machine is moved to a new axis position and another hole is pierced.

When finished with one punch and die combination, the programmer can command that another be placed in the punching position. It is common for the turret of this kind of machine to be able to hold from 30 to 50 punch-and-die combinations. This makes the machine ex-

Figure 2.13   Turret punch press. (*Courtesy Trumpf, Inc.*)

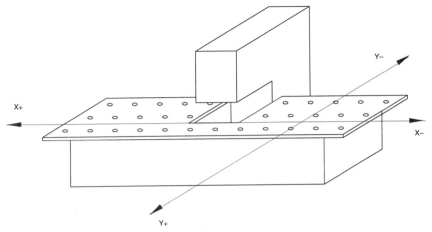

Figure 2.14   Axis motions of turret punch press.

tremely flexible, allowing a wide variety of shapes to be punched in a workpiece with the tooling available in the turret.

Turret punch presses are commonly found in metal fabrication shops. Some people would argue that the operation being performed by the turret punch press is not truly a machining operation but a fabricating operation. However, since this book is addressing a variety of CNC equipment, we classify the operations being performed by the turret punch press as machining operations.

Directions of motion

The part being held in the finger clamps can be moved in two directions. The table, along with the clamps, can move left and right as well as in and out. If the machine is viewed from the front, the left and right motion is considered the X axis. The motion in and out is the Y axis. Figure 2.14 shows this.

Flame Cutting Machines (Also Called *Gas Plasma*)

The flame cutting machine incorporates an oxyacetylene gas torch that is used for cutting. As with a hand-held cutting torch, the surface finish generated by the flame cutting machine is very poor. For this reason, the flame cutting machine is generally used to make the rough stock shape for a workpiece to be machined. That is, the flame cutting machine does not usually produce a finished workpiece. To the contrary, the flame cutting operation is usually the very first operation, cutting the part to a crude shape. This crude shape is close to the finished size, but allows a small amount of stock to be removed by subsequent and more precise machining operations.

Once the flame is ignited, the workpiece is passed under the flame. As it is brought into contact with the flame, the material is melted, and the shape to be generated is machined. The material and thickness of the part to be machined will determine how quick the motion can be during the cutting operation.

Axis directions

The flame cutting machine will have two directions of motion, left and right, as well as in and out. If the machine is viewed from the front, the left and right motion direction is the X axis. The in and out direction is the Y axis.

CNC Laser Equipment

For the most part, laser equipment is a newcomer to the field of CNC. While laser technology has been around for a while, only now is the technology filtering its way into the form of CNC equipment. CNC laser equipment is used in a similar way to the CNC flame cutting machine. Instead of a gas flame, the machine tool uses a laser to generate the heat to melt the workpiece to be machined. Also, since the size of the laser beam itself is much smaller and much more controllable than the gas flame, the accuracy that can be expected is much better than for the flame cutting machine.

Currently, the biggest limitation to CNC laser equipment is the thickness of the part that can be machined. The power available in current laser equipment is capable of cutting up to about 0.250-in-thick steel. As time goes on, the laser manufacturers are expected to improve on this number.

This workpiece thickness constraint limits the possible applications for the CNC laser machine.

Axis directions

CNC laser equipment has two directions of motion. The workpiece can be moved left and right, as well as in and out. If viewed from the front of the machine, the left and right motion is the X axis. The in and out motion is the Y axis.

# Chapter 3

Key Concept No. 2: Preparation for Programming

Any complex project can be simplified by breaking it down into small pieces. This can make seemingly insurmountable tasks much easier to handle. CNC programming is no exception. CNC can be quite intimidating to the beginning programmer viewing a complexly contoured workpiece requiring numerous machining operations. Learning how to break up a complex programming task will be the main thrust of key concept no. 2.

Writing a CNC program is only part of what the CNC programmer must do. Here we intend to explore those things that help the programmer get ready to write a CNC program. We will show you many techniques that will help you prepare to write programs.

Though this key concept has nothing whatsoever to do with actual programming commands, it is among the most important. The preparation that goes into getting ready to write a CNC program is directly related to the success of the program. You know the old saying, "Garbage in, garbage out!" It truly applies to CNC programming. The better prepared the programmer, the easier the programming task will be.

Preparation for programming is especially important to beginning programmers. The beginning programmer will have problems with formatting a CNC program. For the first few programs, the beginner will have trouble enough remembering the various command words, structuring the program correctly, and in general, just getting familiar with the programming process. This problem is made infinitely more complicated if the beginner is not truly ready to write the program in the first place.

Without the proper preparation, writing and verifying a CNC program can be equated with doing a jigsaw puzzle. The person doing the puzzle has no idea as to where each piece will eventually fit. Every piece must be assembled by trial and error. The worker makes a guess as to whether two pieces will fit together and then tries it. Maybe the pieces will fit and maybe they will not. Since the worker has no idea as to whether pieces will truly fit together until assembly is attempted, it is also next to impossible to tell how long it will take to finish the puzzle.

If a CNC program is done without preparation, the programmer will have a tendency to put the program together piecemeal, like a jigsaw puzzle. The programmer will not be sure that anything will work until it is tried. The program may be half finished before it is obvious that something is wrong. Or worse, the program may be finished and running on the CNC machine before some basic error is found. Always remember that CNC machine time is much more expensive than the programmer's time. There is no excuse to waste this precious machine time for something as basic as a lack of preparation.

You can liken the preparation that goes into writing a CNC program to the preparation needed for a speech. The better prepared the speaker, the easier it is to make the presentation, and the more effective the speech will be. Truly, the speaker must think through the entire presentation (probably several times) *before* the speech can be presented. Similarly, the CNC programmer must think through the entire CNC program before the program can be written.

The better prepared the beginner can be, the easier it will be to write the program. Most experienced programmers would agree that writing the program is actually the easy part of the programming process. The real work is done in the preparation stages. If preparation is done properly, writing the program will be a simple matter of translating from what you want the machine to do (from English) to the language the CNC machine can understand. Of course, this translation into a language that the control can understand is called *manual programming*.

Though this preparation is so very important, you would be surprised at how many expert programmers muddle through the writing of a program with no previous preparation. While an experienced person may be able to write programs for simple applications without preparation, even the so-called expert programmer will be lost on complex programs if preparation is not done. To think that you are saving time by not preparing to write a program can be a grave mistake. In reality, you usually add time to the programming process if you do not prepare. The short period of time saved will be quickly lost when you consider the problems created from lack of preparation.

Without preparation, the programmer will be constantly backtracking to repair problems during programming and rewriting programming commands. When the program is finally finished, the program must be verified at the machine. If the programmer was ill-prepared to write the program, chances are there will be many problems yet to present themselves at the machine. We hope you agree that when an experienced programmer would have problems programming without preparation, the beginner is truly doomed without it.

Preparation and Safety

When a CNC program is executed, you can rest assured that the CNC machine tool will follow your program's instructions to the letter. While the control may go into an alarm state if it cannot recognize one command or another, it will give absolutely no special consideration to mistakes made related to motion commands. The level of problems encountered because of motion mistakes ranges from minor to catastrophic.

Minor motion mistakes usually do not result in any damage to the machine or tooling, and the operator is not exposed to a dangerous situation. However, the workpiece may not come out correctly.

For example, say the programmer intended to drill a hole at 5 IPM with a feed rate word of

F5.

in the drilling command. But, in the actual program, the decimal point was placed in the wrong place. Say the actual program command included an

F.5

instead of

F5.

In this case, the control would be told to run the drill at a much slower feed rate (0.5 IPM) than intended. Probably no damage to the tool or machine would result, but at the very least, the machining cycle time would be much longer than necessary.

Catastrophic mistakes, on the other hand, can result in damage to the machine and possible injury to the operator. For example, if the programmer intended to position a tool, at the machine's rapid rate, to a position of 0.1 in above the part, and if program zero was set at the top surface of the part, the correct Z axis positioning command would be

Z.1

However, if the programmer made a mistake and included a minus sign for this axis departure (Z-.1), the tool would be told to position *into* the part (at rapid) by 0.1 in. Depending on what kind of tooling is being used, this would at the very least cause the tool to break. Or the part could be pushed out of its setup. Possibly, if the setup is very sturdy and the tool is very rigid, damage to the machine's way system and/or axis drive system could result. If the tool breaks and parts come flying out of the work area, the operator could also be injured.

All of this is not being stated to scare you. There are several verification procedures that, if followed, almost guarantee that no crash can occur. (These verification techniques will be described in Part 3.) However, as an operator or programmer of a CNC machine tool of any kind, you must recognize the potential for dangerous situations when working with this kind of equipment. Because there is this possibility for injury to the operator and/or damage to the machine tool, *everyone* involved with CNC equipment *must* treat the machine tool with respect. That is, you must be constantly aware of this possible danger and do everything you can to avoid dangerous situations. Being properly prepared to write the program is the single most important thing you can do to achieve this end.

In a joking way, we say there are two kinds of CNC operators, those that have had a crash and those that are going to! It seems inevitable that someday every programmer or operator will have a mishap of one sort or another, just as it seems inevitable that anyone who drives an automobile will eventually be involved in an accident.

All joking aside, *there is no excuse for a crash on any CNC machine.* In Part 3, we will discuss several verification procedures that will help the operator find mistakes in the CNC program. If these procedures are followed, we can assure you that you will not have problems. It has been our experience that beginners are quite cautious with the CNC equipment until they start becoming overly confident. As long as you do not start trying to shortcut the recommended verification procedures, you should have no problems.

So why all this talk about safety during a key concept on program preparation? Just as the well-prepared speaker is less apt to make mistakes during the presentation, the well-prepared CNC programmer will be less apt to make mistakes while writing the CNC program. We cannot emphasize enough the importance of this preparation. Now that you know why you must prepare to write CNC programs, here are the actual steps in this important preparation.

How to Prepare for CNC Programming

As you now know, preparation will make it easier and safer to write CNC programs. Now let's look at the specific preparation techniques themselves. While there are many different types of CNC equipment being discussed in this book, the needed preparation will remain very much the same from machine to machine.

Prepare the machining process

Process sheets, also called *routing sheets,* are used by most manufacturers to specify the sequence of machining operations to be performed on a workpiece during the manufacturing process. The person who actually prepares the process sheet must have a good understanding of machining practice, and must be well-acquainted with the various machine tools owned by the company. This person determines, to the best of his or her ability, the best way to produce the workpiece to the required specifications in the most efficient and inexpensive way.

In most manufacturing companies, this involves routing the workpiece through a series of different machine tools. Each machine tool along the way will perform only those operations the process planner intended, as specified on the process sheet.

When a CNC machine is involved in the process plan, many times the CNC machine will be required to perform several machining operations on the part. Most forms of CNC machines have automatic tool-changing systems that allow many different tools to be used in any one machining cycle. In some companies, the process sheet will clearly specify the order of machining operations to be performed on the CNC machine. Other companies leave it to the programmer's discretion as to how the part should be machined. If the programmer is left to determine the order in which the CNC machine performs its operations, of course the programmer must have a good knowledge of basic machining practice as it relates to that CNC machine.

In any event, the process required to machine a workpiece on the CNC machine must be developed *before* the CNC program can be developed. That is, the programmer must know the order of machining operations required of the CNC machine to be programmed. With a simple process, the experienced programmer may elect to develop the process as the program is being written. While some experienced programmers have the ability to do this, most beginning programmers will find it mandatory to prepare the process first.

For most forms of CNC equipment, the process to machine the part will have a dramatic impact on the success of the program. If the process is correct, the workpiece will come out correctly. If the process is poor, the workpiece will not come out correctly. This can be frustrating for the programmer.

While executing the program, the machine may be doing exactly what the programmer intended it to do. But, if what the programmer intended the machine to do was incorrect, the part will not come out correctly.

This is why basic machining practice is so important to the success of the program. The beginner should always seek help whenever there is a question as to whether the intended process will work correctly.

Here's an example of how the process can affect the success of a program: A programmer is developing a process for a workpiece to be run on a machining center. The programmer elects to machine the part in this order:

1. Rough face-mill top of part

2. Finish face-mill top of part

3. Drill hole to 1.485 diameter

4. Finish bore hole to 1.5 diameter

While you may see nothing wrong with this process, one of the rules of basic machining practice states that you should "rough everything before you finish anything." In the above process, the top of the part is finished before the hole is drilled to 1.485. It is possible that the part may shift in the setup during the drilling operation because of the extreme pressure exerted during this powerful operation. If the part shifts, this will cause inaccuracy in the subsequent operations in the program. For this reason, it would be wiser to follow this machining plan:

1. Rough face-mill top of part

2. Drill hole to 1.485 diameter

3. Finish bore hole to 1.5 diameter

4. Finish face-mill top of part

Notice the difference. In the second process, the part is completely rough-machined before any finishing starts. Most experienced machinists will say this is better machining practice.

Whether you agree with our machining process in this example is of no importance. What is important is that you understand the importance of developing a workable process before the program is written. In the above example, a programmer who had decided to develop the process while writing the program would probably have been at least three tools into the program before discovering the mistake, if it was discovered at all. In this case, the entire program would have had to be rewritten. On the other hand, if the process had been developed first, it would be much more likely that the problem would be spotted before the programming process even begins.

There are times when the order of machining has no effect on the success of the job. In some cases, the sequence of operations can be done in any order, and the workpiece will still be produced to the required specifications. For example, say several different small holes of various diameters must be pierced in thin plate on a turret punch press. It is likely the order by which the holes are pierced will have no bearing on the correctness of the program. However, we still recommend that the programmer prepare a process plan. In this case, the process plan will only serve to help the programmer keep track of the holes to be machined, assure that all holes are pierced, and provide documentation that could be used by others in the future.

Developing your machining process before the program is written will serve several purposes. First, it will allow the programmer to check the process for errors in basic machining practice. The programmer will be forced to think through the entire process before the first CNC command is written. As you have seen, many times while developing a process "up front," the programmer will spot a problem with the process that would have been difficult to repair if the process had been developed while writing the program.

Developing the machining process prior to writing the program also allows you concentrate on your machining practice skills and your programming skills separately. While developing the machining process, you concentrate on machining practice in order to develop a workable process. Your mind is occupied only with the task at hand. While programming, you simply translate this developed process into a language that the machine can understand.

Figure 3.1 shows an example of a planning form that can help you organize your process. As you read on, you will see how this form can really help you document the machining process as it relates to the CNC operation.

Yet another reason to do the process plan first is for future documentation. Months, and sometimes years, after a CNC program is developed, there may be a need to revise it. If the person who does the revision can view the process planning form developed for the program, it will be much easier to make the necessary changes.

The last reason we will give to plan the process first is simply that planning will help you remember the operations to perform during programming. Beginning programmers will have enough problems remembering the various commands needed in the program. The process planning form can be used as a series of step-by-step instructions by which to machine the part, a form of checklist. Without this form, the beginning programmer will be prone to leaving out machining operations in the CNC program. You would look pretty silly if you forgot to drill a hole before you tapped it!

Develop the needed cutting conditions

While developing the machining process, it is wise to consider the cutting conditions you will use with each tool in the program. For metal

Part Number:	Date:		Sequence of Operations Planning Form					
Part Name:	Page:							
Machine:	Prgmr:							
Seq.	Operation		Tool	Station	Feed	Speed	Min. Length	Block No.

Figure 3.1   Example of planning form for developing machining process.

cutting operations (milling, drilling, turning, grooving, etc.), these cutting conditions involve factors like speed, feed, depth of cut, coolant, tool material, and workpiece material. For wire EDM machining, these cutting conditions involve the various factors that contribute to a good machining condition (on time, off time, current, flushing, etc.). For CNC turret punch presses, these conditions involve the workpiece material and thickness which determine the die clearance necessary for the various punch and die combinations to be used.

It is convenient to come up with the various cutting conditions needed in the program while developing the machining process. This will keep you from breaking your train of thought while programming. If you use a planning form similar to the one in Fig. 3.1, you will be able to calculate the speeds and feeds needed while programming and write them right on the form.

Do the required math and mark up the print

The word *numerical* in computer numerical control implies a strong emphasis on numbers. Most college curriculums related to CNC require a strong math background. However, most forms of CNC equipment require less math than you might expect. Believe it or not, most common CNC programs can be completely prepared by applying simple addition, subtraction, multiplication, and division techniques. A basic knowledge of right angle trigonometry is also helpful, but not even mandatory.

While there are times when the manual programmer must apply trigonometry, there are many reference books that give the formulas related to trigonometry in very simple format. This makes it easy for a person who knows little about trigonometry to solve trigonometry problems. While the serious programmer will eventually want to learn trigonometry, it is not mandatory at the start.

The basic reasons for you to do the math related to a workpiece before attempting to write the program are similar to the reasons why you should first come up with a machining process. With the numbers going into your CNC program already calculated, you will not have to break out of your train of thought while programming. Whenever you need a coordinate value for the program, you will have it available. It will be a simple matter of copying the previously calculated number into your program.

Documenting your math also helps if mistakes are made. When you go back to check the mistake, if you have the math documented, you may be able to determine how the mistake was made. This will help you keep from duplicating the mistake in the future.

Marking up the print

The programmer assigned to prepare a program must be given a good copy of the print. The programmer should be allowed to do whatever is desired with the print to help with programming. This is because the programmer's copy of the print will be kept in the programming department as part of the documentation for the program.

Depending on the complexity of the workpiece to be produced, interpreting the print can range from quite simple to very difficult. Once you study the print and understand the machining operations to be performed, you can mark up the print in a way that will make programming easier. The first thing we suggest is that you take a highlighting pen of a bright color and mark those surfaces on the print that require machining by your program. For complicated workpieces, this helps to narrow down what needs to be done.

By marking up the print, we mean the programmer will make indications on the print related to critical information required during programming. The location of program zero, the placement of workholding devices (clamps, fixtures, etc.), as well as actual calculated values (coordinates) to be used in the program should be included on the marked-up print.

There are various techniques used to mark up a print in an understandable way. We will show two ways. Whatever feels the most comfortable to you as a beginning programmer is the method you should choose. Once the print has been marked up, it will provide the programmer with important information during programming and provide for future documentation on the program.

Doing the math

How dimensions are described on the part print will have a great deal to do with how much math is required. Progressive companies have their design engineering departments supply all dimensions in *datum surface dimensioning*. When this technique is used, each dimension on the print will be taken from only one position in each axis. This reduces the amount of math the programmer, not to mention everyone else involved in the manufacturing process, must do.

Unfortunately, not all companies use this technique. The programmer may be expected to do a great deal of math in order to locate the positions required in the CNC program.

When doing the required math for a program, the programmer will be calculating the coordinates to be used in the program. A coordinate is the distance from program zero to a location on the print that must be included in the program. Depending on the number of machine axes, any one position on the print can have two or three related coordinates (one for each involved axis). As stated, most coordinates can be calculated by using simple arithmetic while others require some basic trigonometry to be done.

How the programmer organizes the various values for coordinates depends on the print and what kind of machine is to be programmed. If the print is large and roomy, and if there are only a few coordinates to be calculated, often the programmer can simply write the coordinate values on the print close to the location required in the program. Figure 3.2 shows an example that includes all coordinates right on the print. As you can see, drawings tend to get a little crowded.

On the other hand, if the print is small and crowded, and if many coordinates are required, the programmer may elect to use a *coordinate sheet* on which to write the needed coordinates. On the print, the programmer will place small numbers near the location to be included in the program. This number represents the *point number* to be filled in on the coordinate sheet. The columns on the coordinate sheet will include the various involved axis addresses (X, Y, Z, etc.), and leave room for the programmer to fill in the actual calculated coordinates. Figure 3.3 shows an example.

Whether you elect to choose one of the two given methods or develop a method of your own is totally up to you. We give these two methods

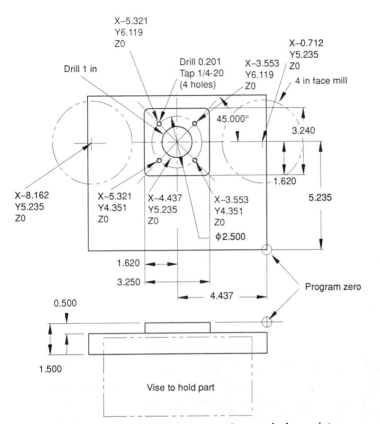

Figure 3.2 Example showing math developed on marked-up print.

only as suggestions. The main point is that you must, by one means or another, derive the needed coordinate values before you start writing a CNC program.

Check the Required Tooling

The next step for preparing to write a CNC program has to do with the tooling used by the program. There are several tooling-related problems that can cause even a perfectly coded program to fail. The programmer can avoid delays at the machine during a program's verification by doing a little checking during preparation for programming.

First, the programmer will want to be sure the various tools to be used by the program are available. Many times the programmer will assume that commonly used tools are in the company's inventory. While this may be the case, it is wise to double-check that all tooling is available.

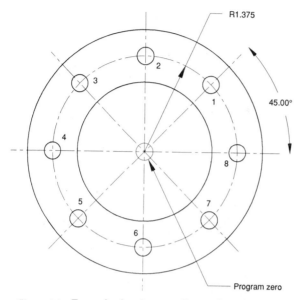

#	X	Y
1	0.9723	0.9723
2	0	1.375
3	-0.9723	0.9723
4	-1.375	0
5	-0.9723	-0.9723
6	0	-1.375
7	0.9723	-0.9723
8	1.375	0

Figure 3.3 Example showing coordinate sheet for developing math.

Second, there are many times when the tool-holding device can present interference problems of which the programmer must be aware. For example, a programmer intends to drill a 0.500-in-diameter hole into the part. However, the hole is very close to a wall that extends above the surface where the hole is to be drilled by about 5 in. In this case, the shank (tool-holding device) must clear the 5-in-tall wall. If it does not, the shank of the tool will collide with the wall before the tip of the tool even reaches the surface of the part to be drilled. Figure 3.4 shows an example of this kind of problem.

This is but one example of the kind of tool-interference checking of which we are speaking. The programmer must be constantly aware of possible tooling limitations and instruct the person assembling the tools accordingly. Notice that the setup sheet shown in Fig. 3.5 has room for *special comments.* When this form is given to the operator, it makes an excellent way of alerting the setup person that some special tooling problem exists.

Third, and especially relevant for machining centers, the programmer must assure that the tool is long enough to reach the surface to be machined. When a machining center is positioned to the extreme minus end of its Z-axis travel, there will usually be some distance from the nose of the spindle to the surface of the part to be machined. To even touch this surface, the length of the tool must be at least the distance from the nose of the spindle to the surface of the part to be ma-

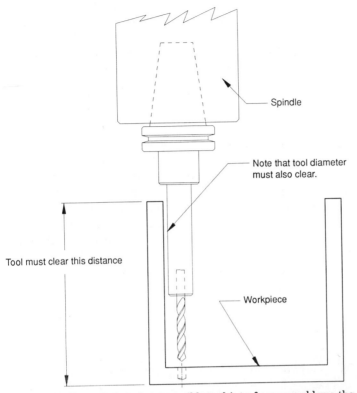

Spindle

Note that tool diameter
must also clear.

Tool must clear this distance

Workpiece

Figure 3.4   This drawing shows possible tool interference problems the
programmer must be aware of.

chined. Add to this distance, the depth to which the tool is intended to
machine. This is the tool's minimum length. Notice there is also a column on the process planning form for minimum length.

Plan the setup

Depending on the type of CNC machine and the company's policies,
the programmer may also have to develop the work-holding setup required to hold the workpiece during machining. If required, the programmer must also understand basic machining practice as it applies
to work holding setups for the machine to be programmed.

For simple work-holding setups, the programmer will usually make
a simple drawing or sketch indicating how the setup is to be made. For
example, if programming a turning center, the operator or setup person may have to be told by a sketch how to bore soft jaws to hold the
part. For a machining center, the sketch may show the location of
clamps, parallels, and stop blocks.

For extremely complicated fixtures, the programmer is usually not involved in the fixture design, though there are companies that consider this to also be part of the programmer's responsibility. In this case a tool designer will design the fixture and supervise its construction. Once the fixture drawing is made, the programmer will need it to see how the part will be held and to plan the program accordingly.

Other documentation

In almost all cases, the programmer is responsible for supplying all documentation related to the program. This means the programmer must provide instructions for the operator on how to get the program up and running. One common technique to do this involves a *setup sheet* that includes a sketch of the setup, the location of program zero, a list of tools related to the job and their related station numbers, and, in general, any other instructions necessary for getting the job up and running. Figure 3.5 shows an example of a setup sheet. While each CNC user will strive to develop setup sheets in a way that best suits individual needs, this example shows many of the techniques commonly used.

Part Number:	Date:	Setup Sheet		
Part Name:	Page:			
Machine:	Prgmr:			
Station	Tool Description	Offset	Insert	Setup Instructions and Notes
				Sketch of Setup

Figure 3.5   Example of setup sheet.

Notice that the setup sheet allows tools and tool stations to be well-described. Also notice that there is room to make a setup sketch and give the operator written instructions. The less a programmer leaves to the operator's imagination, the less the chance of a mistake being made during the setup.

We recommend that the programmer make this set of instructions *before* the program is written. This will help the programmer become more familiar with the job and provide information that can be used when the program is written. With the setup documentation prepared, the programmer can easily reference clamp locations, program zero location, and the correct tool station numbers.

Program listing. Most companies supply a printed copy of the program with the documentation that goes out to the operator and setup person. This will help the operator make minor changes in the program if they are required.

Is It All Worth It?

You have seen that there is a great amount of work involved in preparing to write a CNC program. Once this work is done, the programmer will have a clear and concise idea as to what the program must do. There will be no questions left to ponder while programming, and the programmer can concentrate efforts on the task at hand. The work done in preparation for writing the program will pay dividends in the writing and verification of the program.

As we stated earlier, and as you may be starting to see, the actual writing of the program is really the easy part of the programming. The preparation is where all the real work lies. Once you have written a few programs and understand the language of CNC programming, it will be a relatively simple task to translate what you want the CNC machine to do into a language that the machine can understand.

4

Key Concept No. 3: Types of Motion Commands

In the introductory chapters, you were acquainted with the components of an axis of motion. We discussed how the drive motor, ball screw, and way system work together to provide a precise linear or rotary axis of motion.

Every CNC machine discussed in this book possesses more than one axis of motion. More often than not, it will be necessary for the programmer to command that two or more axes move simultaneously in a controlled manner. For example, an end milling cutter is to be used on a machining center to mill a contour around the outside of a shape. This contour involves machining straight surfaces, angular surfaces, and round surfaces. While some of the movements in this example may involve only one axis, the angular and circular motions must involve at least two axes.

In the early days of NC, this presented real problems. Whenever the program was required to produce an angular or circular surface, the motion had to be broken down into a long series of very small one-axis motions to form the angle or circle as closely as possible to the desired shape. This kind of motion normally required the help of a computer to produce. With the advent of a feature called *motion interpolation,* programming common complex movements became much simpler. With today's current CNC controls, it is relatively easy to command angular and circular motion.

What Is Interpolation?

To truly understand how combined axis motion occurs, you must understand the word interpolation. *Webster's New World Dictionary* (third edition) has several definitions for the word *interpolate*. The one that most applies to CNC is the math-related definition: "To estimate a missing functional value by taking a weighted average of known functional values at neighboring points." When the control interpolates a motion, it is estimating very precisely the programmed path based on a very small amount of input data.

For example, when the control makes a straight motion in two axes, called linear interpolation, all that is required is the start point and end point of the motion. The control automatically and instantaneously fills in the missing points between the start point and end point. What really happens is that the control makes a series of very small one-axis movements from the start point to the end point. This series of motions resembles a stairway. Each step along the way is very small, and the end result will appear to be a perfectly straight line. Figure 4.1 shows a picture of what actually happens during linear interpolation.

As you can see from Fig. 4.1, when two or more axes are programmed, the control forms a series of small one-axis movements. The size of each step determines the *resolution* of the axis. The smaller the step, the better the resolution. The drawing in Figure 4.1 intentionally exaggerates the situation. While there was a time when CNC axis resolution was quite poor, today's current CNC machines will make straight-line motions almost perfectly.

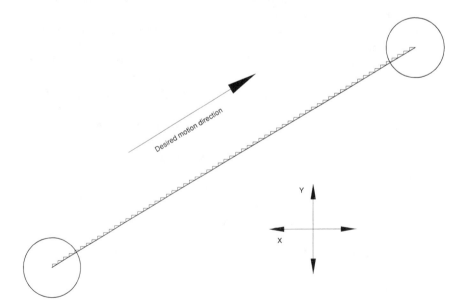

Figure 4.1 Example showing what happens during linear interpolation.

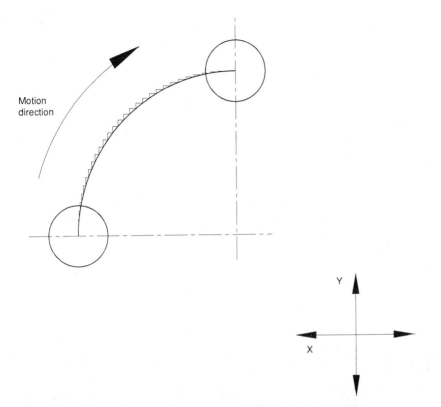

Figure 4.2 Example showing what happens during circular interpolation.

Circular interpolation is performed in much the same way. Note the similarity in Fig. 4.2.

Knowing how a CNC control interpolates motion is not of primary concern to the programmer, though it is nice to know what is going on. What is more important to the programmer is understanding the various interpolation types for the particular CNC machine being programmed and having an understanding of how to make the required motion commands.

CNC control manufacturers have designed various types of interpolation commands around the most common motions the machine tool will be expected to make. The two most commonly used interpolation types are *linear interpolation* (straight-line motion) and *circular interpolation* (circular motion). Because they are so commonly used, these two motion types are equipped as standard features on almost every kind of CNC machine in existence today.

Other kinds of interpolation have much more specific applications and are only provided as options when required. For example, *helical interpolation* is used on machining centers to allow two axes to be commanded in a circular motion while a third axis is moving in a linear motion. The main application for helical motion is thread milling. If the user has no application requiring thread milling, they will have no real need for helical interpolation.

Another kind of interpolation is *parabolic interpolation*. The motion generated in this interpolation is in the form of a parabola. Yet another form of interpolation, *hypothetical interpolation,* allows users to define their own special form of motion, determined by the kind of motion they require. These two types of interpolation are seldom used or required.

Again, the two most commonly used forms of interpolation are linear interpolation and circular interpolation. There is also another type of motion called *rapid motion.* Some control builders refer to this motion type as *positioning.* Rapid motion allows the axis motion to take place very quickly, and is generally used to minimize noncutting time in the program. However, on most CNC machines, the rapid motion will not occur along a straight line, unless only one axis is commanded, so we do not consider it a true interpolation type.

It should be refreshing for beginners to know that there are only three commonly used methods to cause axis motion. Every motion the typical CNC machine makes can be divided into one of these three categories. Once you master these motion commands, you will be able to generate the required motions to machine a workpiece. The usage of these commands remains remarkably similar for the various forms of CNC equipment. While there are some minor variations, mostly related to circular commands, once you truly understand the information presented here, you will be able to apply what you know to any form of CNC equipment.

All motion types share five things in common. First, they are all *modal,* meaning they remain in effect until changed or canceled. If more than one command of the same type is to be made, the programmer need only include the motion type in the first motion.

Second, each of these commands requires that the programmer include the end point of the motion. The control will assume the axes are at the beginning point of the motion prior to the given motion command. This allows the programmer to think of motion commands as a series of connect-the-dots or point-to-point motions.

Third, all motion commands will be affected by whether or not the programmer decides to work from program zero. In the absolute mode (G90 on most controls), the motion commanded will be relative to the program zero point. In the incremental mode (G91 on most controls), the motion will be taken from the current position.

Fourth, each of these motion commands requires movement only of the axes stated in the command. If commanding a motion in only one axis, you need include only the moving axis letter address (X, Y, Z, etc.) and departure value in the motion command. All other axes' letter addresses can be left out. The control will not attempt to move an axis unless an axis departure is included in the motion command.

Fifth, current CNC controls allow leading zeros to be left out of all types of commands. This means the actual G codes used to instate the motion types can be programmed in one of two ways. G00 and G0 mean essentially the same thing to the control, as do G02 and G2, and G03 and G3. Our text will always include the leading zero to maintain compatibility with older controls.

At this point, we will specifically discuss each motion type in detail. During each type, we will address the specific requirements of each motion type as they relate to the various kinds of CNC equipment. When finished, we will give a lengthy example.

G00 Rapid Motion (Positioning)

This type of command is used to position the CNC machine to a location where some action is to occur. Under normal operation, the G00 command will cause the machine to move at its fastest possible rate. The rapid rate will vary from machine to machine. Several current CNC machines boast rapid rates—well over 500 IPM. Because the machine will be moving very quickly while under the influence of a rapid motion command, most control manufacturers allow the operator to override the machine's rapid rate during a program's verification.

The specific programming format of the rapid motion command remains consistent from one CNC machine to the next.

Figure 4.3 shows a sketch of how rapid motion will occur. In this example, we intend the tool to move rapidly from the starting position to the ending position. In the absolute mode, this would be commanded by the following command:

N005 G00 X5. Y5.

Note that the above command will be correct only in the absolute mode, commanded by G90 on most controls. In the incremental mode (G91), the axis departures (X, Y, Z, etc.) would be taken from the tool's current position.

Depending on what the programmer intends to do in the command, any number of axis departures can be included in the G00 command. The programmer can make the machine move in one axis only, or a motion can be made in two or more axes simultaneously. If the programmer wishes the machine to move in only one axis, only the letter

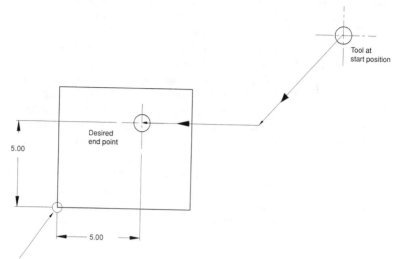

Figure 4.3 Example of rapid motion command.

address (X, Y, Z, etc.) and the value of the departure need to be included in the G00 command. Axes of letter addresses not included in the G00 command will not move during the execution of a G00 command.

The format for this command is virtually the same from one control to the next. The above example command would be interpreted and executed by almost all CNC controls in the same manner. The only thing that may vary from one machine to the next is the actual method by which the motion will be made.

Most CNC controls will turn loose all axes included in the G00 command at the machine's rapid rate. Each axis will be allowed to go as fast as the machine allows. Since, on most machines, all axes move at the same rapid rate, the axis with the shortest departure distance will arrive at its destination first and stop. The machine will continue on in any other included axes until the next shortest departure reaches its destination, and so on. Eventually, all axes will reach their destination and the G00 command is completed.

With this kind of rapid command (as shown in Fig. 4.3), if more than one axis is commanded in the G00 command, the motion will appear to "dog-leg" into position. In this case, the programmer must be concerned with any obstructions between the starting point and ending point of the G00 command. There will be times when the programmer must break up a rapid motion command that includes more than one axis departure into two or more commands to avoid obstructions.

A second and somewhat less popular form of rapid command will make the rapid motion along a straight line. While this makes life a little easier for the programmer because it is easier to plot the true motion, most control builders do not go to the trouble of making the rapid commands behave this way.

A third, seldom-used, form of rapid command will move in only one axis at a time. Even if more than one axis departure is included in the rapid command, the machine will move only in one axis at a time. In this case, some previously defined order of motion will occur. For example, first X will move, then Y, then Z, and so on. The only form of CNC machine we have seen with this type of rapid command is the wire EDM machine. However, only a few wire EDM machine builders use this technique for rapid movement.

While the method by which the machine will move during rapid commands is not of critical importance, the programmer should know how rapid motion will occur on the machine being programmed to avoid problems with obstructions. For beginners, it should be easy enough to ask someone currently working with the machine to give you this basic information.

G01 Linear Interpolation (Straight-Line Motion)

The second type of motion we will discuss causes the machine to move along a perfectly straight path in one or more axes. The control will calculate the path between the start point and the end point of the motion automatically, no matter what angle is involved.

Also of note regarding the linear interpolation command is the motion rate at which the axes will move. This motion rate is programmed differently, depending on the style and application of the machine. Metal cutting machines, like machining centers and turning centers, use an F word to specify the desired feed rate for the command. Even with an F word, it may be possible to specify the feed rate in one of two ways. One is feed per minute and the other is feed per revolution. In the inch mode, this equates to inches per minute and inches per revolution. In the metric mode, this equates to millimeters per minute and millimeters per revolution.

While there are exceptions, most machining centers allow feed rate to be programmed only in feed per minute (IPM or MPM). Since the cutting speed for a tool in RPM remains constant during machining with any one tool, only one feed-per-minute calculation need be made per operation.

On the other hand, turning centers are constantly changing spindle speed in RPM during machining. To program this type of CNC equipment exclusively in feed per minute would be a real headache for the programmer. For this reason, turning center controls also allow feed rate to be programmed in feed-per-revolution mode (IPR or MPR). For machines that allow this, G98 usually specifies the feed-per-minute mode and G99 usually specifies the feed-per-revolution mode.

With controls that allow an F word to control feed rate, the feed rate is modal. This means that if a series of cutting motions are to occur at a desired feed rate, the F word need only be included in the first cutting motion command.

Certain types of CNC machines have different methods to control the motion rate. Most CNC wire EDM machines will not actually have a special word to specify the motion rate, but a series of machining parameters that determine how fast the motion will occur. With wire EDM, it is currently impossible to specify a constant motion rate due to the nature of the EDM process. The adaptive control system of the EDM machine is constantly adjusting the movement rate to allow for the current environment of the cutting condition.

The G01 command is used primarily for machining that occurs along a straight line. Examples of when a G01 command can be used include the drilling of a hole on a machining center and the turning of a straight or tapered diameter on a turning center.

Figure 4.4 illustrates what happens during a linear interpolation movement. Remember that the control will assume the tool to be at the starting point of the motion when the G01 command is given. Here is the command that would make the desired motion in the absolute mode:

    N005 G01 X5. Y5. F10.

As with G00, if the machine is currently in the absolute mode (G90), the above command tells the control to position the machine to an end point of 5 in in X and 5 in in Y from program zero. If in the incremental mode (G91), the control will position the machine 5 in plus in X and five inches plus in Y from the beginning point of the motion.

This motion will occur along a perfectly straight line beginning at the starting point of the motion. Note again that the control is also being told to maintain a constant feed rate of 10 inches per minute or millimeters per minute, depending on whether inch or metric mode has been selected.

G02 and G03 Circular Interpolation

There are many times when surfaces to be machined include radii. Circular interpolation allows the programmer to specify circular mo-

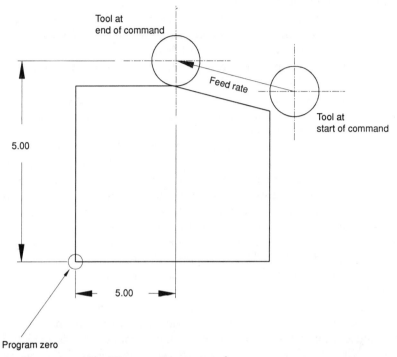

Program zero

Figure 4.4 Example of linear motion command.

tion as the cutting path. When circular interpolation is commanded, two axes will *always* be moving together.

Current CNC controls that are intended for contouring operations will always have circular interpolation as a standard feature (turning centers, machining centers, wire EDM machines, etc.). However, there are some kinds of CNC machines that are not intended for contouring operations. For example, turret punch presses will have no use for this feature. Also some forms of CNC drilling and tapping centers will have no use for circular interpolation.

By contouring, we mean making motions around the periphery of a shape. Contouring can include straight surfaces, angles, and circular motions.

Circular interpolation involves two G words. The reason why there are two G words is to allow the programmer to specify the direction of the circular motion to be machined, clockwise motion or counterclockwise motion. To determine whether a particular motion is clockwise or counterclockwise, the programmer must look at the motion from the plus side of the uninvolved axis. If making a circular movement in *X*-

Y, look at the motion from the *Z* plus side. If making a circular move in *X-Z,* look at the motion from the *Y* plus side, and so on. While this may sound a little complicated, for most applications this is as simple as looking at the plan view of the print from above.

For most CNC controls, G02 represents clockwise motion and G03 represents counterclockwise motion. We must point out that manufacturers of certain controls, especially some turning center controls, reverse the G02 and G03. You must check the programming manual for the machine tool with which you will be working to determine which direction is which. Because almost all CNC control manufacturers use G02 for clockwise circular motion and G03 for counterclockwise circular motion, all examples given in this text will reflect this method.

Like the G01 linear interpolation, the rate of motion can be controlled during a circular movement. All information presented about feed rate control for G01 still applies to G02 and G03.

Also like G01, the G02 and G03 commands require that the end point of the circular motion be included in the circular command. The control assumes that the current position of the tool is the starting point of the circular motion.

However, circular commands do require a bit more information related to the size of the circular arc to be generated. The method of specifying this information varies with the control builder, as does the level of difficulty related to programming circular motion. To specify circular motion on older controls was quite difficult. Current controls make it quite a bit easier. We will begin with the most common and simple way for specifying circular motion on current controls and work our way back to the older, more difficult methods.

Specifying a circular movement with a radius

Many current CNC controls allow the programmer to command circular motion by simply specifying the radius of the arc along with the end point. The control will automatically figure out how to make the circular motion from this very small amount of input data.

Figure 4.5 shows an example of the desired circular command. Here is an example of a series of movements, including a circular command, assuming the absolute mode of programming is used:

```
N055 G01 X7. Y3. F5.
N060 G03 X5. Y5. R2.
N065 G01 X0
```

In this simple example, block number N055 commands a motion to 7

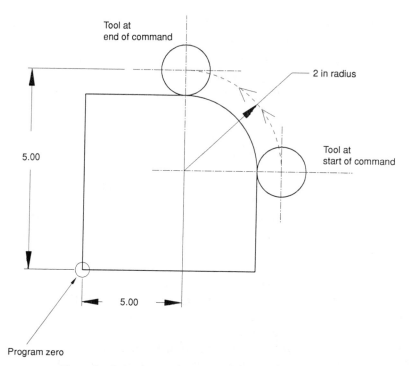

Tool at
end of command

2 in radius

Tool at
start of command

5.00

5.00

Program zero

Figure 4.5 Example of circular motion command.

in in X and 3 in in Y along a straight-line path. This will be the starting point for the subsequent circular command. In block number N060, the counterclockwise circular motion commanded will end at 5 in in X and 5 in in Y. The R word is telling the control that the desired radius is 2 in. In block N065, the movement will continue on in a straight line to an X position of zero.

You can see by this simple example that the R word allows the programmer to easily define the radius of the circular command.

This looks simple, we hope. Remember that the numbers used for circular motions must work out correctly. It is wise to double-check every circular command you make until you feel quite comfortable with circular commands.

Using the R word to specify the radius of the arc is our recommended method of programming circular commands *if* the CNC control allows circular motion to be commanded in this manner. Note that most CNC controls that allow the radius of a circular command to be designated by an R word also allow an older technique using I, J, and K. If R is allowed, it is *much* easier to use than I, J, and K.

Specifying circular motion with I, J, and K

Unfortunately, not all CNC controls allow the programmer to specify circular motion with a simple radius command. Some, especially older CNC controls, require that the programmer do a little more work. They require the programmer to tell the control the location of the center point of the arc. The method by which this is done also varies from control builder to builder.

When the R word is not allowed, the programmer will use the words I, J, and K to help designate the center point of the arc. There are two possible ways in which I, J, and K can be used. Note that no single CNC control will allow both methods of using I, J, and K. You will be using one method or the other. It will be up to you to check in the programming manual that came with the machine tool to determine which use of I, J, and K is applicable to your particular machine.

I, J, and K as directional vectors. The most common method of using I, J, and K in circular commands involves pointing from the starting point of the arc to the center of the arc. To do this involves knowing the distance and direction from the start point to the center of the arc along the two axes involved with the circular movement.

I is related to the X axis, J is related to the Y axis, and K is related to the Z axis. Here is a more formal definition of I, J, and K:

I is the distance and direction from the starting point of the arc to the center of the arc along the X axis.

J is the distance and direction from the starting point of the arc to the center of the arc along the Y axis.

K is the distance and direction from the starting point of the arc to the center of the arc along the Z axis.

The word *direction* implies that the sign of the value can be plus or minus. That is, I, J, and K can have minus values. To determine whether they should be plus or minus, simply draw an arrow on the print from the start point of the arc to the center along each axis. Note that there can be more than one arrow. Now ask yourself if the arrows are pointing in the plus or minus direction. This determines whether I, J, and/or K will be plus or minus.

If any of the arrows you drew is along the X axis, you will use an I in the circular command; if along the Y axis, a J will be used in the circular command; if along the Z axis, a K will be used.

The distance is the actual calculated value from the start point of

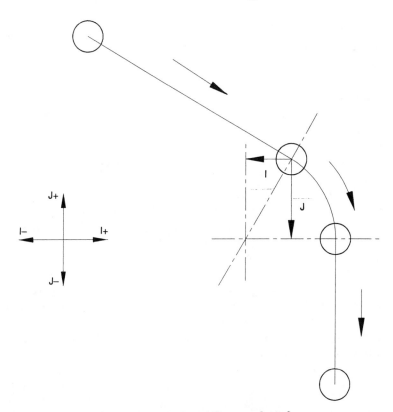

Figure 4.6 Example showing how I, J, and K are evaluated.

the arc to the center of the arc. If the distance from the start point to the center of the arc is zero, it can be left out of the circular command, or included with a zero value.

Figure 4.6 shows an example of how I, J, and K are evaluated. This particular example shows an *X-Y* circular motion coming from an angular surface. Note that both I and J will be required in the circular command and both will be negative.

Here are a series of commands using I and J as the distance and direction from the start point to the center of the arc for the example shown in Fig. 4.5:

```
N055 G01 X7. Y3. F5.
N060 G03 X5. Y5. I-2.
N065 G01 X0.
```

In this rather simple example, block number N055 feeds the tool along a straight line to 7 in in *X* and 3 in in Y at 5 IPM. Block number N060 makes the counterclockwise circular motion to an end point of 5 in in X and 5 in in Y. The word

I-2.

is saying that there is a distance of 2 in along the X axis from the start point of the arc to the center of the arc. The minus sign is saying the direction from the start point to the center of the arc is pointing in the negative direction along X.

In the above example, the only axis involved in the pointing was the X axis. For this reason, only the I word was required. A J0 could have been included, but it would have had no effect. Most programmers like to get into the habit of leaving out information that is not required.

In some cases, the movement will involve angular linear movements into arcs. In this case, there will be times when both an I and J are required to correctly describe the subsequent circular motion.

Figure 4.7 shows an example of a more complex series of motions. Notice the cutter is moving around this part in a counterclockwise general direction. This will help you determine the start point for each movement. In each circular motion, study the I and J to be sure you understand. As you can see, this part requires some trigonometry to determine the values of I and J.

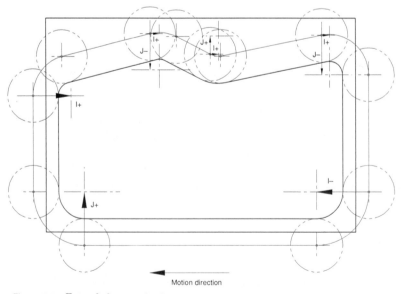

Figure 4.7   Extended example showing how I and J are evaluated.

Using I, J, and K as the absolute arc center position

Yet other forms of CNC controls handle the programming of circular motions somewhat differently. They still require the I, J, and K words in circular commands, but instead of being the distance and direction from the start point of the arc to the arc center, I, J, and K are input as the absolute center position of the arc in X, Y, and Z respectively. While this is much simpler than the other way to use I and J, it is still not as easy as using the R word. For Fig. 4.5, here are example commands showing how I and J would be programmed as the absolute center position in X and Y of the arc:

```
N055 G01 X7. Y3. F5.
N060 G03 X5. Y5. I5. J3.
N065 G01 X0.
```

In this case, the I and J in block number N060 tell the control the absolute position of the arc center in X and Y respectively.

Limitations related to crossing quadrant lines

Each CNC control type will have its limitations as to how much of a complete circle can be commanded in one circular motion. While some current controls have no limitation in this regard, most have definite rules that cannot be broken. A quadrant line is the arc's centerline. Each arc has two centerlines that form an X-Y cross hair at the circle's center. Figure 4.8 shows an example including quadrant lines.

Older CNC controls would not allow a quadrant line to be crossed during a circular command. So, if the programmer needed to program the motion shown in Fig. 4.8, it would have been broken into two separate commands. In this case, the first circular motion command would be programmed to end at the first quadrant line the circular motion comes across. The second circular motion command would continue on to the true ending point of the arc.

Note that the CNC control would execute these two commands in a flowing motion. That is, there would be no pause or stopping at the end of the first circular command.

Some newer CNC controls allow quadrant lines to be crossed. However, there may still be some limitation as to how many quadrant lines can be crossed. Most allow at least one crossing, meaning the motion shown in Fig. 4.8 could be commanded in one circular motion.

Figure 4.9 shows another example requiring two quadrant lines to be crossed. Most current CNC controls require that this motion be bro-

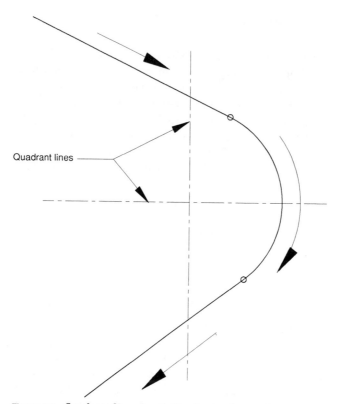

Quadrant lines

Figure 4.8   Quadrant lines in relation to circular motions.

ken into two commands. The first motion can be to either quadrant line. The second motion continues on to the end point of the arc.

Making a full circle in one command.   Even some controls with quadrant crossing limitations allow a full circle to be commanded. This technique is especially handy when a round inside shape is machined with an end mill on a machining center, as would be required to simulate a counterboring operation. In this case, usually the I and J technique must be used. Figure 4.10 shows an example.

In this case, notice the motion begins at the top of the circle in Y. In block N010, the G02 simply includes the

   J-1.

to inform the control as to the position of the arc center. Since the end point is the same point in X-Y as the start point, no X or Y was required in this command.

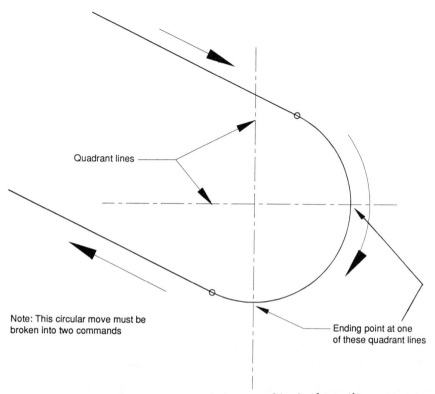

Quadrant lines

Note: This circular move must be
broken into two commands

Ending point at one
of these quadrant lines

Figure 4.9   Example of two quadrant lines being crossed in circular motion.

Helical Interpolation

This kind of motion command is used exclusively on machining centers. While some far-fetched applications are conceivable for this command, it is used primarily for thread milling operations. There are times when a threaded hole is too large to tap on a machining center. And there are also times when male threads must be machined on a machining center. These are two examples of when the user can benefit from thread milling.

There are two basic types of thread milling cutters. One resembles a slot milling cutter, but has the profile of the thread machined on the outside diameter of the cutter. This form of cutter must be programmed to machine around the thread several times to form the entire thread, meaning one movement around the thread for each pitch. For this reason, this kind of thread milling cutter is limited to low production quantities.

The other form of thread milling cutter has several thread profiles machined into the cutter. This allows the entire thread to be ma-

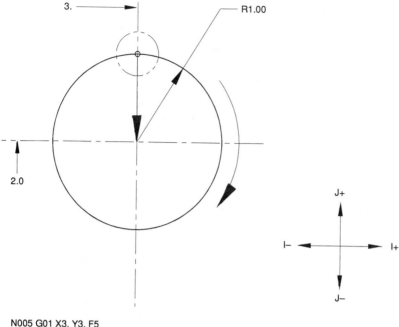

N005 G01 X3. Y3. F5
N010 G02 J – 1.

Figure 4.10   Example of making a full circle in one command.

chined in one pass. Though this type of thread milling cutter is more expensive, because of the improved capabilities of this cutter, this second type of thread milling cutter is more popular for medium to high production.

A helical motion involves three axes moving together. Two of the axes, usually X and Y, move in a circular manner. The third axis, usually Z, moves in a linear manner. Though the diameter of the motion remains constant, the best way to describe this kind of motion is as a spiraling motion.

While some control manufacturers use differing methods for commanding helical motion, most use the same commands as for circular motion, the G02 and G03 commands. With this format, the only difference is the addition of the third axis of motion, usually Z. To perform a thread milling operation correctly, the Z axis departure must reflect the pitch (lead) of the thread to be machined. Also, the method by which the thread milling cutter arcs in and arcs out of the motion around the thread must also take into consideration the pitch of the thread. A full example of thread milling is shown in key concept no. 6.

Figure 4.11 shows a helical motion. Unfortunately, you must study both views to see what is happening. As the motion takes place in X-Y

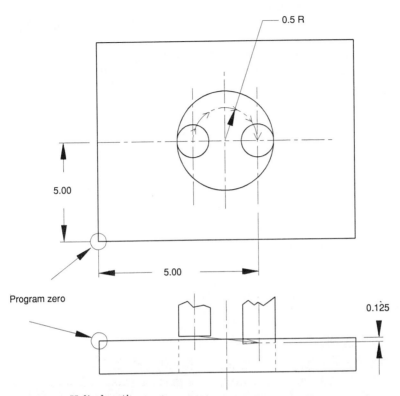

0.5 R

5.00

5.00

Program zero

0.125

Figure 4.11   Helical motion.

in the form of an arc, the Z axis moves in a linear motion, forming a helix. Assuming the current position prior to the helical motion command is

    X4. Y5. Z.1

here is an example command:

    N050 G02 X5. Y5. Z-.125 F5.

Example Program Showing Three Kinds of Motion

Before going any further, we want to illustrate how the three most common motion types—rapid motion, linear motion, and circular motion—are utilized in a program. What follows is an example program for a machining center that stresses all three types of motion. We know that the entire structure of the program cannot be familiar to

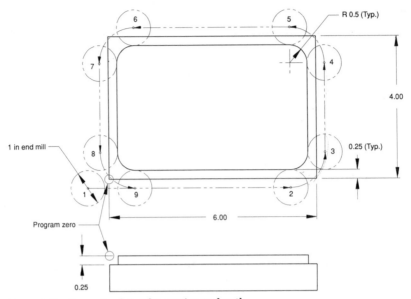

Figure 4.12   Example of circular motion tool path.

you at this time, so pay close attention to only the G00, G01, and G03 commands. The balance of the program is basic format and is included only to make the program complete and correct. Program formatting is discussed at length in key concept no. 5 of programming. While this example illustrates motion types in a machinging center application, if you understand this example, you will be able to relate what you know to other types of CNC controls.

Figure 4.12 is a drawing that shows the workpiece to be machined and the tool path. Notice that the movements (tool path) are the motion of the tool's center. We call this *programming the tool's centerline coordinates*. To plot the centerline coordinates requires that you take into consideration the tool radius as you write the program. This requires more work, especially on angular surfaces. For now, this method is sufficient. We are only stressing the movement types. Later, in key concept no. 4 of programming, we will introduce a feature called cutter radius compensation that will allow you to program the *part* coordinates, and forget about the radius of the cutter while programming.

Since this program requires calculations to be made, you should do all math prior to writing the program. In our case, Fig. 4.13 is the coordinate sheet that corresponds to the numbered points on the previously shown print. As an example, we calculated the X coordinate of point number three by subtracting the 0.25 dimension from the 6.00 in overall length and adding the 0.5 cutter radius (6.00 − 0.25

#	X	Y
1	−0.625	−0.25
2	5.25	−0.25
3	6.25	0.75
4	6.25	3.25
5	5.25	4.25
6	0.75	4.25
7	−0.25	3.25
8	−0.25	0.75
9	0.75	−0.25

Figure 4.13 Coordinate sheet for example program for circular motion.

+ 0.5 = 6.25). For the Y coordinate of point 3, we took the 0.25 dimension and added the 0.5 part radius (0.25 + 0.5 = 0.75). Study this coordinate sheet and make sure you understand how these coordinates were determined for the balance of the tool path.

Finally, we are ready to look at the program for the drawing shown in Fig. 4.12. Remember, this is a complete program, and at this point, we are only stressing the use of motion commands, so you can ignore the commands that do not have a G00, G01, G02, or G03 if you wish. For your reference, each line is documented with information in parentheses.

```
O0002 (program number)
N005 G92 X10. Y10. Z10. (assigns program zero)
N010 G90 S300 M03 (select ABS mode, turn spindle on at 300 RPM clockwise)
N015 G00 X-.625 Y-.25 (rapid to point 1)
N020 G43 H01 Z.1 (instate tool length compensation, rapid to 0.1 above part)
N025 G01 Z-.25 F30. M08 (fast feed to depth, turn on coolant)
N030 X5.25 F5. (straight line cut to point 2)
N035 G03 X6.25 Y.75 R1. (form radius to point 3)
N040 G01 Y3.25 (straight line cut to point 4)
N045 G03 X5.25 Y4.25 R1. (form radius to point 5)
N050 G01 X.75 (straight line cut to point 6)
N055 G03 X-.25 Y3.25 R1. (form radius to point 7)
N060 G01 Y.75 (straight line cut to point 8)
N065 G03 X.75 Y-.25 R1. (form radius to point 9)
N070 G00 Z.1 M09 (rapid up to clear part, turn off coolant)
N075 G91 G28 Z0 (go to zero return position in Z)
N080 G28 X0 Y0 (go to zero return position in X and Y)
N085 M30 (end of program)
```

Notes about the program follow:

1. Notice the modal commands. In block N020 the control will continue rapid motion, even though the G00 is not specified again. In all cutting blocks after N030, the control will continue to feed the tool at 5 IPM. The feed rate command is modal. The R word in all circular movements is *not* modal. It *must* be programmed in each circular move even though it stayed at

R1.

and didn't change.

2. Because we were programming the tool's centerline, the R word had to reflect the radius of the tool. If the cutter radius changed for any reason, this whole program would have to be changed.

3. Notice once again that *all* movement commands specified the *ending point* of the current motion.

4. Notice that, between every cutting command, we changed from G01 to G03 and vice versa. These commands are necessary. One common mistake that beginners often make is to forget to change motion types. For example, they may program a circular movement but forget to include the G01 in the subsequent straight-line cutting command. If this happens, the control will think that the subsequent command is supposed to be a circular command. An alarm will be generated because no R, I, J, and/or K words were included.

5. Whenever an axis is not moving in a command, it can be left out. For example, in block N025, only the Z axis is moving. So X and Y can be left out of the command. In the same way, in block N040 only *Y* is moving, so X and Z are left out. It doesn't hurt to include the uninvolved axes, but it opens the door for typing mistakes. Also, these redundant words take up memory space, meaning fewer programs can be stored in the control at any one time when many redundant words are programmed.

We hope this example program has driven home the concept of the three motion types. If not, please review this information. We will be building on it as the book continues.

# Chapter 5

Key Concept No. 4:
Types of Compensation

Just about every form of CNC machine tool uses some form of compensation. Machining centers have tool length compensation and cutter radius compensation. Turning centers have tool nose radius compensation and offset compensation. Wire EDM machines have wire radius compensation.

Since this feature is so common in the various forms of CNC equipment, it is mandatory that you understand it thoroughly. The descriptions we give in this chapter should get you well on your way to understanding this important feature.

What Is Compensation?

Before a marksman fires a rifle, the distance to the target must be judged. If the target is estimated to be at 50 yards, the sight of the rifle is adjusted accordingly. When adjusting the sight, the marksman is compensating for the distance to the target. But even after this preliminary compensation and before the first shot is fired, the marksman still cannot be sure whether the compensation has been made correctly. After the first shot, the marksman can fine-tune the sight of the rifle more accurately. The second shot should be considerably more accurate than the first. This process can be repeated until the sight is aligned perfectly. This rifle analogy is amazingly similar to the use of compensation with a CNC machine.

The uses for compensation

Compensation is used for several different purposes on different types of CNC machines. In almost all cases, compensation will let the programmer forget about tooling in one way or another. It is designed to make the use of CNC equipment much easier.

This brings up a good point. In almost every case, it is not mandatory that compensation be used. There are usually ways to circumvent compensation, but not using compensation is usually the old way of handling the programming problem and, in most cases, makes programming more difficult.

Understanding offsets

All forms of compensation for all kinds of CNC controls work with *offsets*. Offsets are used with compensation to tell the control the numerical value to be compensated. In the marksman analogy discussed earlier, the 50-yard distance needed compensating. The amount of adjustment made on the sight of the rifle can be related to the value being stored in a CNC offset.

Offsets to the CNC machine are like memories of an electronic calculator. Each offset is a storage location in which a value, or values, needed by the program can be stored for use by the program. The value stored in the offset, by itself, has absolutely no meaning whatsoever until it is referenced by the program.

Current CNC controls allow multiple offsets. Most controls have from as few as 16 offsets to as many as 999, depending on the machine's application. The operator has complete control of the offset table in the control's memory. The operator can change the value of an offset at any time.

On most CNC controls, the offset value is invoked in the program by an offset number. For example, offset 1 may have a value of 0.500, offset 2 may have a value of 0.625, offset 3 a value of 1.000, and so on.

Certain CNC controls may have one offset number referencing several values. For example, each offset of a turning center usually includes at least two values, the X offset and the Z offset. Both are referenced in the program by the same offset number.

Yet other CNC controls organize the tool offsets by the way they will be used in the program. They may have designated a series of offsets for cutter diameter or radius, and another series of offsets for tool length. This technique makes it easy for the operator to find the offset in question.

Figures 5.1, 5.2, and 5.3 show examples of the three offset tables just mentioned. As you can see, the offset table in Fig. 5.1 allows only one value per offset. Many machining center and wire EDM controls use this technique. However, this form of offset table makes it somewhat confusing as to what the usage of the various offsets will be. The operator will have to be told explicitly what each offset represents.

Figure 5.2 shows another method used for machining centers. In

#	Value
1	
2	
3	
4	
5	
6	
7	
8	
9	
10	
11	
.	
.	
.	
.	
99	

Figure 5.1 Offset table used by many types of machine tools.

#	Length	Diameter
1		
2		
3		
4		
5		
6		
7		
8		
9		
10		
11		
.		
.		
.		
99		

Figure 5.2 Offset table used by many machining center controls.

this table, each offset number represents two values, the tool length value and the tool diameter value. With this method, it is more clear to the operator what each offset represents.

Figure 5.3 shows an offset table used by most CNC turning center controls. Notice that there are two values stored in each offset, the X offset value and the Z offset value.

#	X	Z
1		
2		
3		
4		
5		
6		
7		
8		
9		
10		
11		
.		
.		
.		
99		

Figure 5.3 Offset table used by most turning center controls.

A program command of some kind is always required to instate or access the offset. Again, the kind of application determines the required program command. We will give more information on the programming commands related to offsets a little later. Common words used to specify the desired offset number to be used by the program include D, H, and T.

For use in specific applications, the value stored in each offset being used will have some special logical meaning. The offset value may represent the length of a tool, the radius of a milling cutter, the radius of the EDM wire plus overburn, or some deviation from a tool's correct and desired position.

It is mandatory that the programmer and operator are in sync when it comes to the meaning of offsets to be used in a program. That is, the operator must know what values are to be stored in the offsets required in the program. If the operator inputs offsets incorrectly, or forgets to input them at all, the results could be disastrous. For this reason, the programmer must include offset information on the setup sheet to be included with the programmer's documentation to the operator.

However, in some shops, the meaning of each offset is left quite vague. Many shops have unwritten rules relating to the meaning of offset numbers. For example, on a machining center that uses the offset table shown in Figure 5.1, it may be an unwritten rule that the length value for each tool be stored in the same offset number as the tool station number. For tools that require a tool radius or diameter offset, the operator may be expected to add a constant number to the tool station number to come up with the offset number in which to store the tool's radius or diameter value. The length value for tool 1 is stored in offset 1 and the radius or diameter value is stored in offset 31. For tool 2, offset 2 is the length offset value and offset 32 is the radius (or diameter) value, and so on. As you start working in any new company, you should always check to find out how offsets are handled for the various CNC machines the company uses.

Forms of Compensation

Now we begin our discussions of the specific types of compensation available for different CNC machine tools. If you are a beginner to CNC, it is necessary that you understand the reasons *why* the various forms of compensation work as they do. In fact, we say it is as important to understand why you need compensation for the various purposes, at least at first, as it is to know the commands to use it. Knowing the reasons why compensation is needed will help you understand *how* to use it.

Also along these lines, you will find that the actual programming commands related to the various forms of compensation will vary dramatically from one CNC control to the next, yet the basic reason for its use will not. That is, *why* tool length compensation is needed will stay the same from one machining center to the next, but its usage will vary. If you understand *why* you need the compensation type, you are well on your way to understanding *how* to use it. Also, knowing why compensation is used for its various purposes will allow you to easily adapt to any one version of its use.

Here is a list of the compensation types we will discuss in order of presentation and the CNC machine tools related to the compensation:

Compensation	Related machine
Tool length compensation	Machining centers
Cutter radius compensation	Machining centers
Fixture offsets	Machining centers
Dimensional tool offsets	Turning centers
Tool nose radius compensation	Turning centers
Wire radius compensation	Wire EDM machines
Wire taper compensation	Wire EDM machines

Tool Length Compensation

This form of compensation is used on current CNC machining centers and similar CNC machines (CNC milling machines, CNC drilling machines, CNC drilling and tapping machines, etc.). Tool length compensation allows the programmer to forget about each tool's length as the program is being prepared. In fact, the main benefit of using tool length compensation is its ability to allow the length of each tool to vary with absolutely no change to the program. It is very important that you understand this feature, since it is used with *every tool in every program* written for machining centers.

Figure 5.4 shows a tool being held in the spindle. A typical machining center tool is made up of the shank or adaptor, extension, and the cutting tool itself. The tool shown in Fig. 5.4 is a drill, but all machining center tools will have a similar makeup.

The tool shown in Fig. 5.4 happens to be for a taper shank machining center spindle. Notice that the tool length is measured from the tip of the cutting tool to the nose or face of the spindle. Depending on the style of tool shank used by the machining center, this configuration may vary. However, the length of the tool will *always* be the distance from the tip of the cutting tool to the nose of the spindle.

With any machining center using taper shank tooling, there will al-

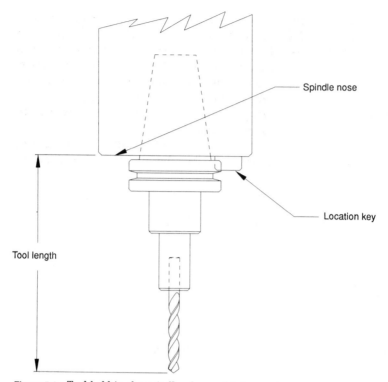

Figure 5.4   Tool held in the spindle of a machining center.

ways be a small gap between the nose of the spindle and the butt of the tool adapter flange. Figure 5.5 shows this gap. If measuring a tool's length away from the machine, you must take this gap into consideration. Fortunately, tooling manufacturers maintain a high level of consistency from one tool shank to the next. This means that, when the gap distance is measured one time, it will remain consistent for all tools.

Figure 5.6 shows a simple tool length measuring fixture that can be used to measure tool lengths away from the machine. Note the shim placed beneath the butt of the tool shank flange. With this arrangement, a simple and inexpensive height gauge can be used to measure each tool's length.

The style of cutting tools used in machining centers will vary dramatically. A CNC machining center can utilize face mills, end mills, drills, taps, reamers, and boring bars. Almost any form of rotating cutting tool can be used in this kind of machine. Most CNC machining centers also utilize an automatic tool changer, meaning that several tools can be used in the same CNC program. These rotating tools will vary dramatically in configuration as well as length. For example, a face mill is usually quite short to ensure rigidity. On the other hand, a

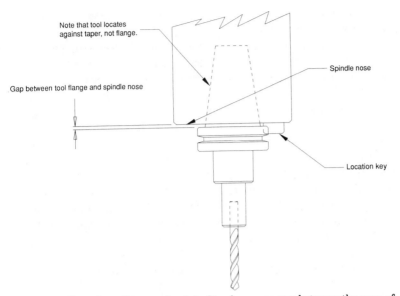

Note that tool locates against taper, not flange.

Spindle nose

Gap between tool flange and spindle nose

Location key

Figure 5.5 Note that all taper shank tooling has some gap between the nose of the spindle and the flange of the tool.

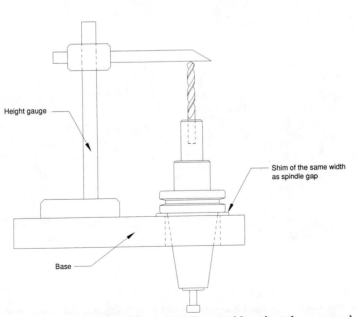

Height gauge

Shim of the same width as spindle gap

Base

Figure 5.6 Tool length compensation allows tool lengths to be measured off line, away from the machine tool.

drill or reamer may have to be very long in order to reach the surface of the workpiece to be machined.

Since the length of tools used in a machining center will vary, the distance the Z axis must travel with each tool to reach the same surface of the workpiece will also vary from tool to tool. For example, the Z axis must travel farther from its Z-axis tool change position to reach the same part surface with a short tool than with a longer one. Figure 5.7 illustrates this by superimposing two tools in the spindle.

In the early days of NC, the programmer had to keep track of this variance in tool length from tool to tool precisely, which made programming quite difficult. In those early days, when there was no such thing as tool length compensation, every tool used on the machining center had to be preset to a known length. The programmer had to know this preset length for each tool before the program could even be written. Note that some tools are easier to adjust to a given length than others. For example, a straight-shank drill held in a collet chuck allows quite a nice range of tool lengths, so that it is possible to predict

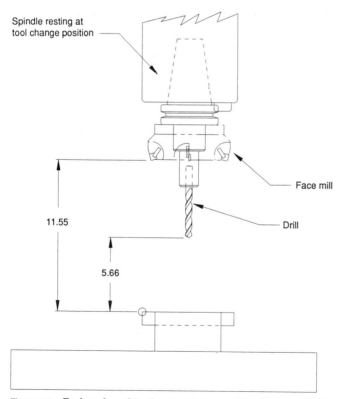

Figure 5.7   Each tool used in the machining center will have a different distance from the tip of the tool, at the tool change position, to the work surface to be machined.

its length. On the other hand, a face mill held in a face mill adaptor allows absolutely *no* adjustment. It can be assembled only to one length. With certain tools the programmer actually had to set up the tool and measure its length before the program could be written.

With tool length compensation, the programmer need not be overly concerned with each tool's length, at least as to how it relates to the machine's motion. The program will behave properly no matter how long the tools in the program are, within the limits of the machine's Z-axis travel. The programmer inputs a command for each tool that tells the control to use tool length compensation. This command will take into consideration the tool's length, subtracting this length value from the programmed motion distance in Z. The resulting motion is shorter by the tool length amount. The tool length value for each tool is stored in the control separately from the program, in the table of offsets. This is what keeps the program independent of the length of each tool. The tool length compensation command points to a tool offset in which the length value for each tool is stored. The operator or setup person must correctly input a length value for each tool to be used in the program.

Earlier, we talked about tool presetting. You must know the difference between tool presetting and simply measuring the length of the tool. With tool presetting, the setup person must adjust the length of the tool to a precise value. Tool presetting is time-consuming and difficult. On the other hand, when using tool length compensation, the operator or setup person simply assembles the tool with no concern for the tool's length. Once assembled, the tool length is easily measured.

At least one offset is set aside for use with each tool. In this offset, the operator or setup person will store the length of the tool. Normally the offset numbers chosen will correspond to the tool station numbers. The length of tool 1 will be stored in offset 1, the length of tool 2 in offset 2, and so on.

Just because tool length compensation is being used does *not* mean the programmer can forget about tool lengths completely. There are several things about the tool's length, aside from Z-axis motion commands, that can affect the success of the program. The rigidity of the tool, the ability of the tool to reach the surface to be machined without interference, and the speeds and feeds for the tool are among the things with which a programmer must still be concerned. In key concept no. 2, we discussed many of the things that a programmer must be concerned with in regard to tooling.

Using tool length compensation in a program

Now that you have an idea of the benefits and features of tool length compensation, let's look at how it is used. As with many features in programming, tool length compensation can be applied in several ways. Add to this the fact that the actual commands related to tool length compensation differ among control manufacturers, and you can see why we cannot be truly specific and accurate for every CNC machining center control being used today. For this reason, it is of primary importance that you understand what has been presented about tool length compensation to this point. If you do, you should be able to understand the example to follow and apply what you know to just about any variation that comes along.

What follows is one way tool length compensation can be applied. It is the most popular way, but you may find this technique varies from control builder to builder. The technique we show allows the benefits of tool length compensation to be maximized.

Most CNC controls use three CNC words related to tool length compensation. G43 is the G code that instates or initializes tool length compensation. There will be one and only one G43 command for each tool in the program. Typically this is the tool's first Z-axis motion command for the tool.

Included in the G43 command is an H word that tells the control which tool offset number is being used as the tool length. You can remember the letter H easily, since it stands for the *height* of the tool. This H code is usually kept as the same number as the tool station number. For example, tool station 1 will use H code 1 (H01), tool station 2 will use H code 2 (H02), and so on.

Also included in the G43 command is a Z-axis departure. It is the location along the Z axis at which you want the *tip* of the tool to stop. It is programmed relative to program zero. For example, if you wanted the tip of the tool to stop at a distance of 0.100 in above program zero, the Z axis would be commanded as

Z.1

The third word related to tool length compensation on most CNC machining center controls is a G49 command. G49 is the cancellation command for tool length compensation. Depending on control model and programming format, it may or may not be necessary to cancel tool length compensation. More on this will follow futher on in the book.

Example program showing tool length compensation. Before looking at a full example program, let's look at the actual single command to instate tool length compensation. Note that this command assumes the control to be in the absolute mode (G90):

N015 G43 H01 Z.1

In this command, the control is being told to bring the *tip* of the tool to an absolute position in the Z axis of 0.1 in above the program zero point.

Let's dissect this command a little more. Remember, this command is the tool's first Z-axis positioning command for the tool. At the time when the control reads this command, it is resting at the machine's reference point, very close to the plus limit of the Z axis. The G43 command tells the control to instate tool length compensation. The H01 word tells the control to look in offset 1 to find the tool length value. The tool length value would be the length of the tool currently in the spindle. The Z.1 tells the control the position at which to stop the *tip* of the tool. Prior to actually moving, the control automatically calculates the required Z-axis departure by subtracting the tool length stored in offset 1 from the distance between the Z point programmed (Z.1 in this case) and the nose of the spindle, currently resting at the reference point. In essence, the tip of the tool comes to rest at a position 0.1 in above the program zero point.

Once instated, tool length compensation remains in effect until canceled. The control will continue to keep the tip of the tool as the programmed point. Therefore, subsequent Z-axis commands will reflect the tip of the tool. Depending on the programming format, it may not be necessary to cancel the tool length compensation. If the machine is commanded to go to its reference point in the Z axis for a tool change, tool length compensation need not be canceled. But if no special reference point command is given for the Z axis at the end of the tool, and a simple positioning command is used, it is necessary to cancel tool length compensation (with a G49 command on most controls) on the tool's return for a tool change.

Here is a full program that uses three tools. At this time, do not worry about the format of this program. Program formatting is discussed in key concept no. 5. For now, just concentrate on the tool length compensation commands. Figure 5.8 shows the workpiece to be machined. Here is the sequence of operations for this simple program:

Sequence	Description	Tool	Station	Feed, IPM	Speed, RPM
1	Drill ½-in hole	½-in drill	1	5	500
2	Drill ⅜-in hole	⅜-in drill	2	4	750
3	Drill ¼-in hole	¼-in drill	3	3	875

Remember that this program allows the tool length of each tool to be kept separate from the program. After the program is written, the operator or setup person can assemble the tools and measure their lengths. Then the offsets can be entered accordingly. For example, if

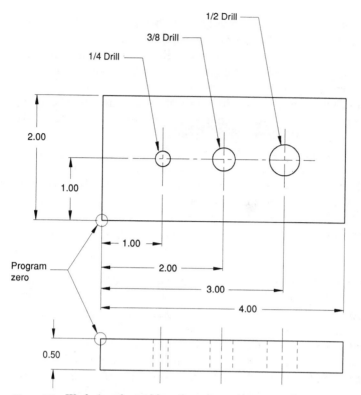

Figure 5.8   Workpiece for tool length compensation example.

the ½-in drill was measured to be 5.4353 in long, the ⅜ drill was measured to be 6.8503 in long, and the ¼ drill was measured to be 5.4086 in long, the offsets would be entered as follows:

Offset	Value, in
1	5.4353
2	6.8503
3	5.4086

Note that this program assumes tool 1 (½-in drill) to be in the spindle when the program is activated.

```
O0001 (program number)
(½ drill)
N005 G54 G90 S500 M03 T02 (select coordinate system, ABS mode, turn spindle
on at 500 RPM, get tool 2 ready)
N010 G00 X3. Y1. (move to first X and Y position)
N015 G43 H01 Z.1 (instate tool length compensation)
N020 M08 (turn coolant on)
N025 G01 Z-.8 F5.0 (feed to hole bottom at 5 IPM)
```

N030 G00 Z.1 M09 (rapid out of hole, turn coolant off)
N035 G49 G91 G28 Z0 M19 (cancel tool length compensation, send tool to tool change position, orient spindle on the way)
N040 M01 (optional stop)
N045 T02 M06 (place second tool in spindle)
(⅜ drill)
N050 G54 G90 S750 M03 T03 (select coordinate system, ABS mode, turn spindle on at 750 RPM clockwise, get tool 3 ready)
N055 G00 X2. Y1. (move to first tool position)
N060 G43 H02 Z.1 (instate tool length compensation)
N065 M08 (turn coolant on)
N070 G01 Z-.8 F4. (feed to hole bottom at 4 IPM)
N075 G00 Z.1 M09 (rapid out of hole, turn coolant off)
N080 G49 G91 G28 Z0 M19 (cancel tool length compensation, send tool to tool change position, orient spindle on the way)
N085 M01 (optional stop)
N090 T03 M06 (place tool 3 in spindle)
(¼ drill)
N095 G54 G90 S875 M03 T01 (select coordinate system, ABS mode, turn spindle on clockwise at 875 RPM, get tool 1 ready)
N100 G00 X1. Y1. (move to first X-Y position)
N105 G43 H03 Z.1 (instate tool length compensation)
N110 M08 (turn coolant on)
N115 G01 Z-.8 F3. (feed to hole bottom at 3 IPM)
N120 G00 Z.1 M09 (rapid out of hole, turn coolant off)
N125 G49 G91 G28 Z0 M19 (cancel tool length compensation, send tool to tool change position, orient spindle on the way)
N130 G91 G28 X0 Y0 (send X and Y axes to reference position)
N135 M30 (end of program command)

This program requires some explanation. Again, remember that it is *not* our intention at this point to discuss the formatting of a CNC program. That will be done in the next key concept. Our intention here is to discuss the commands related to tool length compensation.

Notice first and foremost that the tool length compensation initialization commands (the G43 commands) are always included in each tool's first Z-axis motion. In blocks N015, N060, and N105, the tool currently in the spindle has not made any prior Z motion.

Notice also that only one G43 command is required per tool. Subsequent Z-axis positioning commands for each tool are given to the *tip* of the tool relative to program zero in Z. So once instated, tool length compensation remains in effect until canceled.

Lastly, notice that tool length compensation was canceled (by G49) at the end of each tool, during the tool's retraction to tool change position.

One point the above program does not stress is that tool length compensation can be instated within a motion command including more than just the Z axis. For time-saving reasons, the programmer may elect to instate tool length compensation in a three-axis movement to the work. As long as no obstructions are in the way, this is just fine.

Here is an example command that would move all three axes at the same time while also instating tool length compensation:

N010 G00 X3. Y1. G43 H01 Z.1

Remember that most controls would allow the words in the above command to be in any order.

While the above program may not be entirely clear for the simple reason that you have not yet been exposed to the various programming words, we hope it is clear in regard to tool length compensation.

Cutter Radius Compensation

Cutter radius compensation (also called *cutter diameter compensation*) is used on machining centers and similar CNC machines. This feature allows the programmer to forget about the cutting tool's radius or diameter during programming. Like all forms of compensation, it makes programming easier, since the programmer need not be concerned with the exact cutter diameter while the program is being prepared. Cutter radius compensation also allows the radius of the cutting tool to vary without modification to the program.

Cutter radius compensation is *not* applied to all forms of cutting tools. It is needed only for cutting tools that have the ability to machine on the periphery of the cutter, and only *when* machining on the periphery of the cutter. Tools like end mills, shell mills, and some face mills have this ability. Drills, reamers, taps, boring bars, and other center cutting tools have absolutely *no use* for cutter radius compensation.

There are many times during milling operations when the programmer wishes to machine the edge of a workpiece. This edge can be in the form of a straight surface or in the form of a contoured surface. Like a woodworking router, a milling cutter is driven along the edge of the workpiece. Figure 5.9 shows an example of this.

One way to program the milling cutter's path is to program the motions by the centerline of the milling cutter. This technique was demonstrated in key concept no. 3 for the three kinds of motion commands. As you saw during key concept no. 3, this means the programmer must take into consideration the diameter of the milling cutter. For example, if the milling cutter is 1 in in diameter, all motions programmed must be kept precisely ½ in away from the surfaces to be milled. Even this assumes that there is no tool pressure pushing the cutting tool away from its programmed path. Tool pressure will be discussed further later.

While using the centerline coordinates of the tool path is a popular way of programming, it has several disadvantages and limitations.

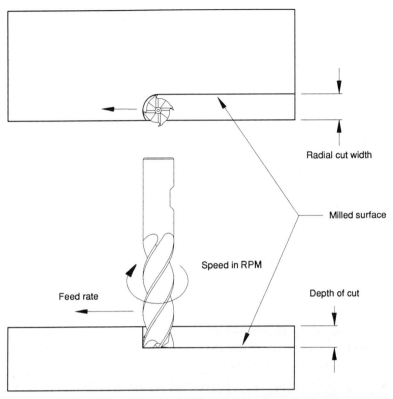

Radial cut width

Milled surface

Speed in RPM

Feed rate

Depth of cut

Figure 5.9 Side milling with an end mill.

Here we list and explain them. Each of these limitations presented real problems before cutter radius compensation became available.

Figure 5.10 shows the difference between the tool's centerline coordinates and part coordinates. Notice that the tool's centerline coordinates require at least one extra calculation to be made for each axis in the coordinate system. On the other hand, part coordinates are usually taken right from the print.

Reasons for cutter radius compensation

There are several reasons why cutter radius compensation is so helpful. It relieves several programming burdens. Here we list and explain why it is so important to use this feature when you have reason to do so.

Changes in tool diameter. Using a 1-in diameter end mill to machine the right side of a rectangular workpiece being held in a vise would be

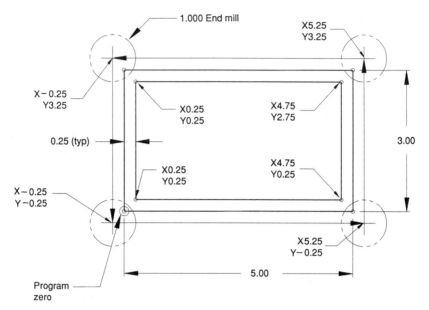

Figure 5.10 Cutter centerline coordinates require more calculations to be made than part coordinates.

considered a simple operation by most experienced programmers. If a programmer prepares the program on the basis of the tool's centerline coordinates—*not* using cutter radius compensation—the end mill must be kept away from the right side of the rectangular workpiece by precisely 0.500 in throughout its motions.

Imagine that the operator is making the setup and discovers the company is out of 1-in end mills. There are 0.875-in-diameter end mills and 1.25-in-diameter end mills, but no end mills left in 1-in diameter. In this case, the programmer would have to change the program in order to use an end mill diameter other than the one programmed.

Tool pressure. Most machinists will agree that the cutting tool will seldom machine the workpiece as desired on the first try. The cutting tool, workpiece, and even the machine tool itself are under a great deal of pressure during machining. The more powerful the machining operation, the greater the pressure. Even when a milling cutter is kept quite rigid (sturdy end mill holder and short overall length), there will be some deflection of the tool during machining. This is because the cutting edge of the tool will have a tendency to push away from the surface being machined.

If you have ever scraped paint from the wall of a room, you have experienced this kind of deflection. When scraping paint, you do your best to push the scraper in a way that will remove the paint, but many times your scraper will be deflected from the surface to be scraped. In machining terms, this tendency to deflect is called *tool pressure.*

Generally speaking, the weaker the machine tool and cutting tool, the more potential for deflection. While the small deflection may not be substantial enough to cause problems, there are times when it will. This is especially true when the accuracy required of the part is demanding.

In the previously discussed example related to milling the right side of a workpiece, even if an end mill precisely 1 in in diameter is used, there is still the potential for deflection of the tool during machining. Depending on the expected tolerance for this surface, the amount of deflection may be enough to cause the part to be out of tolerance. If this were the case, and if fixed centerline coordinates were used, it would mean having to reprogram the milling cutter to allow for deflection.

Also note that, as the cutter dulls, deflection will increase. This means a sharp cutter will have less deflection than a dull one. This change in deflection amount during the life of a cutting tool can present real headaches while machining if fixed centerline coordinates are used in the program.

Complex contours

Even with the previously discussed possible problems, a case can be made for using fixed centerline coordinates for simple parts. With the previously given simple example of milling the right side of a workpiece, it would be relatively easy to simply add or subtract the radius of the cutter to the square surface of the part to be machined. However, as the surface to be milled becomes more complicated, calculating the centerline coordinates for the path of the end mill also becomes more complicated. If angular surfaces and radii must be machined, many times it can be difficult enough to calculate coordinates on the workpiece, let alone the coordinates for the centerline of the cutter.

Figure 5.11 shows an example of when calculating the tool's centerline coordinates for the tool path would be very difficult indeed. In this part's case, it would be hard enough to calculate the actual surfaces of the workpiece itself, let alone the tool's centerline coordinates.

Roughing. The last reason we present for using cutter radius compensation has to do with roughing operations. We have already stated that, when a complex contour must be machined, it can be difficult

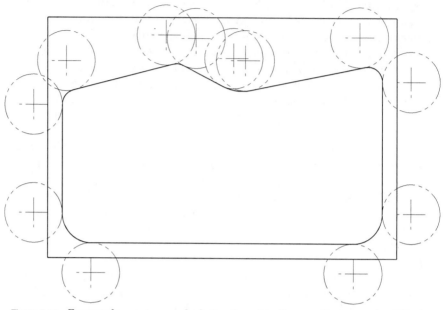

Figure 5.11   For complex contours, calculating the centerline coordinates can be difficult indeed.

enough to come up with workpiece coordinates, let alone cutter centerline coordinates. Add to this the complication of having to allow for a constant amount of finishing stock throughout the surface to be machined during roughing. In essence, this doubles the amount of work the programmer must do. The programmer not only must calculate the centerline coordinates of the end mill during finishing, but also must calculate the centerline coordinates during the roughing operation.

The dotted lines and arcs in Fig. 5.12 show how the part must be machined prior to finishing. Note how much additional work is required in order to calculate the tool's centerline coordinates for this roughing pass.

Advantages to using cutter radius compensation

As you may have guessed, using cutter radius compensation allows the programmer to easily overcome the previously mentioned limitations of programming fixed centerline coordinates.

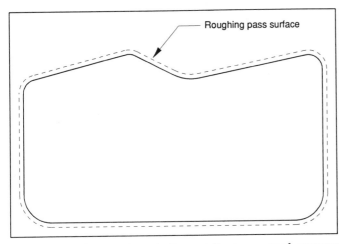

Figure 5.12   If the cutter's centerline coordinates are used, programming roughing passes would double the necessary calculations.

A range of cutter sizes can be used.   When cutter radius compensation is used, the operator is not forced to use the precise cutter diameter programmed. While there are limitations with regard to minimum and maximum cutter sizes, a range of cutter sizes is possible. This allows the operator much more flexibility than if fixed centerline coordinates are used in the program. Also, no program modification is necessary if a cutter size must change. Figure 5.13 shows how various cutter diameters can be used.

Sizing the workpiece becomes easier.   When cutter radius compensation is used, it is both possible and easy to overcome any tool pressure the cutting tool may be influenced by during machining. Once the workpiece has been machined and measured, it will be obvious if there has been any deflection caused by tool pressure. If there has been, and once the amount of deflection is known, it will be easy for the operator to compensate for the next workpiece.

Machining complex contours becomes easier.   When cutter radius compensation is used, it is no longer necessary that the programmer calculate the centerline coordinates of the tool. Only the workpiece surfaces need be calculated. While calculating the desired cutting path on

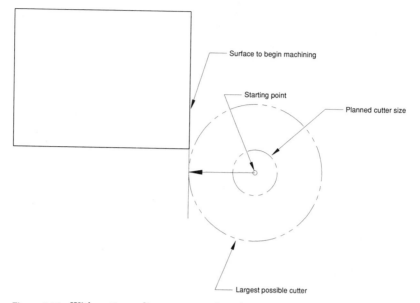

Figure 5.13   With cutter radius compensation, the operator will be allowed a wide range of cutter sizes.

the workpiece itself may be difficult, depending on how the print was dimensioned, at least the programmer can forget about the tool's centerline coordinates.

Roughing becomes much easier.   When cutter radius compensation is used, it becomes very easy to make roughing passes. In fact, the same series of coordinates used for the finishing pass can be used for roughing. How this is done involves understanding the offsets related to cutter radius compensation (discussed a little later). For now, suffice it to say that there is a way to use the actual finishing coordinates and trick the control into allowing the desired finishing stock for a subsequent finishing tool.

Commands related to cutter radius compensation

As you have seen, using cutter radius compensation has *many* advantages. Even so, you would be surprised at the number of programmers who elect not to use it. Though not overly complicated to use, cutter radius compensation does have certain specific rules that cannot be broken. In addition, the method by which cutter radius compensation is programmed does vary from control builder to builder. Also, early versions of cutter radius compensation were cumbersome and difficult to program. For these reasons, some experienced programmers have elected to not use cutter compensation at all. They persist in programming the tool's fixed centerline coordinates.

We strongly recommend the beginner to stick with it until the use of cutter radius compensation is mastered for the particular CNC controls to be programmed. Over the long haul, it will save you countless programming hours and make programming much easier.

As stated, the techniques required to use cutter radius compensation vary from control to control. The method we will show is the most popular and incorporates techniques that will work on a great number of CNC controls. Though this is the case, you must be prepared for CNC controls which vary the method to use cutter radius compensation.

There are three G codes used with cutter radius compensation. Two of these G codes are used for instating or initializing cutter radius compensation. Note that either G41 or G42 is used to instate cutter radius compensation. The third G code, G40, is used to cancel cutter radius compensation. Along with these three G codes, a programming word is also used to specify the offset number to be used with radius compensation. Usually a D word (sometimes an H word, depending on control model) is used to specify the offset number.

The steps to programming cutter radius compensation

There are three distinct steps to programming cutter radius compensation. First, it must be instated. Second, the motion commands must be given. Third, it must be canceled. Let's look at programming cutter radius compensation each step of the way.

Instating cutter radius compensation. The first thing a beginner must be able to do is decide how the cutter will be related to the surface being machined during the cutting. While cutting, the cutter will either be on the left side or right side of the surface being machined. To evaluate this relationship, the programmer must look in the direction the tool will be moving during the cut. While looking at the print, the programmer may have to rotate the print to look in the cutting direction. While looking in this direction, the programmer must ask the question, "What side of the workpiece is the cutter on?" If the cutter is on the left side of the workpiece, the programmer must include a G41 in the block to instate cutter radius compensation. If the cutter is on the right side of the workpiece during machining, the programmer

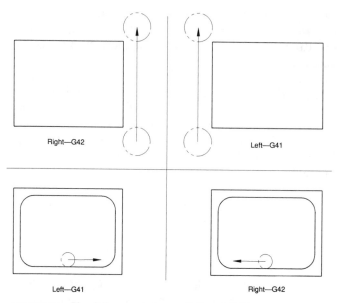

Figure 5.14   The difference between G41 and G42.

must include a G42 in the block to instate cutter radius compensation. The G41 word represents a cutter-left condition and the G42 word represents a cutter-right condition. See Figure 5.14 to get the idea.

If you understand basic machining practice, there may be an easier way to remember and determine G41 or G42. When using a right-hand cutting tool, as you normally will, if the machining operation is to be a climb milling technique, G41 is used. If conventional milling is to be done, then G42 is used. Figure 5.15 shows the difference between climb and conventional milling.

Prior to the command to instate cutter radius compensation, the programmer must position the tool close to the surface to be machined. Usually a distance of slightly more than the desired cutter radius from the surface to machine is sufficient. Of course, these prior movements must be programmed relative to the centerline of the tool. This requires that the programmer has a general idea as to what size cutter is to be used. As stated, cutter radius compensation allows a range of cutter sizes to be used. We recommend that these prior clearance approach movements allow for the maximum cutter size the programmer wishes the operator to use. Also, most CNC controls require that the tool be positioned in a way such that, when the tool comes into contact with the workpiece and begins machining, a right angle is formed.

Figure 5.16 shows this. Notice that the tool is first moved to an X

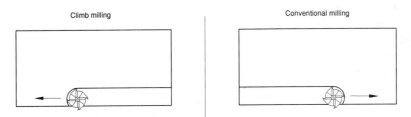

Figure 5.15 If you understand basic machining practice, you can easily tell the difference between G41 and G42. Climb milling is G41 and conventional milling is G42.

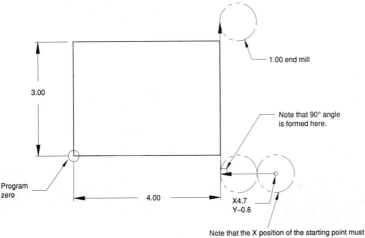

Figure 5.16 When cutter radius compensation is instated, most controls require a movement at a right angle to the first direction of cut.

position of 4.7 in and a Y position of −0.6 in. This is *still* based on the tool's centerline. Between the movement when cutter radius compensation is instated during the move to

X4.

and the first machining motion along the Y axis, a right angle is formed.

The last thing you need to know about instating cutter radius compensation has to do with the offset used with the compensation. A D word (H on some controls) is used to point to the offset number where the cutter radius compensation value (normally the radius of the tool) is to be stored.

There are two possible ways to use the offset. One way (and our suggested way for manual programmers) is to program the actual surface of the workpiece. By this method, the coordinates for the program often can be taken right from the print. With this technique, the offset value must be the actual radius of the cutter to be used. For example, if the operator uses a 1-in end mill, the value stored in the offset will be 0.500 in. Another benefit of using this method, aside from the ease of coming up with coordinates, is the fact that the control will be more likely to give an alarm if some workpiece violation is going to occur.

The other way of using the cutter radius compensation tool offset is to actually program the centerline coordinates for a desired tool diameter. This method is an older way to program. However, it is popular with companies that have computer-aided manufacturing (CAM) systems to help them prepare the program. With the CAM system, it is just as easy to have the CAM system generate centerline coordinates as it is to generate workpiece coordinates.

In this case, the offset is equal to the radius deviation from the intended cutter size. If the cutter being used is larger than the intended cutter, the offset value will be positive, and if the cutter being used is smaller than the intended cutter, the offset value is minus.

For example, imagine a program is written around the use of a 1-in end mill. If the 1-in end mill is used, the value stored in the offset will be zero. If a 0.875-in-diameter end mill is used, the value stored in the offset will be *minus* 0.0625 in. If a 1.25-in-diameter end mill is used, the value stored in the offset would be *plus* 0.125 in.

Here is one last point about the offset number. Some CNC machining center controls designate a specific series of offsets for cutter radius compensation. With these controls, it is relatively easy for the operator to tell which offset to use for any given tool station, but other controls do not designate the difference between the various types of tool offsets (tool length, tool radius, etc.). With this type of control, it is left to the programmer's discretion to decide on all offset numbers. With tool length compensation, we recommended using the tool station number as the tool length compensation offset number. For cutter radius compensation, we recommend simply adding a constant number to the tool station number to calculate the offset number to be used for cutter radius compensation. For example, if the constant number is 30, tool 1's length is stored in offset 1 and its radius is stored in offset 31. This allows for a nice, logical way for the programmer and operator to stay in sync with the various offsets to be used with each program.

Motion under the influence of cutter radius compensation.   Once instated, cutter radius compensation remains in effect until canceled. During these movements, the control will be keeping the tool on the left side or right side, depending on the condition related to G41 or G42, of a series of lines and circles determined by the subsequent straight-line motion and circular motion commands given. It is during these movements that some odd things can happen if the user does not understand what is happening when cutter radius compensation is used.

There are some basic rules that govern these movements. If you keep in mind that the control is keeping the tool compensated during the movements on the left side or the right side of the programmed motions, it will be relatively easy to assure success.

The first general rule has to do with a CNC control feature called *buffer storage*. As mentioned during the introduction to CNC, the CNC control is constantly looking ahead to see what will be happening next. While executing one command, the control has already looked ahead to at least the next command to see what will be occurring next. For general-purpose machining, this look-ahead buffer is designed to keep the motion from coming to a halt between commands. Thanks to this buffer, the control can flow from one command to the next without pausing.

When cutter radius compensation is used, the look-ahead buffer is also used for another purpose. It is used to allow the control to calculate the correct stopping point for the *current* command. If the next movement is a circular motion, tangent to the surface being machined, the control will calculate the end point in one fashion. If the next movement is along a line that intersects the current direction of motion, the control will calculate the end point in another fashion. This makes the *next* motion command *very* important to the control when under the influence of cutter radius compensation.

It is this very fact that made early versions of cutter radius compensation so difficult to work with. In the early days of cutter radius compensation, there was no such thing as a look-ahead buffer. In those days, the programmer was forced to include a directional vector that pointed in the direction of the next motion with each motion command to be made under the influence of cutter radius compensation.

Even with today's CNC controls, there are some limitations related to the look-ahead buffer. First of all, the look-ahead buffer can hold only a limited number of commands. This number varies from control builder to builder, but rest assured there is some limitation in this regard.

There are two points to make about this limitation. First, when under the influence of cutter radius compensation, the programmer will want to keep the number of nonmotion commands to a minimum. That is, things that have nothing to do with motion should be kept out of this area of the program. These things include spindle commands, unrelated M codes, feed rates, and even motion commands in uncompensated axes (usually the Z axis). Especially when the look-ahead buffer is very small, there is no sense wasting its capacity with information unrelated to cutter radius compensation. Usually, when this limitation is exceeded, the control will alert the operator with some form of alarm.

Second, if the movements are very small and if the feed rates are very fast, the control may not be able to keep up with the desired motions. Even though the control is doing its best to compensate as quickly as possible, the motion may be occurring faster than the control can react. In this case, usually the control will bog down and react at a slower rate than usual. Generally, this means that the desired feed rate cannot be reached.

Another rule related to cutter radius compensation has to do with the size of the cutter. When using cutter radius compensation, the cutter must be able to fit into the surfaces to be machined. This is another area where basic mistakes are commonly made. Remember that the control will be trying to keep the tool constantly compensated to the right side or left side of the programmed surfaces. This means that the control will be trying to bring the tool tangent to each surface being machined. There are times when, because of the size of the cutter, it is impossible for this tangency to occur without the workpiece being violated (gouged). It can be difficult to tell when this condition will occur prior to running the program. Figure 5.17 shows an example of when the cutter will not fit into the surface to be machined.

With Fig. 5.17 in front of you, you may think it impossible to make such a silly mistake. However, while programming, it can be difficult to envision having such a problem without actually drawing a circle the diameter of the tool in all machining positions. Add to this the fact that not all drawings are made to scale, and you can see why this is a common problem.

Because of this cutter size limitation, there are certain types of motions that simply cannot be made without reducing cutter size. Another example of this is in machining of inside radii. The part radius *must* be larger than the radius of the cutter. Otherwise the tool will violate the radius. Most CNC machining center controls will generate an alarm if this condition exists. Sometimes it can be difficult to diagnose this kind of problem. However, if you constantly try to visualize the cutting tool being compensated to the left or right side of the programmed shape, it should become easier.

Canceling cutter radius compensation

The G40 word is used to cancel cutter radius compensation. While the rules governing how cutter radius compensation must be canceled will

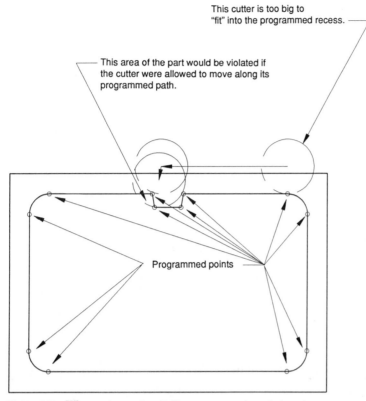

This cutter is too big to
"fit" into the programmed recess.

This area of the part would be violated if
the cutter were allowed to move along its
programmed path.

Programmed points

Figure 5.17   When using cutter radius compensation, the programmer must
be careful not to violate the surfaces of the workpiece being machined.

vary from control to control, some general statements will apply. First
and foremost, cutter radius compensation once instated *must* be canceled. You must remember that the control will continue to keep the cutter on the left or right side of the programmed path until the compensation is canceled. If the programmer forgets to cancel cutter radius
compensation, the compensation will remain in effect throughout the
balance of the program, even for subsequent tools. Depending on the
types of motions to be made, it is possible that an alarm will sound if the
subsequent motion commands happen to break the rules of cutter radius
compensation. Even if no alarm sounds, movements will not be correct.

Second, it is wise to retract the tool from the surface being machined, if
possible, before canceling cutter radius compensation. On some controls,
the cancellation command will actually cause a motion equal to the
value being stored in the offset as the control brings the tool back on center. This motion will be in the direction of the surface just machined,
meaning the tool could gouge the shape by the radius of the cutter.

In some cases, it may be impossible to move the tool above the surface because a witness mark would be left on the workpiece by the milling cutter. In this case, the G40 cancellation word can be included in a motion command to a clearance point away from the surface just machined. If this is done, remember that the coordinates given in the G40 command must reflect the centerline of the tool.

Third, one basic mistake beginners make when attempting to use cutter radius compensation is to try to do too much after instating cutter radius compensation. If more than one contour is to be machined, it is wisest to cancel the cutter radius compensation before going on to the next contour. Beginners have a tendency to leave the cutter radius compensation on as they rapid the tool to the second contour to be machined. While there is nothing wrong with making rapid motion commands under the influence of cutter radius compensation, remember that the motions to the next contour will be affected by the previous cutter compensation commands. Our suggestion when multiple surfaces must be machined is to handle each flowing contour independently. Instate, cut, and then cancel. Then move into position to machine the second contour. Instate, cut, and then cancel, and so on. This will keep the control from becoming confused between contours.

Other techniques with cutter radius compensation

At this point, you should see the advantages of using cutter radius compensation. While you may not yet know every detail, you do know the three steps that are required to use it. Before we give some extended examples of its use, we want to introduce some other techniques that make cutter radius compensation especially helpful.

Sizing workpieces with offset adjustments. Earlier in this section, we stated that cutter radius compensation allows for easy sizing of the workpiece. Now we want to explain further. Imagine that the programmer is machining an outside 3-in-square shape. A 1-in end mill is used and the programmer wants to use workpiece coordinates in the program. The operator measures the end mill and finds it to be precisely 1 in in diameter. In the corresponding tool offset, a value of 0.500 in is stored.

Now the operator runs the cycle and then measures the workpiece. Instead of finding the outside square shape to be precisely 3 by 3 in, the operator measures 3.001 by 3.001 inch. In this case, the end mill probably experienced some deflection during machining. This kind of problem can also occur from a poor tool holder. In any event, the workpiece is 0.001 in oversize. However, it is oversize only 0.0005 in per side. That is, all sides of the part are (in this case) 0.0005 in too big. What should the operator do?

In this case, the operator could *reduce* the offset by 0.0005 in, making the new value 0.4995 in. The reason to reduce the offset is to keep the centerline of the cutter 0.0005 in *closer* to the surface being machined. Think about it. It is the kind of technique an operator will use on a daily basis.

When trying to decide which way to adjust an offset, think about what would happen if the value of the offset were zero. If the programmer is programming workpiece coordinates, the centerline of the spindle will follow the programmed path, and the workpiece will be violated (gouged) by the radius of the cutter. If you keep this in mind, it should help you decide in which direction, plus or minus, an offset should be adjusted when the workpiece is not coming out to size.

Allowing for finishing stock. We stated earlier that cutter radius compensation allows the programmer to use the same set of coordinates needed for finishing for the roughing pass as well. If this technique is used, it can save countless calculations for the roughing pass coordinates, especially on complicated contours.

Here's the trick. For the roughing cutter, the operator will lie to the control about the cutter size. The control will be told that the cutter is *larger* than its actual size by the amount of stock to be left for finishing. This will keep the cutter away from all surfaces to be machined by the difference between the offset value and the actual size of the cutter. Here is an example.

A programmer wishes to use a 1-in end mill to rough a contour, leaving 0.030-in finishing stock for the finishing end mill, also to be 1 in in diameter. The operator is told to make the offset for the roughing end mill 0.030 in larger than its actual size. In this case, with a 1-in roughing end mill, the operator makes the offset 0.530 in. For the finish end mill, the operator tells the truth about the tool, entering a 0.500-in value. When the program is activated and when the rough end mill does its machining, it stays away from the finished surface by, in this case, 0.030 in. That is, the control thinks that this tool has 0.530-in radius, and it compensates accordingly. The 0.030 stock is evenly left on all surfaces. When the finish end mill machines the workpiece, it machines to size, since the offset value for this tool is accurate. Figure 5.18 illustrates how you can trick the control in this manner.

Example programs showing cutter radius compensation

Now you should be ready to look at some example programs that show the use of cutter radius compensation in different applications. As we

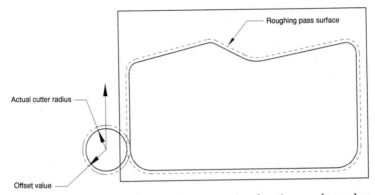

Figure 5.18   To rough-mill the part, the same series of motions can be used as for finishing if the operator makes the offset used for roughing larger than the milling cutter's actual size.

said with tool length compensation, we cannot hope that every detail of these programs will be understood. As you learn more about program formatting during key concept no. 5, you will be able to recognize more of the commands shown here. For now, simply concentrate on understanding the commands related to cutter radius compensation.

You have been exposed to the raw tools of cutter radius compensation. The examples below should drive home your understanding of cutter radius compensation.

We will limit each program to just the one tool required to show cutter radius compensation, making each example as simple as possible.

Finish milling one side of the part.   Figure 5.19 shows the print to be used for this example. Planned cutter size is 1.000 in in diameter.

Offset no.	Value, in
1	5.4545
31	0.5000

Program:

O0004 (program number)
N005 G54 (select work coordinate system no. 1)
N010 G90 S350 M03 (select ABS mode, turn spindle on at 350 RPM)
N015 G00 X6.6 Y-.6 (move to point 1)
N020 G43 H01 Z.1 (rapid down to just above part)
N025 G01 Z-1.1 F30. M08 (fast feed to cutting surface, turn on coolant)
N030 G42 D31 X6. F4. (instate compensation, feed to point 2)

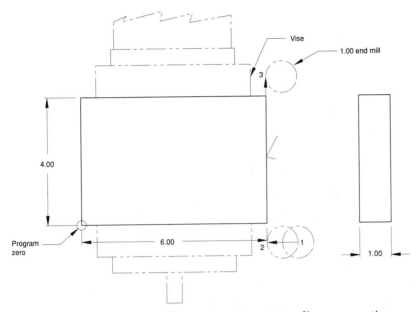

Figure 5.19   Print for first example program using cutter radius compensation.

N035 Y4.2 (feed to point 3)
N040 G00 Z.1 M09 (rapid above part, turn off coolant)
N045 G40 (cancel cutter radius compensation)
N050 G49 G91 G28 Z0 M19 (go home in Z, orient spindle)
N055 28 X0 Y0 (go home in X and Y)
N060 M30 (end of program)

This program is relatively simple. One point we will make here is that you notice in blocks N030 and N035 that the *part surface* was programmed. Also, note that the cutter could have been 1.200-in in diameter (6.6 minus 6.0 and the result, 0.6, multiplied by 2) and still work. Any larger diameter, and the cutter would be violating the workpiece at the approach point (in block N030), prior to instating cutter radius compensation. On most controls, you will receive the "overcutting will occur" alarm.

Example showing the milling of an outside contour.   Figure 5.20 shows the print for this example. Planned cutter size is 1.000 in in diameter.

Offset no.	Value, in
1	5.4056
31	0.5000

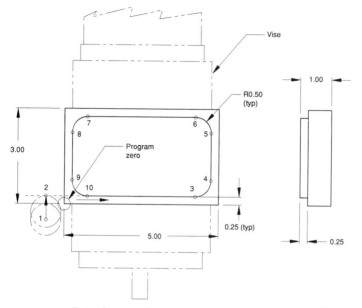

Figure 5.20   Print for second example program using cutter radius compensation.

Program:

```
O0005 (program number)
N005 G54 (select coordinate system no. 1)
N010 G90 S350 M03 (select ABS mode, turn spindle on at 350 RPM)
N015 G00 X-.6 Y-.6 (rapid to point 1)
N020 G43 H01 Z.1 (instate length comp, rapid above part)
N025 G01 Z-.25 F30. M08 (fast feed to surface, turn on coolant)
N030 G42 D31 Y.25 F4. (instate cutter comp, feed to point 2)
N035 X4.25 (feed to point 3)
N040 G03 X4.75 Y.75 R.5 (circular move to point 4)
N045 G01 Y2.25 (feed to point 5)
N050 G03 X4.25 Y2.75 R.5 (circular move to point 6)
N055 G01 X.75 (feed to point 7)
N060 G03 X.25 Y2.25 R.5 (circular move to point 8)
N065 G01 Y.75 (feed to point 9)
N070 G03 X.75 Y.25 R.5 (circular move to point 10)
N080 G00 Z.1 M09 (rapid to clearance point above part, turn off coolant)
N085 G40 (cancel cutter compensation)
N090 G49 G91 G28 Z0 M19 (go home in Z, orient spindle)
N095 G28 X0 Y0 (go home in X and Y)
N100 M30 (end of program)
```

In this program, again note that all coordinates reflected the part itself. Even the R word in the circular movement commands was specified as the radius of the workpiece. Maximum cutter size for this program: 1.200 in.

Example showing milling of inside circle with ramp-in/ramp-out. Figure 5.21 shows the print to be used for this example. Planned cutter size is 1.000 in in diameter.

Offset no.	Value, in.
1	5.3343
31	0.5000

Program:

O0006 (program number)
N005 G54 (select coordinate system no. 1)

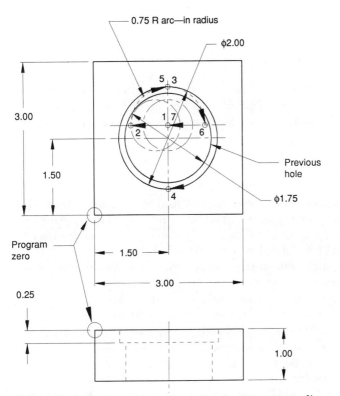

Figure 5.21   Print for third example program using cutter radius compensation.

N010 G90 S350 M03 (select ABS mode, turn spindle on at 350 RPM)
N015 G00 X1.5. Y1.75 (move to point 1)
N020 G43 H01 Z.1 (instate length comp, move just above part)
N025 G01 Z-.25 F30. M08 (fast feed to surface, turn on coolant)
N030 G42 D31 X.75 F3. (instate radius comp, feed to point 2)
N035 G02 X1.5 Y2.5 R.75 (circular move to point 3)
N040 G02 Y.5 R1. (circular move halfway around to point 4)
N045 G02 Y2.5 R1. (circular move back to start point 5)
N050 G02 X2.25 Y1.75 R.75 (ramp off to point 6)
N055 G01 G40 X1.5 M09 (cancel compensation on the way back to point 7)
N060 G00 Z.1 (move away in Z)
N065 G49 G91 G28 Z0 M19 (go home in Z, orient spindle)
N070 G28 X0 Y0 (go home in X and Y)
N075 M30 (end of program)

Note that this program has shown you an example of how you can cancel cutter radius compensation during a movement command. Notice in block N055, the compensation is being canceled during a movement command back to the center of the circle in X.

We hope these examples have given you a good way to become familiar with, as well as to test your knowledge of, cutter radius compensation. While it is not mandatory that you fully understand this information the very first time through, it is important that you stick with it until you are comfortable with this information. Please review this information until you feel confident. As stated in the beginning of this presentation, many programmers give up on cutter radius compensation at the first sign of problems. You should be able to see by now that it will be well worth the time it takes to learn this feature thoroughly.

Fixture offsets

This compensation type is used on machining centers, wire EDM equipment, turret punch presses, and laser equipment. Fixture offsets, also called *work coordinate system multiple settings,* allow the user to work with several coordinate systems from within the same program.

Fixture offsets also relieve the programmer from the burden of assigning the program zero point in the program.

During key concept no. 1, we introduced program zero. We said that the programmer chooses the program zero point for each program wisely, based on the methods by which the print is dimensioned. We also stated that one way to assign program zero is to use the G92 command. There was a time when the G92 command was the only way available to the programmer to assign program zero. As you saw during key concept no. 1, the G92 command is quite limiting because only one program zero point per program can be easily assigned.

There are times when it is convenient, if not mandatory, that the programmer be able to set up more than one coordinate system in the program. Here are some examples:

First, there are times when it is necessary to machine more than one workpiece in a setup. The degree of similarity of workpieces to be machined can range from identical to completely different. There may also be differing operations on the same workpiece to be performed while several workpieces are being held at different attitudes in the setup. Having the ability to assign more than one coordinate system makes it easy to go from one workpiece to the next and still be able to refer to print dimensions.

Second, there are times when a workpiece to be machined must be held in a rotary device of some kind, especially on horizontal machining centers. It may be necessary to work on two or more sides of the workpiece in the same setup. Fixture offsets allow the programmer to assign one program zero point for each side of the workpiece to be machined.

Third, there are times when a dedicated workpiece must be run over and over again. By dedicated, we mean the CNC machine tool was purchased for the sole purpose of running a small series of workpieces, maybe only one workpiece. Maybe these workpieces use a fixture that can be easily mounted on the machine in a consistent manner. Possibly several dedicated workpieces are run this way. In this case, it can be very convenient to place the fixture offsets in the control's memory for the next time the workpiece is run, saving the operator from having to input the program zero values over and over again.

Comparison to G92. Fixture offsets even have advantages over G92 commands that are not related to multiple program zero points. As stated, the G92 command is the older of the two methods of assigning program zero. Fixture offsets, being newer, have overcome some of the limitations of the G92 command.

As you know, a G92 command includes a letter address for each axis of the machine. The value for each letter address represents the distance *from* the program zero point to the *current* location of each axis. Generally speaking, this current position is the machine's reference point. As long as the machine is at its proper starting point when the G92 command is executed, everything will be just fine. But, as mentioned, if the machine is not where it is supposed to be, the control will simply *assume* the machine is where it should be. This is the greatest cause of crashes on CNC machines.

On most forms of CNC equipment, this basic problem has been overcome with fixture offsets. On most controls, the fixture offset values are *not* taken from the current position of the machine. They are taken from the machine's *reference point*. The control will keep constant and automatic track of the machine position relative to its reference point. If the machine is out of position when the fixture offset command is executed, the control will take this distance from the reference point into consideration before moving. In essence, the machine cannot be out of position. On most controls, this is the major advantage of using fixture offsets over G92 even when only one coordinate system is required.

Possible compatibility problems.   Many CNC controls with fixture offsets allow the use of G92 and fixture offsets in the same program. While there are not many applications where this is helpful, it is possible. However, there are CNC controls that do not allow fixture offsets to be used with the G92 command. If, for some reason, you intend to use the two together, you must check the programming manual for the particular control to find out whether or not it is possible.

Fixture offsets from the operator's viewpoint.   On most controls, as with G92, the operator or setup person must take a measurement. This measurement taken for fixture offsets will differ from the G92 measurement in two ways. First, the measurement for fixture offsets will be to find the distance between program zero and the machine's reference point. While this may also be the case with G92, remember that the G92 command is the distance from program zero to the *current* machine position. If the programmer elects to start from some position other than the machine's reference point, the measured distance will *not* be to the machine's reference point.

Second, the direction of measurement is usually reversed with fixture offsets. That is, the sign (plus or minus) of the value going into the fixture offset is usually opposite that of the values that would have gone into the G92 command. With G92, the distance is measured *from* program zero *to* the starting position. With fixture offsets, the distance is usually measured *from* the machine's reference position *to* the program zero point. This means that values that would have been positive for a G92 command will be negative for fixture offsets.

How the actual measurement is taken will vary, depending on the operator's or setup person's preferred technique and the setup being made. Once the measured values are known for each axis, the operator must store the values in the proper fixture offsets. When the control has fixture offsets, there will be a table of offsets similar to the tool offsets used with tool length and cutter radius compensation. However, this table of offsets will include several values for each offset. If the machine has three axes (X, Y, and Z), there will be three values for each offset. The operator must now enter one value for each axis being assigned (*X, Y, Z,* etc.).

This brings up a good point. The more fixture offsets being used, the more measurements the operator must take. For example, if the program is to use three coordinate systems, and if the machine has three axes, *nine* measurements must be made (*X, Y,* and *Z* for three-coordinate systems). Also, these values must be correctly entered into the proper offsets. When more than one coordinate system is involved, the use of fixture offsets makes programming easier, but makes more work for the operator during setup.

Fixture offsets from the programmer's viewpoint

As with all forms of compensation, programming with fixture offsets is much easier. Depending on how many fixture offsets are available, there will be a series of G codes, one for each fixture offset. While the techniques to program fixture offsets will vary dramatically from control to control, one very common configuration on CNC controls allows six fixture offsets. This means six G codes will be involved. This common configuration uses the series of G codes from G54 through G59 to command the various fixture offsets to be used. G54 will represent fixture offset 1, G55 will represent fixture offset 2, and so on through G59.

In a program, all that the programmer needs to do is include the proper G code (G54 to G59) on or before the desired motion commands. From there, if in the absolute mode, the control will automatically move the axes relative to the current program zero point, making it very easy for the programmer to switch from one program zero point to another from within the same program.

Example program showing fixture offsets

Figure 5.22 shows a simple drawing of the workpiece to be machined. Figure 5.23 shows how four of these parts are mounted on the table of a vertical machining center in different places.

Imagine you had to machine four identical parts on the table. Even though these four parts are identical, the same principles would apply if they were not.

Note that the program zero points for these four parts are in the lower left-hand corner. Before the program can be run, these four program zero points will have to be measured and corresponding values must be entered into the proper fixture offsets.

Now let's look at the program. Notice how simple it is to change from one coordinate system to another within the program.

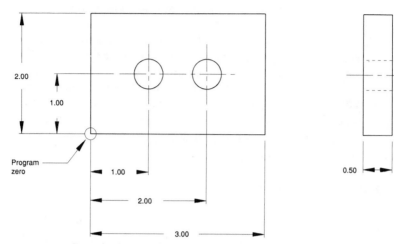

Figure 5.22   Print for fixture offset example program.

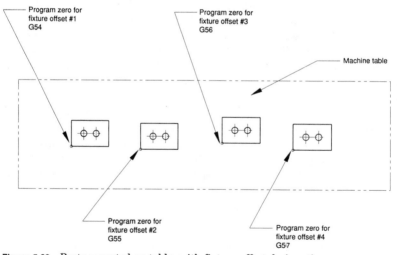

Figure 5.23   Parts mounted on table, with fixture offset designations.

O0007 (program number)
N005 G54 (select coordinate system no. 1 to machine leftmost part)
N010 G90 S700 M03 (select ABS mode, turn spindle on at 700 RPM)
N015 G00 X1. Y1. (move over first hole in leftmost part)
N020 G43 H01 Z.1 (rapid above part)
N025 M08 (turn coolant on)
N030 G01 Z-.75 F3. (drill first hole)
N035 G00 Z.1 (rapid out of hole)
N040 G00 X2. (move over second hole)
N045 G01 Z-.75 (drill second hole in first part)

```
N050 G00 Z.1 (rapid out of hole)
N055 G55 X1. Y1. (select fixture offset 2 and move over first hole)
N060 G01 Z-.75 (drill first hole in second part)
N065 G00 Z.1
N070 X2. (move over second hole in second part)
N075 G01 Z-.75 (drill second hole in second part)
N080 G00 Z.1
N085 G56 X1. Y1. (select fixture offset 3 and move over first hole)
N090 G01 Z.75 (drill first hole in third part)
N095 G00 Z.1
N100 G00 X2. (move over second hole in third part)
N105 G01 Z-.75 (drill second hole in third part)
N110 G00 Z.1
N115 G57 X1. Y1. (select fixture offset 4 and move over first hole)
N120 G01 Z-.75 (drill first hole in fourth part)
N125 G00 Z.1
N130 G00 X2. (move over second hole in fourth part)
N135 G01 Z-.75 (drill second hole in fourth part)
N140 G00 Z.1 M09
N145 G91 G49 G28 Z0 M19
N150 G28 X0 Y0
N155 M30
```

Notes about the program.

1. For this program, all parts have the same *Z* height above the table. If you are working on parts of different thicknesses, you must be careful with your *Z*-axis commands, especially when moving from one coordinate system to another.

2. Though this is a very simple program, it showed much of what you need to know about using multiple fixture offsets.

3. If you are changing tools, and continuing to machine on the parts, we recommend including the current fixture offset command at the beginning of each tool. This will help if you have to pick up in the middle of the program to rerun a tool.

Dimensional tool offsets

This kind of compensation is used on all forms of turning centers. Like other forms of compensation, it makes life easier for the programmer. This form of compensation also helps the operator as well.

Among turning center programmers and operators, this form of compensation is usually referred to as *offsets* or *tool offsets*, not as *dimensional tool offsets*. We give it this descriptive name to keep you from becoming confused between this form of compensation and the offset table used on the various CNC controls.

Turning centers have the ability to perform a wide variety of operations on a workpiece. Among the common operations performed are rough turning, rough facing, finish turning, finish facing, drilling, rough boring, finish boring, threading, grooving, and knurling. Dimensional tool offsets are required for *every* kind of machining operation performed on turning centers. They are required for three reasons.

First, it is almost impossible for an operator or setup person to set every tool into position perfectly. There will almost always be some misalignment of the tools being held. Note that this misalignment may be very small, sometimes no more than 0.0005 in, but probably enough to throw the workpiece out of tolerance. Dimensional tool offsets allow the operator to fine-tune the positions of the tools without physically moving the tools in the setup.

Second, dimensional tool offsets are required to allow for tool wear. Even if the operator or setup person *could* perfectly set each tool into position, as the series of workpieces is being run, each tool would experience tool wear. For example, imagine one of the tools is used to finish-turn a 3-in diameter. The operator is lucky and the tool is placed perfectly into position. When the very first workpiece is run, the 3-in diameter measured precisely 3.0000 in. Now imagine the operator has 1000 of these parts to run. The second part is run, then the third, and so on. Eventually the "3-in" diameter will no longer measure 3.0000 in because of tool wear. When the operator gets 50 parts into the job, the diameter is found to be 3.0006 in. This means that 0.0003 in of material has worn from the cutting edge of the tool. This small amount of tool wear is probably not sufficient to warrant changing tools. The tool is not severely damaged, it is just slightly worn. Dimensional tool offsets can be used to bring the workpiece back on size. This process can be repeated until the tool *is* worn sufficiently to warrant replacement. At this time, the tool will be changed and the dimensional tool offset is set back to its original value.

Note that the amount of tool wear will vary dramatically, depending on the material being machined, the configuration of the tool, the cutting conditions set up by the programmer, and the material of the cutting tool itself. But rest assured that there will be some tool wear in *every* tool used in *every* job run.

Third, dimensional tool offsets are designed to help the operator machine the first workpiece on size. Because of the two previously mentioned reasons for using dimensional tool offsets, it should be obvious to you that the operator will not know the status of a tool until a workpiece is machined. That is, until a tool cuts something, the operator will have no idea as to whether or not the tool is adjusted correctly.

You can relate this to a radio in an automobile. With the new digital technology, you do not know what station is tuned in until you turn the radio on. Once on, you can adjust the radio to the desired station. In a similar way, the operator of a CNC turning center will not know whether a tool will machine on size until the tool cuts something. Once it does, it will be easy for the operator to adjust it to machine the proper size.

It would be foolish for the operator of a CNC turning center to allow a tool to cut a critical surface without first adjusting the offset for the tool. This adjustment can be made in the direction to force the tool to cut so that it leaves extra stock. If this technique is used, the operator can rest assured that there will be stock left to machine. Also, once the tool has machined, the operator can easily measure the surface to tell how much more stock needs to be machined. The offset for the tool can then be adjusted accordingly. While all of this may sound a little complicated, it is something the operator of a CNC turning center will do on a regular basis. Whenever a new tool is loaded into the machine, the operator will adjust the offset to force the tool to machine with extra stock, allow the program to cut with the tool, measure the machined surface, adjust the offset to bring the tool on size, and finally, rerun the tool to machine the surface to size. If this technique is used for each tool in the program, it almost guarantees that the first part to be machined will come out correctly.

The use of dimensional tool offsets. Now that you have a basic understanding as to *why* dimensional tool offsets are so important, let's look at how they are used. The use of dimensional tool offsets remains quite consistent from one turning center control to the next. From the programmer's standpoint, a T word will be included in the program to specify which offset number is to be used. Actually, the T word does two things on most turning center CNC controls. It is usually a four-digit number. The first two digits represent the tool station number of the turret being used, and actually activate the machine's turret to rotate to the commanded tool station. The second two digits of the T word represent the offset number to be used with the tool and instate the offset.

It is wise to use some logical approach to assigning tool offset numbers. We recommend using the same offset number as the tool station number. For example, use offset 1 for tool station 1, offset 2 for tool station 2, and so on. This allows the programmer and operator to remain in sync.

Here's an example of a turret command:

```
N015 G00 T0101
```

In this command, the T0101 tells the control to index to station 1 and, at the same time, instate offset 1. For most CNC turning center controls, if there is a value in offset 1, the turret will actually move by the amount in the offset. For this reason, we recommend including a

G00 word with the turret command to tell the control how to make this possible movement (at rapid). Other CNC turning centers are a little smarter. The offset will not be instated until the next motion command. This means that no motion will occur in the above command, therefore no G00 will be needed.

Each dimensional tool offset will have two values in the actual offset table. One of the values represents the X-axis offset, and the other the Z-axis offset. The control screen will make it clear on the offset page as to which is which. This allows the operator to adjust for problems in both the X and the Z axes. If a diameter is coming out incorrectly, the X-axis offset will need adjusting. If the length of the workpiece (or one of the shoulders) is coming out incorrectly, the Z-axis offset would need adjusting.

Deciding what needs to be done with tool offsets to make the necessary adjustments can sometimes be tricky. For this reason, we recommend that you adjust the offset for each tool in the program and make it machine on size *before* going on to the next tool.

Tool nose radius compensation

All forms of "single-point" turning and boring tools have some form of radius on the very cutting edge of the tool. In the inch system, there are four standard sizes for the *nose radius* of turning and boring tools. They are: 1/64, 1/32, 3/64, and 1/16 (0.0156, 0.0316, 0.046, and 0.0625) in. Though you may consider these radii to be quite small, when a cutting tool is used for machining contours on a workpiece, the small nose radius may be sufficient to cause a deviation from the required shape of the workpiece. Figure 5.24 shows you what this radius looks like on a single-point turning tool.

When a program is written for any single-point turning or boring tool, it is the extreme surface of the tool in X and Z that is actually being programmed. Figure 5.25 shows this.

For straight turns and straight faces, the point of the tool will actually come into contact with the workpiece, machining it correctly (look at Fig. 5.26). When the surface is at 90° to the cutting tool (turned diameters and faces), the tool nose radius will not affect the surface contour being machined.

However, when angular and circular surfaces are machined, the point of the tool being programmed will *not* actually come into contact with the workpiece, and the surface being machined will not match print dimensions. Though this deviation is quite small, there are times when it is sufficient to cause the part to be scrapped.

Figure 5.27 shows the imperfection caused by the radius of the tool on angular and circular surfaces. As you can see, angular surfaces will

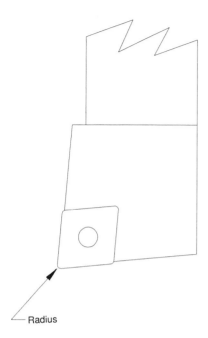

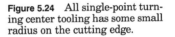
Radius

Figure 5.24 All single-point turning center tooling has some small radius on the cutting edge.

come out with excess stock and circular surfaces will not be precisely to size. Though Fig. 5.27 intentionally exaggerates this condition, if no concern is given to this radius, the part may not come out to size.

Tool nose radius compensation gives the programmer an easy method of ensuring that the radius of the cutting tool remains in constant and correct contact with the surface to be machined. It forces the radius of the cutting tool to remain tangent to the surfaces of the workpiece being machined.

However, we must point out that there are two times, even with single-point tools, when we recommend that tool nose radius compensation *not* be used. The first is in roughing. The small deviation from the correct tool path will not be sufficient to cause problems. The finishing tool (using tool nose radius compensation) will sweep the small amount of excess stock from the surface as if it were not even there. Also, while roughing, the tool is constantly changing direction, making a roughing pass in one direction, then reversing direction to make the next pass, and so on. This constant reversal in direction during roughing will make tool nose radius compensation more difficult to program.

The second time that we recommend *not* using tool nose radius compensation for single-point turning tools is when the angular and circular surfaces generated by the finishing operation are not critical at

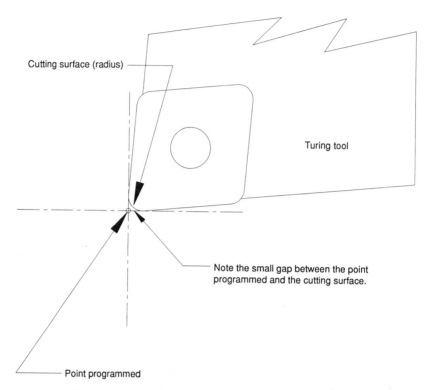

Cutting surface (radius)

Turing tool

Note the small gap between the point
programmed and the cutting surface.

Point programmed

Figure 5.25 This drawing illustrates that the point being programmed is not the actual cutting surface of the tool.

all. If simple chamfers and radii are simply being made to break sharp edges, there is no real need for tool nose radius compensation. It does not hurt to use it, but it is not necessary. On the other hand, if the surfaces to be machined *are* critical, tool nose radius compensation must be used.

Steps to using tool nose radius compensation

From the programmer's standpoint, programming tool nose radius compensation is quite easy. It is actually quite similar to programming cutter radius compensation for a machining center, but much easier. The three steps to using both forms of compensation are identical. First, you must instate compensation; second, you must give the motion commands; and third, you must cancel compensation.

Instating tool nose radius compensation.   For most CNC turning center controls, it is quite simple to instate tool nose radius compensation. As

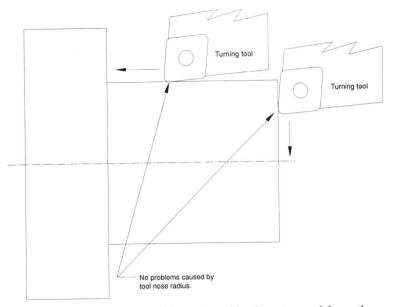

Figure 5.26 For machining straight surfaces like diameters and faces, the radius of the cutting tool has no ill effect on the workpiece.

with cutter radius compensation, you must be able to determine how the tool will be related to the workpiece during the machining operation. To do so, you look in the direction the tool will be moving during the cut, rotating the print if necessary. Looking in this direction, ask the question, "What side of the surface to be machined is the tool on, the right side or the left side?" If the tool is on the left side of the surface to be machined, you will use a G41 to instate tool nose radius compensation. If the tool is on the right side of the surface to be machined, you will use a G42 to instate tool nose radius compensation. See Fig. 5.28 to see how this is done.

Once you know which side of the surface the tool is on, left or right, all you must do to instate tool nose radius compensation is to include the corresponding G code (G41 for left or G42 for right) in the tool's first motion command to the workpiece.

Motion commands under the influence of tool nose radius compensation.
The same basic limitations that apply to cutter radius compensation also apply to tool nose radius compensation. However, since the radius of the turning tool is *much* smaller than the radius of the typical tool used with cutter radius compensation, there seem to be fewer application problems with tool nose radius compensation.

The biggest problem seems to be in machining narrow recesses. The

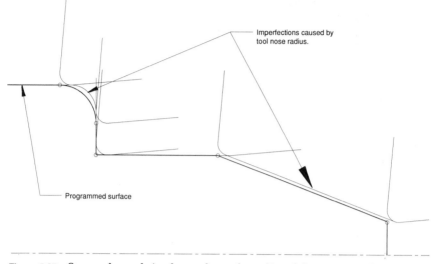

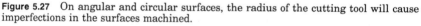

Figure 5.27 On angular and circular surfaces, the radius of the cutting tool will cause imperfections in the surfaces machined.

recess must be at least wide enough to allow twice the tool radius into the recess. If it is not, the tool will actually violate one side of the recess before the depth of the recess is reached. Figure 5.29 demonstrates this possible problem.

Once tool nose radius compensation is instated, the control will continue to keep the cutting tool properly aligned during the machining operation.

Canceling tool nose radius compensation. As with cutter radius compensation, the programmer must remember to cancel tool nose radius compensation. If the compensation is not canceled, the control will remain under the effect of tool nose radius compensation even on subsequent tools, possibly causing them to machine incorrectly.

A G40 word is used to cancel tool nose radius compensation. The G40 can be simply included in the tool's command to retract to the tool change position.

Offset used with tool nose radius compensation

As with all forms of compensation, tool nose radius compensation requires an offset to tell the control the radius of the tool. While the actual method used to assign offsets varies from one turning center control to the next, most use the same tool offset in which dimensional

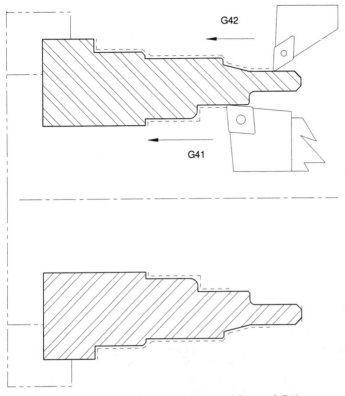

Figure 5.28 This drawing illustrates the use of G41 and G42.

offset values are stored. As you know, dimensional offsets use two
storage positions in the same offset, one for X and one for Z. When tool
nose radius compensation is used, a third position in the offset is used.
Usually it is designated with the letter address R. In this position, the
operator stores the radius of the tool being used. As mentioned earlier,
the radius of the turning tool will not vary nearly as much as the radius of milling cutters used with cutter radius compensation. The four
typical choices will be: 0.0156 (1/64), 0.0316 (1/32), 0.0456 (3/64), and
0.0625 (1/16) in. Also, when a cutting tool's insert dulls, it will be replaced by another with an *identical* nose radius. This means many of
the advantages discussed during cutter radius compensation do not
apply to tool nose radius compensation. With tool nose radius compensation, the benefit is limited to keeping the profile of the workpiece
accurate.

Figure 5.30 shows the revised offset table for turning center controls. The T in the offset table is also required on most turning center
controls when using tool nose radius compensation. This position al-

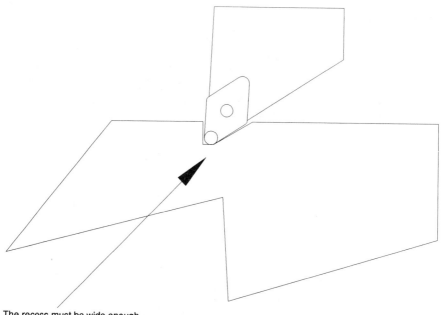

The recess must be wide enough
to allow the tool nose radius in.

Figure 5.29 When using tool nose radius compensation, the programmer must be careful not to allow the tool radius to violate the surface of the part.

lows the user to tell the control the *type* of tool being used with a code number. This helps the control to make decisions during motion commands as to which way to compensate the tool. For example, the control will have to know whether the tool is designed to machine on the outside diameters (turning tool) or the inside diameters (boring bar) in order to make compensations correctly. If the T offset technique is required, the programming manual for the control will show the code numbers relative to the T offset. While not all CNC turning center controls require this T word, it is a common technique.

Example program showing tool nose radius compensation

In this example, we will stress only the use of tool nose radius compensation. While the various programming words may still seem a little foreign, concentrate on what is happening relative to tool nose radius compensation. Figure 5.31 shows the workpiece to be machined. Notice that the part has already been rough-machined. All that is left is to finish-turn and finish-bore the workpiece.

#	X	Z	R	T
1				
2				
3				
4				
5				
6				
7				
8				
9				
10				
11				
.				
.				
.				
99				

Figure 5.30 This revised turning center offset table shows the R and T designation many turning center controls require.

Offset no.	X	Z	R	T
1	00.0000	00.0000	0.0316	3
2	00.0000	00.0000	0.0316	2

Program:

N005 G50 X _____ Z _____ (set program zero)
N010 G00 T0101 M41 (index to tool 1, select low spindle range)
N015 G96 S300 M03 (turn spindle on at 300 SFM clockwise)
N020 G00 G42 G00 X4.55 Z.1 M08 (instate tool nose radius compensation, rapid to first position, coolant on)
N025 G01 Z0 F.007
N030 X4.75 Z-.1
N035 Z-1.
N040 X5. Z-1.5
N045 X5.15
N050 G03 X5.25 Z-1.55 R.05
N055 G01 Z-3.
N060 X5.36
N065 X5.5 Z-3.07
N070 Z-4.
N075 X6.1
N080 G00 G40 X _____ Z _____ T0100 (rapid back to start point, cancel tool nose radius compensation)
N050 M01 (optional stop)
N055 G50 X _____ Z _____ (set program zero for second tool)
N060 G00 T0202 M42 (index to tool 2, select spindle range)

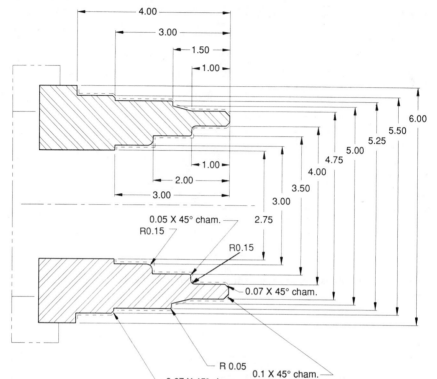

Figure 5.31 Print used for tool nose radius compensation example program.

N065 G96 S300 M03 (turn spindle on at 300 SFM clockwise)
N070 G00 G41 X4.14 Z.1 (instate tool nose radius compensation, rapid to first position)
N075 G01 Z0 F.006
N080 X4. Z-.07
N085 Z-.85
N090 G03 X3.7 Z-1. R.15
N095 G01 X3.6
N100 X3.5 Z-1.05
N105 Z-2.
N110 X3.3
N115 G02 X3. Z-2.15 R.15
N120 G01 Z-3.
N125 X2.65
N130 G00 Z.1 (rapid out of hole)
N135 G40 X _____ Z _____ T0200 (cancel tool nose radius compensation, rapid back to starting point)
N140 M30 (end of program)

Wire Radius Compensation

This section will discuss one of the compensation types for wire EDM machines called *wire radius compensation*. For wire EDM machines, this feature allows the programmer to input coordinates into the program with no concern for the wire diameter. In most cases, the coordinates can be taken right from the print, making programming much simpler. As with cutter radius compensation, the control will automatically keep the wire a specified distance (the offset value) away from the surfaces of the workpiece during the cutting motions. This generates the motions necessary to make the part to the proper size.

One advantage of using wire radius compensation has to do with multiple passes. Depending on the accuracy and finish required of the wire EDM process, it is possible that more than one pass may be made around the workpiece. In fact, for critical workpieces, it is common to make four or five total passes around the shape, each contributing to better accuracy and finish. This is similar to metal cutting operations when roughing and finishing passes are made.

With wire radius compensation, the same coordinates used to machine the final pass can be used for all passes around the workpiece. This means that the programmer will not have to develop multiple sets of coordinate values for the various trim passes required to make the part to the size and finish requirements you need. For all passes, the actual coordinates taken for the finished workpiece can be used.

Another advantage to wire radius compensation has to do with machining punches and dies. When wire radius compensation is used to machine punches and dies, it allows the programmer to easily incorporate die clearance in the program with the same set of coordinates that machined the part to size. *Die clearance* is a term used to describe the clearance between a punch and die. Of course, for a punch and die to function properly, there must be some clearance between them. The thickness of the part to be punched determines how much die clearance is required. The configuration of the part to be punched determines whether the punch or the die must incorporate the die clearance. If the hole must be punched to a given size, the *punch* must be made to that size, and the die must be made bigger by the die clearance amount. If the blank is to be made to size, the *die* must be made to the same size and the punch must be made undersized by the die clearance amount. Also note that if wire radius compensation is used, the same set of coordinates used to machine the punch can be used to machine the die. Only minor program changes are necessary to transform a punch program to a die program.

How wire radius compensation is used

Now that you know some of the advantages to using wire radius compensation, let's look at how you do it. Wire radius compensation requires three steps in the program:

1. Instate wire radius compensation

2. Drive the wire through its motions

3. Cancel wire radius compensation

Initializing wire radius compensation. First, let's look at how you instate wire radius compensation. There are three G words related to
wire radius compensation:

G40 Cancel

G41 Wire left

G42 Wire right

To instate wire radius compensation, you will be choosing from G41
(wire left) and G42 (wire right). You must be able to decide how the
wire will be related to the workpiece during the cutting motions. To
evaluate this relationship, look in the direction that the wire will be
moving during the cut (rotate the print, if necessary). Looking in this
direction, ask yourself, "What side of the workpiece is the *wire* on? Is
the wire on the left side or the right side of the workpiece?"

If the wire is on the left side of the workpiece, you will use G41 to
instate wire radius compensation. If the wire is on the right side of the
workpiece, you will use G42. Note that this technique is common to
cutter radius compensation and tool nose radius compensation. Figure
5.32 shows you how to determine G41 and G42.

You can make a general rule for wire radius compensation. If you
make the arbitrary decision that you will *always* machine workpieces
in a counterclockwise direction around the shape, dies (or any inside
shape) will *always* use G41 and punches will *always* use G42. Of
course, if you decide to machine workpieces in a clockwise direction
around the part, the rule will reverse. Also note that if you are mak-

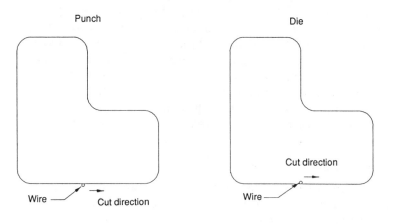

Figure 5.32 These drawings illustrate the difference between G41 and
G42.

ing reverse-direction trim passes, the condition will change for each trim pass (G41 for the first pass will change to G42 for the next reverse-direction pass).

Prior to the command to instate wire radius compensation, it is important to position the wire so that as you initialize compensation and begin cutting, a right angle is formed. As the wire moves to the programmed point after the initialization block, it will automatically be offset so that the edge of the wire is flush (tangent) with the surface to be cut. Figure 5.33 demonstrates this.

Offsets with wire radius compensation can be commanded with one of two words, depending on control builder. Usually a D or H word is used to instate the offset. However, even the D or H word varies from control to control with regard to what the word actually does. Some controls use the D or H word to specify the offset number in which the compensation value is to be stored (like cutter radius compensation). Others use the D or H word to specify the actual value of the offset itself.

In the initialization block, a D or an H word will be included with the G41 or G42 to give the control the offset information. This information tells the control how far to keep the wire away from all surfaces to be cut. For example, if you needed an offset of 0.0062, it can either be programmed directly with the D or H word as

H.0062   or   D.0062

Or the D or H word may point to the offset. D01 or H01 would point to offset 1. In offset 1, a value of 0.0062 will be stored. Remember that no single wire EDM control can use both methods of assigning offsets. Ei-

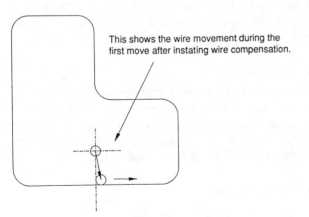

This shows the wire movement during the first move after instating wire compensation.

Figure 5.33   For most wire EDM controls, the movement instating wire radius compensation must form a right angle to the first surface to be machined.

ther you will be assigning offsets directly with D or H, or the offset table will be used.

Note that the actual value of the offset can be somewhat difficult to determine for a wire EDM machine. If using a 0.010-in-diameter wire, you may think a 0.005-in wire radius offset would be used, but this is incorrect. In *all* cases, the offset *must* take into consideration the *overburn* or *gap width* amount. In finishing, this gap width may be very small, so the offset will be close to 0.005 in for a 0.010-in-diameter wire, but not exactly. Another problem with determining the offset values occurs during roughing. In roughing, the programmer has to allow not only for the roughing pass overburn, but also for the amount of stock consumed in trim passes. Beginners can easily get bogged down trying to come up with offset values by themselves. To help the programmer, most wire EDM machine manufacturers give the user a previously tested set of cutting conditions and offsets to be used for machining a variety of workpiece materials. This list of conditions and offsets also takes into consideration part finish, workpiece material, workpiece thickness, and wire diameter.

Driving the wire around the shape. Once wire radius compensation is properly instated, the programmer will program the workpiece coordinates to drive the wire through its programmed path around the shape. Because the wire radius is quite small compared to cutter radius compensation and tool nose radius compensation, most of the problems previously discussed about machining with compensation do not apply to wire radius compensation.

The only problem the programmer must be careful with concerns inside radii. When machining an inside radius, the radius to be machined *must* be bigger than the *largest* offset. This means that a 0.010-in-diameter wire could never machine a 0.005-in inside radius because of the gap width.

Canceling wire radius compensation

When finished with the cutting movements, the programmer *must* remember to cancel the wire radius compensation with G40. It is important that this is done on a straight-line motion command (not G02 or G03). At the end of the cancellation command, the wire's centerline will be on the programmed end point of the command.

Example program showing wire radius compensation

Now let's put it all together. Here is a simple program that shows how to program wire radius compensation correctly. Look at the method by which the compensation is instated and canceled. Also, this example shows just the method of using wire radius compensation and makes no trim passes. Figure 5.34 shows the simple drawing for the punch shape.

Program:

```
N005 (wire radius compensation example)
N010 G92 X1. Y-.15 (set program zero)
N015 G90 (select absolute mode)
N020 G42 H.0055 (select wire right and 0.0055 offset value)
N025 C653 (select machining condition)
N030 G01 Y0 (move to point 2)
N035 X1.5 (move to point 3)
N040 Y.5 (move to point 4)
N045 X1. (move to point 5)
N050 Y1. (move to point 6)
N055 X0 (move to point 7)
N060 Y0 (move to point 8)
N065 X.9 (move to point 9)
N070 G40 Y-.02 (move to point 10 and cancel compensation)
N075 M00 (program stop to apply magnets)
N080 G01 X1. (move to point 11 to cut off)
N085 Y-.15 (move back to start point)
N090 M02 (end of program)
```

Note that in block number N020, we showed the technique whereby the actual *value* of the offset was included in the program. Here are two other ways block number N020 could have been programmed (based on control model):

N020 G42 D01   or   N020 G42 H01

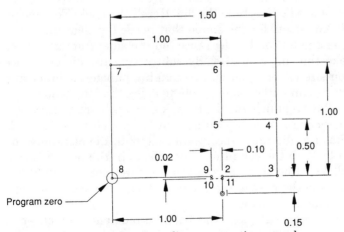

Figure 5.34 Print used for wire radius compensation example.

In either of these second methods, the control expects the offset value to be stored in offset 1.

Incorporating die clearance with offsets. As stated, die clearance is the difference in size between the punch and die. There must be some clearance between punch and die in order for the combination of punch and die to work. The amount of die clearance will vary, depending on the material and thickness to be punched.

It is important to machine test squares to attain the exact values for your on-size offsets. The on-size offsets, if used, will make the part come out to the *exact* size that the part coordinates in the program commanded. As mentioned earlier in this key concept, the programmer can lie to the control about the values of the offsets and force it to machine the part with die clearance.

If the part to be punched requires that the *hole* is of a certain size, the *punch* must be on size, and the die clearance must be incorporated in the die (the die must be made bigger). On the other hand, if the blanked shape being punched must be a certain size, then the *die* must be made on size and the die clearance must be in the punch (the punch must be smaller). Of course, in *all* cases there *must* be some die clearance between the size of the punch and the size of the die.

The amount of die clearance is determined by the material and thickness of the part to be punched. The thicker the part, the more die clearance that is required. The programmer can check in any reference manual for punches and dies to determine the needed die clearance amount for a particular material and thickness.

After determining the amount of die clearance that is necessary, the programmer can easily use wire radius compensation to create the needed die clearance. In fact, the same coordinates used to machine the punch can be used to machine the die. Here's the trick: To come up with the modified offsets for the shape that needs die clearance, simply *subtract* the amount of die clearance (on the side) from *all* offsets needed to make the shape on size. By subtracting the die clearance from all the on-size offsets, you will be making punches smaller and dies bigger. Here's an extended example to drive the idea home.

Imagine you have a punch and die to make, and you are going to use 0.010-in-diameter wire. You determine that the punch must be made on-size and that the die must incorporate 0.002-in die clearance (on the side). You look up the conditions and offsets in the builder's condition manual. You find that you need to use three sets of conditions and offsets to make the part to the finish and accuracy requirements. So, three passes must be made.

For the first (roughing) pass, the book recommends an offset of 0.0085 in. For the second pass, the book recommends 0.0056 in, and for the third pass, the book recommends 0.0053 in. You test the recommended conditions and offsets on a test square and find them to be correct.

Since the punch must be made on-size, the programmer will machine the punch with the recommended on-size offsets (0.0085, 0.0056, and 0.0053 in). When the die is machined, the on-size offsets must be modified by the amount of die clearance. The programmer will take the on-size offsets and *subtract* 0.002 from all one-size offsets to come up with the offsets for a 0.002-in die clearance.

The die offsets will be 0.0065 in for the first (rough) pass, 0.0036 in for the second pass, and 0.0033 in for the third pass. These offsets will keep the wire 0.002 in closer to the programmed path than the on-size offsets, and correctly form the die clearance.

Wire Taper Compensation

This form of compensation applies only to wire EDM machines. The primary need for machining taper is for machining dies. Dies normally require some form of clearance angle on the bottom of the die to allow the blanked slugs from the punched workpiece to fall easily through the die. There are several different ways to machine this taper, depending on the requirements for cycle time, clearance angle, and basic personal preferences.

The method by which the wire EDM machine produces taper is unique. To machine taper, the wire EDM machine must have four axes of motion. The X and Y axes are for the basic table motion, and allow the desired shapes to be programmed. The U and V axes control motion of the upper guide. (See drawings presented during key concept no. 1.) The wire can be tilted from above with the U and V axes. At first glance, it may seem that taper programming would be quite difficult. But, rest assured, taper programming is as easy as programming wire radius compensation.

We hope to discuss the most commonly used methods of taper machining in detail. You may have to read carefully to catch the implications of what is being presented.

First, let's discuss the most basic considerations:

Do you want die land? *Die land* is the amount of nontapered area of the die. The purpose of die land is to allow the die to be sharpened (ground on the top surface) with absolutely no deviation in the size of the die opening. Depending on the taper angle, die land may not be completely necessary. Many companies machine only 0.25° (1/4°) taper as the taper angle. If your desired taper angle is this small, and if the part to be punched is relatively thick, the very small change in die opening size during resharpening will be minimal. For example, if machining 0.25° as the taper angle, when you sharpen the die and remove 0.007-in stock from the top surface of the die, the change in die opening size will be only 0.0000305 in (on the side). For most applications, this small deviation will not affect the performance of the die.

Knowing this, most companies can save a great deal of machining time by machining taper right up to the top of the die. That is, they machine absolutely no land whatsoever. But remember, as the taper angle grows, the amount of deviation to the die opening during sharpening will also grow. You can easily calculate the amount of growth (on the side) by applying this formula:

Growth (on the side) = tangent of taper angle
× amount of stock to be removed per sharpening

By applying the above formula, you can determine whether the deviation during sharpening will be acceptable.

How much die land do you need?   Of course, the amount of die land will be directly related to the life of the die. The larger the die land, the longer the die life. However, you must also be concerned with the ease of removing the blanked part. You can refer to punch-and-die reference handbooks to get a recommendation for the amount of die land for various blanked-part thicknesses.

How big do you want the taper angle to be?   The taper angle in a die affects the strength of the die. Generally speaking, you will want the taper angle to be at its smallest acceptable angle to keep the die strong. The size of the taper angle will also affect the best and easiest way to machine the die. We will discuss this more later.

Do you want to cut the die with the top up or down?   For most applications, it is best to cut the die with the top of the die up. It makes it easier to measure the die opening, once machined, while it is still mounted in the setup. However, you must also be concerned with getting the slug out of the machine after the first roughing pass! If the wire diameter is very small, and the taper angle is very large, it may be impossible (for large dies that cover the entire work table) to remove the slug after the first roughing passes. If you multiply the tangent of the taper angle times thickness of the part that you are machining, you can easily tell whether the slug can be removed through the die opening. If the value you calculate is *smaller* than the wire diameter you are using, you will be able to remove the slug from the top. If not, you must turn the die over and machine with the top of the die down to the table top (for large dies that fill the table).

Now that you know some of the considerations with which you must be concerned, let's look at some recommended methods of machining dies with taper.

Tapering to the top of the die (no die land). This is the easiest way to machine a die. With this method, you only need one program that will rough and finish the entire die. The only implication relative to taper during programming is that you specify the taper direction (left or right) and the taper angle correctly.

Machining a die with a land. In regard to machining a die with a land, there is a constant argument going on as to whether it is smarter to machine the die land first and then the taper angle, or whether it is smarter to machine the taper angle first and then the land. Much of this is based on personal preference. Depending on the taper angle and thickness of the die land, we recommend that you run the taper angle first and then machine the land. This may not be possible if the taper angle or die land thickness is too large.

We recommend this because the total cycle time to machine the die would be dramatically reduced with this method. If you machine the die land first, the entire shape being machined must be finished to the proper size in the (straight) land area of the die. Then the taper angle must be machined. If several trim passes are necessary, this can almost double the production time for the die. By using our recommended method, the taper angle is cut first with roughing conditions, leaving only a very small area to be machined for the land area of the die. The necessary trim passes may be reduced, but even if not, the surface area for machining this area is reduced to the point that the trim passes will occur much faster.

Imagine, for example, that you are machining a 0.25° taper angle and the thickness of the die land is to be 0.125 in. In this case, if the taper is machined first, the amount of stock left for the land to be machined is only 0.00054 in, and one trim pass will easily remove this stock. To calculate the amount of stock for the die land area, simply multiply the tangent of the taper angle times the thickness of the land.

On the other hand, if the taper angle is 5° and the land thickness is 0.25 in, the amount of stock will be 0.0218, which is greater than the wire diameter. In this case, if you try to use our recommended method, the wire will actually be creating a slug as you machine the die land. This slug will cause problems during cutting, and in this case you will be wiser to machine the die land first.

What we are saying here is that if the stock left in the die land area after machining the clearance angle is greater than about half the wire diameter, then you should cut the die land first. If not, you can save a great deal of time by cutting the taper angle first.

There is one exception to this rule. If you are machining within extremely accurate tolerances (0.0001 in or smaller), you may be better off to machine the die land first in all cases. For extremely close-tolerance work, the cutting conditions will remain more constant and flushing will be better, allowing for an overall better machining environment. This improvement may make the difference between a good die and a scrap die for extremely close-tolerance work, but for general-purpose applications, use our general rule.

How to program taper

Now that you know some of the implications of taper cutting, you can decide for yourself which method of machining taper is best for your particular application. Here we will give you the raw tools to do machine tapering in any way you wish.

As with wire radius compensation, there are three steps to programming taper:

1. Initialize taper (G51 for left, G52 for right)
2. Make machining motions
3. Cancel taper cutting mode (with G50)

(Notice how similar these commands are to wire radius compensation commands. These commands should be easy to remember!)

Initializing taper

First let's discuss initializing taper correctly. To initialize the taper cutting mode, you must look in the direction the wire will be moving during the cut (rotate print if necessary). Looking in this direction, ask yourself, "Which way does the *upper guide* have to move to generate the desired taper, left or right?" If the *upper guide* has to move to the left, you will use G51 to initialize taper. If the *upper guide* has to move to the right, you will use G52 to initialize taper. Figure 5.35 shows how to do this.

As with wire radius compensation, it is wise to make a general rule to help remember which direction is needed. If machining dies with the top of the die *up,* you will use G51 (left) whenever making passes in a general direction counterclockwise around the die. G52 (right) will be used when making clockwise direction passes around the die. Of course, if you cut with the top of the die down, or if machining in the other direction, the rule will reverse.

Now that you can decide between right and left taper (G51 or G52), let's discuss the *angle* you must also include in the initialization command. Most controls use an A word to command the desired taper angle. If you want to machine a ½° taper (per side), you will include the word A.5 in the command to instate the taper cutting mode.

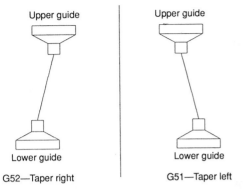

Figure 5.35 captions:

Upper guide

Upper guide

Lower guide

Lower guide

G52—Taper right

G51—Taper left

Figure 5.35 The difference between G51 and G52 for taper machining.

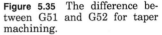

Straight wire is G50 (cancel).

There is a limitation of the maximum possible taper angle on your particular machine, based on part thickness. The thinner the workpiece, the greater the taper angle can be. You must consult the machine-tool builder's machine manual to determine the limitations for maximum taper angle for any one machine tool.

To initialize taper properly, you simply program the G51 (left) or G52 (right) in a command with the A word. Then move to the first surface you wish to machine with taper. The control will instate the taper *during* the movement to the first surface to be cut.

Once initialized correctly, the taper cutting commands will cause the wire to maintain the proper angle around all coordinates programmed.

After all the cutting movements have been made, you *must* remember to cancel taper cutting with a G50 command. The control will bring the wire back to vertical during the *next X* and/or *Y* movement.

The following is an example program that shows how to program taper correctly. Figure 5.36 shows the print for this die shape.

Program:

```
N005 (example program for taper machining)
N010 G92 X.5 Y.2 (set program zero)
N015 G90 (absolute mode)
N020 G41 H.0055 (set radius compensation left with 0.0055-in offset)
N025 G51 A1. (taper left with 1° angle)
N030 C653 (select condition)
N035 G01 X-.485 (move to point 2)
N040 Y.015 (move to point 3)
N045 G03 X.5 Y0 I.015 (arc in approach to point 4)
N050 G01 X1. (move to point 5)
N055 Y1. (move to point 6)
N060 X0 (move to point 7)
```

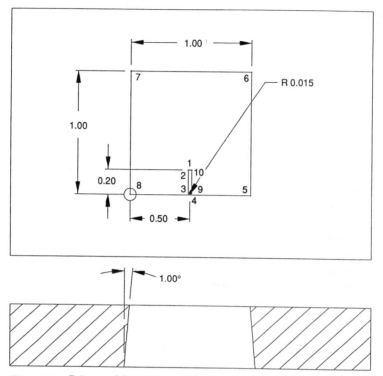

Figure 5.36 Print used for taper machining example program.

```
N065 Y0 (move to point 8)
N067 X.4 (move to glue stop point)
N068 M00 (stop to apply magnets)
N070 X.5 (move back to point 4)
N075 G03 X.515 Y.015 J.015 (arc out to point 9)
N080 G01 Y.2 (move to point 10)
N085 G40 G50 X.5 (cancel radius compensation and taper, move back to point 1)
N090 M02 (end of program)
```

This program needs a little explanation. Blocks N035 to N045 and N075 to N085 allow the wire to instate and cancel compensation and offset on a dummy move. This does two things. First, you know that wire radius compensation, when it is instated, will make an angular move to instate. For punches this is fine, since the approach is in an area that is not actually part of the workpiece, and will eventually be ground off on a surface grinder. But for a die, there will be no subsequent grinding operation. The entire inside shape must be smooth and cannot have any witness marks where the wire approaches the first surface. So, instating offset and taper in the dummy move ensures that the proper taper and offset will be instated when the wire makes its first move to the actual surface of the part to cut.

Second, the *arc in* and *arc out* movements to and from the die surface also ensure that there will be no witness marks in this area. The wire will form a perfectly smooth approach to the first surface to cut.

Some programmers will approach directly the first surface to cut, go all the way around the die, and overtravel the approach point by 0.050 in or so. While this does approximately the same thing, there will be witness marks in the approach area, since the wire will attempt to recut the 0.050-in overtravel area.

Blocks N067 and N068 are necessary, especially for heavier parts, to allow the operator a chance to apply magnets or glued tabs on the slug that is going to fall when the part is cut off.

Taper settings from the operator's viewpoint

As you have seen, there really isn't all that much to programming taper in a die. You simply instate taper, use it, and cancel it. However, there are also some machine settings that must be made by the operator. If these settings are not properly made, the resulting taper will be unpredictable. The control *must* know some basic facts about the setup in order to form the taper correctly. The actual values needed and the method by which they are input will vary from builder to builder. Here are some examples of what may be required:

1. Distance from table top to upper guide
2. Distance from table top to lower guide

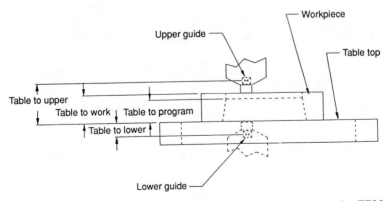

Figure 5.37 This drawing illustrates the kind of information most wire EDM controls require from the operator.

3. Distance from table top to top of workpiece

4. Distance from table top to the land

Figure 5.37 shows the relationship of these numbers. Because of the variety of different CNC wire EDM controls, we cannot be very specific about what the control must know about the setup for taper cutting to work. Suffice it to say that this information is necessary to help the control calculate how much the U and V axes must be moved to maintain the desired taper. Refer to your control's programming manual for more information on how this is done.

# Chapter 6

Key Concept No. 5:
Program Formatting

Until this point, we have been giving you the many building blocks necessary to understand CNC programming for a number of CNC machine types. Knowing about the kinds and usages of CNC machines, the preparation that goes into programming, the types of motion, and the various kinds of compensation is essential to your understanding of CNC.

In this key concept we will put it all together. Here we will show the physical techniques to prepare CNC programs for several types of machines. We must point out the fact that there are *many* ways to format CNC programs, some promising more success than others. This, combined with the fact that there are varying codes used to accomplish the same purpose from one control builder to the next, means that we cannot be totally specific about the formatting of CNC programs for every type of control.

The method we will show stresses the most common way to format the CNC program for the particular machine type being discussed. The highest priority in all cases will be *safety*. If there is some dangerous condition that may exist, we will stress the safe way to handle the problem.

The second priority will be ease of use. There are many ways to format a CNC program. The way we show will be so easy to use that a beginner can apply the techniques given with no difficulty.

Rest assured that this text will give you basics of successful CNC manual programming. With an understanding of the formatting information presented here, you should be able to easily adapt to the machine with which you will be working.

We will begin by giving you the reasons why it is so important to format your programs in a strict and organized manner. Next, we will show you the various formats for specific types of CNC machines. Finally, we will give a list of the most commonly used programming words for the various CNC machine types and describe each word.

Reasons to Strictly Format CNC Programs

As mentioned, there are many, many successful ways to format CNC programs. In fact, the same program could be assigned to five different programmers, and the result would be five different CNC programs, all of which will machine the part properly. We will be showing one way for each type of CNC machine *that works*. We consider the various formats to be safe, allowing you a great deal of flexibility during production. However, if you find yourself in a company currently utilizing CNC machines, and there are programmers who are already writing programs, you may have to adapt to the company's way of doing things. We intend to give you enough background about program formatting to allow you to easily understand different methods from what is shown.

Also, you will find some discrepancy between the various M codes used by different control manufacturers. It is left to the discretion of each machine-tool builder to decide the functions of M codes. While the most basic ones are becoming more standard among the builders, you will want to check to be sure you are using the right ones.

A great part of all CNC programs is nothing more than format information. By format information, we mean the CNC machine requires information in a certain order to function properly. You will find that this format information is redundant. The first time through this information you may be a little confused, but after writing several programs, you will find that you will be writing the same information over and over again, and soon you will have it memorized.

Once again it is as important to know why you are doing things as it is to know how to do them. There are three basic reasons why you must format your programs in a certain way.

Becoming familiar with programming

The first reason has to do with providing beginners a way to become familiar with programming commands. You must understand program formatting well enough to repeat past successes. You will find that program formatting is extremely redundant. If you get in the habit of formatting your programs in the same manner time after time, within a short period of time you will have the program format memorized.

Another point along the same lines has to do with the fact that the program formatting, as we present it, allows you to simply remember what the various programming words do. You will not have to be developing all programming commands off the top of your head. Note the distinction. Without having some kind of format to go by, you would have to memorize each command that goes into your program. With an understanding of program formatting, you will be able to simply look at an example and remember what each command does.

An analogy to this memory jogging is the way a driver interprets the road signs related to driving an automobile. It is doubtful that any driver can recite from memory every road sign used to direct traffic. But most drivers, once they see a sign, are likely to recognize its meaning. The same can be said for program formatting. You can use the various formats we give as a crutch to help you remember the commands required to successfully write programs.

Consistency

Program formatting also allows a high level of consistency from one program to the next. If all programmers in a company use the same basic format to develop their CNC programs, it makes life easier for everyone. Since there are so many possible ways to successfully format CNC programs, this can lead to misunderstandings and arguments between programmers and operators as to whether one format is better than another. We have a saying that applies. We say "You can't argue with success!" If you and others in your company currently feel comfortable with the way you are doing things, by all means keep doing it that way. But we will still ask you to read this presentation carefully because we might give you some ideas about changing your methods. Experienced programmers will agree that actually writing the program is the easy part of the programming process. They will say that coming up with an efficient process to machine the part and doing the related math is much harder than actually writing the program. As time goes on, you will become so familiar with these program-format-related commands that you will be able to write them almost effortlessly.

Picking up in the middle of the program

The third and most important reason to learn how to format CNC programs properly has to do with CNC machines that allow more than one tool to be used in the same program. Machining centers, turning centers, and turret punch presses are among the type of machines that require a strict format to be followed from tool to tool.

The correct formatting from tool to tool will allow the operator to rerun tools from time to time. There will be many times when you wish to run only one of the tools in the program a second or third time. For example, maybe a drilling operation on a machining center is not going deep enough and you want to rerun just the drill to make it go deeper. Of course, some adjustment to the program or tool offset would be necessary before rerunning the tool. If the drill was the fifth tool in sequence, that means four other tools preceded it. In this case, you will want to rerun *only* the drill. You will not want to rerun the entire program just to get to the drill. Rerunning the entire program would be a waste of time and might actually cause some damage to the workpiece being machined.

For the program to allow tools to be rerun, the programmer must make each tool in the program independent of the rest of the program. You can think of the programming commands for each tool as a miniprogram, self-sufficient and capable of activating all necessary functions of the machine. Actually, this way of thinking also helps the programmer break down the major task of programming a complicated workpiece into smaller and easier-to-handle pieces.

Here is a specific example of why program formatting is so important with regard to being able to rerun tools. Say you are programming a machining center and have two tools that run in sequence (maybe tools 1 and 2), both running at 400 RPM. As you write the section of program for the second tool, you may decide to leave out the S400 word in the second tool, since the spindle speed is modal. Everything will work just fine as long as the program is activated in the correct sequential order, and the second tool *always* immediately follows the first tool. But say the operator runs the entire program before discovering that the second tool has done something wrong. If the operator picks up in the middle of the program at the beginning of the second tool, it will run at the same RPM as the *last tool* in the program, which may not be 400 RPM!

This example shows but one of the possible problems that can occur if the programmer does not follow a strict format for each tool in the CNC program. Your goal with program formatting is to make each tool in the program independent of the rest of the program. You will be breaking down the entire program into smaller miniprograms, each making up one tool. There are many instances when you will have to program what you consider to be redundant or repetitious words and commands in order to accomplish this. Also, there may be times when you may deceive yourself and forget to include these very important format-related commands in your program. You must constantly try to envision each tool by itself. Make each tool capable of standing alone, without help from previous tools in the program.

Here is a another, somewhat obscure, example of when formatting is very important. Many programmers make a basic formatting mistake with multiple operations for machining one hole. This mistake keeps each tool from being independent of the rest of the program. Imagine you have a vertical machining center that requires only the Z axis to be at the machine's reference point to make a tool change. That is, the X and Y axes can be at any location for a tool change to be made. With this kind of machine, to save time, you will return only the Z axis to the machine's reference point to make a tool change. Using this machine, say you had to drill and tap one ½-13 hole. In the program, you will call up a $^{27}\!/_{64}$-in tap drill, move it over the hole, and drill the hole. Then you will return the Z axis to reference position and change tools to the tap. Since your tool is at the same location as the previously drilled hole, you would think all you need to do is move the tap straight down (no need to move X and Y, right?). However, if you do *not* include the redundant X and Y locations during the tap's start-up information, what do you think will happen if you rerun only the tap? Don't you agree, if there were no X and Y positioning movement during the tap's operation, the tapped hole will be machined at the X and Y position where the machine happens to be? The tapped hole will *not* be in the correct location! Who knows where the machine may be when it is necessary to rerun just the tap?

This, we hope, is making sense. You must understand the importance of program formatting. Many times, while writing a program, a beginner will be tempted to shorten the program by not programming redundant information from tool to tool. Or, it may not be obvious to the beginner when this information is necessary. If you expect each tool to be independent and you want to be able to run from the beginning of any tool, you must program this redundant information in each tool. Our rule in this regard is: "When in doubt about whether a redundant command is necessary, include it in the program!" No harm will come from programming redundant commands. The control will simply ignore them.

The Four Types of Program Formatting

Depending on the style of the CNC machine, formatting will have either two or four basic forms. For machines that do not have tool changing devices, such as wire EDM machines, there will be only two basic forms of program formatting information, program start-up format and program ending format. For machines with tool changing devices, such as machining centers, turning centers, and turret punch presses, there will be four basic forms of program start-up.

1. Program start-up format
2. Tool ending format
3. Tool start-up format
4. Program ending format

Any time you begin writing a new program, you will be following the program start-up format. This will allow you almost to copy information that goes into your program. While the actual values will change for speeds, feeds, tool station numbers, and coordinates, the basic *format* will remain the same *every time you begin a new program*.

After writing the program start-up format, you will write the portion of the program that machines with the first tool. Eventually, you will be finished with the first tool, and you will follow the format to end the tool's use (tool ending format). Immediately after that, you can follow the tool start-up format for the next tool. From there, you will toggle between tool start-up and tool ending format until you are finished with the program. After the machining information for the very last tool in the program, you will follow the format to end the program (program ending format).

Our basic point is that you will *not* have to memorize anything related to format. You must simply be able to look at the example given in our recommended formats, and almost copy down what you see. It is much easier to look at an example and remember the function of the commands than it is to come up with the format from memory. If you follow the structured formats given on the following pages, you can almost guarantee that you will not make basic mistakes in your program. So, for a while, you can use this format information as a crutch. After you write several programs, you will have it memorized.

Now let's look at the actual format for the various types of machines being discussed. Keep in mind that we are showing four different methods in this order:

Vertical machining centers

Horizontal machining centers

Turning centers

Wire EDM machines

Vertical Machining Centers

Note that this format assumes that the first tool is in the spindle when the program is activated.

Program start-up format

O0001 (program number)
N005 G91 G28 X0 Y0 Z0 (assures machine is at its reference position)
N010 G54 (sets program zero)
N015 G90 S300 M03 T02 (selects ABS mode, turns spindle on at 300 RPM
clockwise, gets tool 2 ready)
N020 G00 X5. Y5. (move to first X-Y position for tool)
N025 G43 H01 Z.1 (instate tool length compensation, move to first Z position)
N030 M08 (turn coolant on)
N035 G01...F3. (in first cutting movement, don't forget the feed rate)

Tool ending format

N075 M09 (coolant off)
N080 G91 G28 G49 Z0 M19 (send Z axis to tool change position, cancel tool length
compensation, orient spindle on the way)
N085 M01 (optional stop)
N090 T02 (ensure that the next station is still ready)
N095 M06 (make the tool change)

Tool start-up format

N135 G54 (set program zero)
N140 G90 S450 M03 T03 (select ABS mode, turn spindle on at 450 RPM clockwise,
get next tool ready)
N145 G00 X4. Y4. (move to this tool's first X-Y position)
N150 G43 H02 Z.1 (instate tool length compensation, move to first Z position)
N155 M08 (turn coolant on)

Program ending format

N310 M09 (turn coolant off)
N315 G91 G28 G49 Z0 M19 (go to tool change position in Z, cancel tool length
compensation, orient spindle)
N320 G28 X0 Y0 (go to zero return position in X and Y)
N325 M01 (optional stop)
N330 T01 (be sure first tool is in waiting position)
N335 M06 (place first tool back in spindle)
N340 M30 (end of program)

Understanding the format

While messages are placed in parentheses to help you understand
what was going on in the format, this format bears further explanation. First we will list the various words involved in the format and
describe them. Some of the words may be well understood by now, but
for the sake of completeness, they are included. After looking at the
various words, we will discuss the format in general.

G words.

G00 This is the rapid positioning command. Whenever you wish motion to
occur at the machine's fastest possible rate, G00 is the command to use. G00 is
discussed at length during key concept no. 3.

G01 F _____   Though it may not be readily apparent, the feed rate in the first cutting command is also part of the program start-up and tool start-up format. To make each tool independent from the rest of the program, the feed rate must be included for each tool. Note that G01 is discussed in key concept no. 3.

G28   This command is used by most controls to send the machine to its reference point. As you know, the reference point is a very accurate location along the machine's travels in each axis. For most axes, this position is very close to the plus limit for the axis. The reference point makes an excellent position from which to start the program. It facilitates part loading and allows the operator to easily check and see whether the machine is in the proper starting position (control panel lights will come on when the machine is at its reference point). We include a G28 as a safety command to assure that the machine is at its proper starting position at the very beginning of the program. While not all controls require this, it is not a bad habit for the programmer to do this. If the machine has an automatic tool changer, most builders use the reference point (at least in Z) as the location where a tool change can occur. The method by which the G28 is commanded will vary from control to control. Most require an intermediate point motion to be included in the G28 command. While somewhat difficult to understand, the purpose for this intermediate point is to allow the programmer to have the tool clear obstructions before the motion to the machine's reference point. With the G28 command, only the axes included in the G28 command will move. If an axis (X, Y, and/or Z) is left out of this command, the machine will not move in those axes. You can think of the G28 command as two commands in one. First the machine will move to the intermediate point, then it will move to the reference point. If you study our given technique, you will find that we pulled a trick on the control. We included a G91 (incremental mode) in *all* G28 commands. For each axis included in the G28 command, we also made the value of axis departure zero. This tells the machine to move nothing, then to go to the reference point. While we rarely recommend using the incremental mode, it can be used to advantage in the G28 command. All other motion commands are made in the absolute mode.

G43   This is the command for tool length compensation. It is important that each tool in a machining center program have a length compensation command. It is *always* included in the very first Z-axis motion command. By using our given technique, the X and Y axes will move over the part *before* the Z axis moves at all. For beginners, this is the safest way since, as the tool comes close to the workpiece, only one axis will be moving. If cycle time becomes of primary concern, it may be necessary to combine all three motions into one command. This can be easily done by combining G43 H __ Z _____ with the previous movement command in X and Y. Tool length compensation is discussed during key concept no. 4.

G49   This command is used to cancel tool length compensation. It is included in each tool's return to tool change position in the Z axis in the G91 G28 G49 Z0 M19 command. In our given format, it is not absolutely required because the first Z-axis motion of the *next* tool will instate tool length compensation again. However it is a good habit to cancel tool length compensation at the end of each tool. Tool length compensation is discussed in key concept no. 4.

G54 (through G59)   This command is used to assign the program zero point. It is a fixture offset command that tells the control which coordinate system to use. Fixture offsets are described in key concept no. 4. Remember, there are two ways to assign program zero: fixture offsets and G92. Our given format uses fixture offsets. If you are using the G92 command to assign program zero you can substitute G92 X _____ Y _____ Z _____ for the G54 given in the program start-up format. The values included in X, Y, and Z must reflect the distances from program zero to the reference point, as discussed in key concept no. 1. If G92 is used, no G54 through G59 commands will appear in the program. When you use fixture offsets and especially when you use more than one, we recommend including the current fixture offset command as part of each tool's start-up format. This will assure that the proper fixture offset is used even if tools are run out of sequence.

G90   This command assigns the absolute mode of programming. It is important to include this command at the beginning of *every* tool, before any motion commands, to assure that the absolute mode is selected, especially if you use our recommended method of commanding G28 (switching to incremental). This command is discussed at length in key concept no. 1.

G91   This command specifies that the incremental mode of programming be used. We have stated several times that the better programming mode is the absolute (G90) programming mode, so you may be questioning this command. Note that the G91 is included only in G28 commands. Note also that whenever an axis (*X, Y,* or *Z*) is included in this command, the value of the axis is *always* zero. On most machining center controls, G28 is a command that tells the machine to return to its reference point. However, the G28 requires a movement to an intermediate position (see G28). The G91 tells the control to make this movement in the incremental mode. Since the value of axis departure is always zero, the tool will move incrementally nothing (that is, it will stay in its current location) and then go to the reference point.

M words.

M01   This word is called an *optional stop.* It works in conjunction with an on/off switch on the control panel labeled *optional stop.* When the control reads the M01 word, it looks to the position of the on/off switch. If the switch is on, the machine will stop executing the program. The spindle, coolant, and anything else still running will be turned off. The machine will remain in this state until the cycle is activated again by pressing the *cycle start* button. Then the control will continue executing the program from the point of the M01. If the toggle switch is off, the control will ignore the M01 command entirely. It will be as if the M01 was not in the program at all. The purpose of including this command in the tool and program ending format is to give the operator a way to stop the machine at the completion of each tool. Many times, especially during a program's verification, it will be necessary for the operator to stop the machine after each tool to confirm that the tool has machined correctly. In some cases, if the tool has not machined as intended, the subsequent tools in the program may be damaged. For example, a hole is to be drilled and tapped. Of course, the drilled hole must be deep enough for the tap to reach its desired depth. After the drill has machined the workpiece, if the optional stop switch is on, the machine will stop when the drill has finished. The operator can check to confirm that the drill has machined the hole deep enough for the tap. Note that there is another program-stopping M code called simply *program stop*. The program stop command is programmed by an M00 word and *forces* the machine to stop no matter what. There is no on/off switch involved and the operator will have no option; the machine will stop.

M03   This M word is used on almost all CNC machining center controls to turn the spindle on in a forward or clockwise direction. When you actually see the spindle running in this direction, it may appear that this statement is wrong. It may appear to you that the spindle is running counterclockwise. You must evaluate the M03 from the perspective of the spindle nose, not from below the spindle. Generally speaking, the M03 is used with right-hand tools. For machining center applications, almost all tooling is right-hand. It is seldom that the machining center programmer comes across a left-hand drill, tap, or milling cutter, though they are available. If using left-hand tools, the spindle must be started in a counterclockwise or reverse direction. The command to start the spindle in the reverse direction is M04. You may have noticed by now that we never had to command the spindle to be stopped. The M19 or M06 commands related to the automatic tool changer will stop the spindle as part of their normal function. However, if you find it necessary to command the spindle to stop, M05 is the command almost all machining center controls use to accomplish this.

M06   This word is used by most machine-tool builders to command that a tool change take place. It exchanges the tool in the waiting station with the tool in the spindle. All mechanical motions necessary to make the tool change are done by the M06 command. Though our given format separated the T word and the M06 into two commands, most CNC controls allow the two words to be combined into one command.

M08   This M code is used on almost all CNC machining centers to turn on the (flood) coolant. If coolant is desired during the cutting operation, we recommend that you wait to turn on coolant in the program until the tool is positioned close to the part. If coolant is turned on while the machine is still a distance from the workpiece, it is possible that it will splash all around and possibly outside the work area, creating quite a mess. Some CNC machining centers also allow another kind of coolant, called *mist coolant*. For certain applications, this combination of air blow and coolant is better for cutting conditions. If the machine has this feature, it is almost always commanded with M07.

M09   This command turns off the coolant (flood or mist). If coolant has been turned on, it is important to turn off the coolant before the motion to the machine's automatic tool change position to keep from drenching the work area and operator.

M19   This command is called the *spindle orient* command. Most CNC machining centers with automatic tool changers require that the spindle be rotated to a precise location before a tool change can occur. This accurate position is called the *orient position*. The spindle is oriented with the tool changing device. Though an M06 command will cause the spindle to orient, it takes time for this orientation to take place. Notice that we include the M19 in the motion command returning the Z axis to its tool change position. This means the spindle orientation will be occurring during the Z-axis motion. When the machine reaches the tool change position, it is likely that the spindle orientation will be complete. The subsequent M06 command will occur faster. The amount of time saved will vary from machine to machine, but will range from about 0.5 to 3 seconds. When you consider the number of tool changes a CNC machining center will make in its period of use, it is a very good idea to get in the habit of including an M19 for every tool of every machining center program you write. Though not absolutely required, this command will save some time every time a tool change command is given.

M30 The program ending format we have shown ends with an M30. This is a program ending word. It tells the machine to turn off anything still running (spindle coolant, etc.) and rewind the memory back to the beginning, then the control stops. While the M30 command we have shown is very popular among control builders, some builders require an M02 as the program ending word. In this case, the M02 will do everything the M30 would.

Other words in the format.

End-of-block word Though it is not shown in or given format, *every* command *must* end with an end-of-block command. If typing the program into the control's memory at the control keyboard, the operator must press a key that represents the end-of-block word at the end of *every* command typed. On some controls, the end-of-block key is represented by a semicolon (;); on others, by an asterisk (*). The end-of-block character usually visually appears on the control screen to let the operator know it has been included. When you type programs on another device, like a computer, many times the end-of-block character will be transparent. On these devices, the carriage return, input, or enter key will automatically register as the end-of-block word.

O word Most controls allow more than one program to be stored in the control's memory. Most controls that allow this require a special word to designate which program is which. The O word is one of the most common ways to do this. The O word assigns the program's number. The operator can easily scan from one program to another by the O word. It is usually the very first word of the program.

S word The S word tells the control the desired RPM for the spindle. The S word, by itself, does not actually turn on the spindle, the M03 does that. The S word just lets the control know what spindle speed is desired. Most current CNC machining center controls allow the programmer to specify the speed directly in RPM, and in 1-RPM increments. It is important that the spindle RPM be included in *every* tool's start-up format.

T word The T word is used for machining centers with automatic tool changers. Usually the T word has a two-place format. That is, two digits are used with the T word. With most machining centers, the T word simply tells the machine to rotate the tool magazine or carousel to the desired position. This position is called the *waiting position* or *next tool position*. The tool does not actually go into the spindle yet, it simply gets put in ready station. Any subsequent M06 command will put it in the spindle. Notice that we included the next tool T word in the program and tool start-up formats. This gets the next tool ready while the tool in the spindle machines the workpiece, saving time later. Also notice that we include the *same* T word again in the tool and program ending format, just before M06 tool change command. This is done for two reasons. First, it is possible that the operator may manually rotate the magazine during the program's cycle to check or replace tooling. If this is done, the operator may forget to rotate the magazine back to the proper position. The T word just before the M06 assures that the proper tool will go into the spindle. Second, it is much easier to pick up in the middle of the program if the T word is close to the M06 command.

H word   This word is part of tool length compensation and is used to tell the control which offset is to be used. Tool length compensation was discussed in key concept no. 4.

Other notes about the format.   We do *not* claim this to be the only way to format programs for a CNC vertical machining center. In fact, some controls will require the given format to be altered. We will be the first to admit that there are *many* ways to successfully format CNC programs. However, this information should truly get you on the right track, giving you the basics of program formatting. You may find that certain CNC controls require special commands we have not discussed. You may also find that more M codes are required for your machine's application. The basic structure we have shown, and the logic that we have given, can be used on *any* vertical CNC machining center.

Example program showing format information for vertical machining centers

Here we show an example program that will stress the use of the format information just presented. Although the program is quite simple, it does show many of the concepts discussed so far. Pay particular attention to the strict format followed for each tool. Note that the whole program can be broken down into miniprograms, each making up a tool. Figure 6.1 shows the part to be machined.

Program start-up:

```
O0003
N005 G91 G28 X0 Y0 Z0
N010 G54 (¼-in drill)
N015 G90 S1300 M03 T02
N020 G00 X1. Y1.
N025 G43 H01 Z.1
N030 M08
```

Cutting:

```
N035 G01 Z-.85 F3.0
N040 G00 Z.1
```

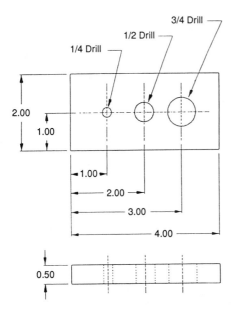

Sequence of Operations:

Seq.	Description	Tool	Station	Feed	Speed
1	Drill to 1/4 (1)	1/4 Drill	1	3.0, in IPM	1300, in RPM
1	Drill to 1/2 (1)	1/2 Drill	2	4.0, in IPM	1000, in RPM
1	Drill to 3/4 (1)	3/4 Drill	3	5.0, in IPM	800, in RPM

Figure 6.1 Drawing for vertical machining center program format example.

Tool ending:

```
N045 M09
N050 G91 G28 G49 Z0 M19
N055 M01
N060 T02
N065 M06
```

Tool start-up:

```
N068 G54 (½-in drill)
N070 G90 S1000 M03 T03
N075 G00 X2. Y1.
N080 G43 H02 Z.1
N085 M08
```

Cutting:

    N090 G01 Z-.85 F4.0
    N095 G00 Z.1

Tool ending:

    N100 M09
    N105 G91 G28 G49 Z0 M19
    N110 M01
    N115 T03
    N120 M06

Tool start-up:

    N130 G54 (¾-in drill)
    N125 G90 S800 M03 T01
    N130 G00 X3. Y1.
    N135 G43 H03 Z.1
    N140 M08

Cutting:

    N145 G01 Z-.85 F5.
    N150 G00 Z.1

Program ending:

    N155 G91 G28 G49 Z0 M19
    N160 G28 X0 Y0
    N165 M01
    N170 T01
    N175 M06
    N180 M30

By studying this program, you should be able to see why we say program formatting is redundant. Notice that blocks N010 through N030 are identical in structure to blocks N068 through N085. In like manner, blocks N045 through N065 are identical in structure to N100 through N120. These formats are repeated one more time for the third tool, and they will be repeated for every tool in every program you will ever write for vertical machining centers. While you still may be a little confused about the meaning of every word, you should agree that there are not that many new words to become familiar with.

Horizontal Machining Centers

The programming format for horizontal machining centers is very similar (almost identical) to the format for vertical machining centers. For this reason, we ask that you read the formatting discussions for vertical machining centers first. We will discuss at length only the differences.

As with vertical machining centers, this format assumes that the first tool is in the spindle when the cycle is activated.

Program start-up format

O0001 (program number)
N005 G91 G28 X0 Y0 Z0 (assures machine is at zero return position)
N010 G54 (sets program zero)
N015 G90 S300 M03 T02 (selects ABS mode, turns spindle on at 300 RPM clockwise, gets tool 2 ready)
N020 G00 X5. Y5. (move to first X-Y position for tool)
N025 G43 H01 Z.1 (instate tool length compensation, move to first Z position)
N030 M08 (turn coolant on)
N035 G01...F3. (in first cutting movement, don't forget the feed rate)

Tool ending format

N075 M09 (coolant off)
N080 G91 G28 G49 Y0 Z0 M19 (send Y and Z axis to tool change position, cancel tool length compensation, orient spindle on the way)
N085 M01 (optional stop)
N090 T02 (ensure that the next station is still ready)
N095 M06 (make the tool change)

Tool start-up format

N135 G54 (set program zero)
N140 G90 S450 M03 T03 (select ABS mode, turn spindle on at 450 RPM clockwise, get next tool ready)
N145 G00 X4. Y4. (move to this tool's first X-Y position)
N150 G43 H02 Z.1 (instate tool length compensation, move to first Z position)
N155 M08 (turn coolant on)

Program ending format

N310 M09 (turn coolant off)
N315 G91 G28 G49 Y0 Z0 M19 (go to tool change position in Y and Z, orient spindle)
N320 G28 X0 (go to zeru return position in X)
N325 M01 (optional stop)
N330 T01 (be sure first tool is in waiting position)
N335 M06 (place first tool back in spindle)
N340 M30 (end of program)

As with the vertical machining center format, this format allows a high level of flexibility for rerunning programs and is quite safe to use. As stated, this format is very similar to the vertical machining center format. If fact, you may have some difficulty finding differences! The changes are, indeed, minor.

The only difference is the method by which the machine is sent to the tool change position. While most vertical machining centers require that only the Z axis be returned to the reference point for a tool change, most horizontal machining centers require that Y and Z be sent to the reference point for a tool change.

Unfortunately, there is one major problem with horizontal machining center programming that our format did not address. Most horizontal machining centers have some kind of rotary device that allows the part to be rotated during the machining cycle. To be able to truly program horizontal machining centers, the programmer must have an idea as to how this rotary device is programmed. After we show an example program using the above format, we will introduce much of what the programmer needs to know about rotary devices.

Example program for horizontal machining centers

Figure 6.2 shows the print for this program.

Program start-up:

```
O0003
N005 G91 G28 X0 Y0 Z0
N010 G54 (¼-in drill)
N015 G90 S1300 M03 T02
N020 G00 X1. Y1.
N025 G43 H01 Z.1
N030 M08
```

Cutting:

```
N035 G01 Z-.85 F3.0
N040 G00 Z.1
```

Tool ending:

```
N045 M09
N050 G91 G28 G49 Y0 Z0 M19
N055 M01
N060 T02
N065 M06
```

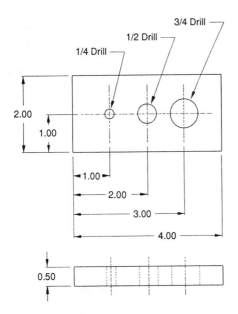

Sequence of Operations:

Seq.	Description	Tool	Station	Feed	Speed
1	Drill to 1/4 (1)	1/4 Drill	1	3.0, in IPM	1300, in RPM
1	Drill to 1/2 (1)	1/2 Drill	2	4.0, in IPM	1000, in RPM
1	Drill to 3/4 (1)	3/4 Drill	3	5.0, in IPM	800, in RPM

Figure 6.2 Drawing for horizontal machining center program format example.

Tool start-up:

```
N068 G54
N070 G90 S1000 M03 T03 (½-in drill)
N075 G00 X2. Y1.
N080 G43 H02 Z.1
N085 M08
```

Cutting:

```
N090 G01 Z-.85 F4.0
N095 G00 Z.1
```

Tool ending:

```
N100 M09
N105 G91 G28 G49 Y0 Z0 M19
N110 M01
N115 T03
N120 M06
```

Tool start-up:

```
N124 G54
N125 G90 S800 M03 T01 (¾-in drill)
N130 G00 X3. Y1.
N135 G43 H03 Z.1
N140 M08
```

Cutting:

```
N145 G01 Z-.85 F5.
N150 G00 Z.1
```

Program ending:

```
N155 G91 G28 G49 Y0 Z0 M19
N160 G28 X0
N165 M01
N170 T01
N175 M06
N180 M30
```

Rotary Tables and Indexers

More and more companies are applying rotary devices on their machining centers. This section will apply to you only if your machining center has some kind of rotary device that allows the workpiece to be rotated during machining. For vertical machining centers, you may have an indexer or rotary table device that rests on top of the table. For horizontal machining centers, your machine may have an indexer or rotary device built into the table of the machine tool itself. If your machine does not have either of these devices, then you can simply skip this section or stay tuned and learn about something you may have to deal with in your future.

We will depend on you to determine what kind of rotary device (if any) your particular machine has. If you are not sure, ask someone currently working with the machine to help you, or check in your machine-tool builder's manual. We will discuss the two most commonly used types of rotary devices. Before we can get started, you *must* know which one you will be working with.

The reasons for rotary devices

These devices can be applied to vertical as well as horizontal machining centers. The most basic purpose of the rotary device is to provide the ability to expose several sides of the workpiece to the spindle during one setup. This means that the user can accomplish two things:

1. The number of setups can be reduced. The total time it takes to produce the parts can be reduced. Fewer programs, less machine down time, and faster throughput of workpieces in the shop are among the advantages experienced by companies using rotary devices on their machining centers.

2. It is easier to hold accuracy from one surface of the workpiece to the next. If multiple setups are made, many times the accuracy suffers, since it is difficult to locate the part perfectly in each setup. With an indexer or rotary table, since the part is not being removed from index to index, the overall accuracy from surface to surface on the part will improve.

Rotary tables vs. indexers—what is the difference?

There are two types of rotary device that are commonly applied to machining centers. We will discuss both of these types in this section. As mentioned earlier, it is important that you know which one applies to your machine.

Indexers

An indexer is a device that allows the part to be quickly rotated a certain number of angular degrees. The rotation is usually very fast, meaning that it is impossible to be machining the workpiece while the indexer is rotating. Also, there is no control of the rotation rate at which the index takes place. Generally speaking, most indexers can only rotate in one direction, meaning that the rotation to the desired angle may not always be the in most efficient direction.

There are many kinds of indexers. Usually the designation on the indexer has to do with the smallest angular increment the indexer is capable of rotating. This angular increment also determines the method by which the indexer is programmed.

Common indexers include:

90° indexers

45° indexers

5° indexers

1° indexers

If the machine you will be working with has an indexer, it will be necessary for you to know which kind you have. If you are unsure, ask someone currently working with the machine or check the machine manual.

The methods of programming indexers. Indexers for 90° and 45° are usually activated by a single M code. The M code number is determined by the machine tool builder or by the company that supplies the indexer. It should be listed in the series of M codes that come with the machine or indexer.

With this kind of indexer, when the programmer wants an index to occur, the M code is commanded, and *one* index of the specified angle occurs. This can be somewhat cumbersome. Imagine you have a 45° indexer, and you want to index 180°. In this case, four successive M codes, one per command, must be programmed.

Indexers for 5° vary yet further. Some builders give you a series of M codes for their 5° indexers, based on the most common angles of index. For example, M71 may be a 5° index, M72 a 15° index, M73 a 45° index, and M74 a 90° index. This would minimize the number of M codes you would have to string together for an odd index angle.

Other 5° indexers force the user to mechanically set the angle of index required, and only one M code is used to activate the indexer. This kind of indexer can be quite cumbersome to work with, since several M codes may have to be strung together to index to an odd angle.

Indexers for 1° are usually the easiest indexers with which to work. Most 1° indexers are programmed with a special word. Usually the B command is used. With the B word, a programmer can easily specify the exact angle of index desired. For a 27° index, the programmer simply specifies B27. Also, many 1° indexers have two M codes to specify the direction of rotation.

As you can see, there are several possibilities for how a particular machine's indexer is programmed. We cannot be very specific for your particular machine. We leave it up to you to ask someone or to check the machine-tool builder's manual or indexer manual to find the method by which your particular indexer is programmed.

Rotary tables

A rotary table can also be applied to either a vertical or horizontal machining center. On a vertical machining center, it is usually mounted on the top of the table, like an indexer. On a horizontal machining center, it is mounted inside the machine as part of the table mechanism.

The rotary table is much more flexible than an indexer. That is, the method by which rotation is controlled is much more effective. The programmer can control more precisely the rotation angle desired as well as the direction in which the indexer rotates. The motion rate (feed rate) at which the rotary table turns can even be controlled. When it comes to the rotation angle, the programmer is allowed to program to within 0.001°. This means that the rotary table has 360,000 positions!

The rotary motion of a rotary table is actually an axis of the machine. The designation for what letter address is used depends on the machine-tool builder and how the rotary axis is related to the machine tool. On horizontal machining centers, where the rotary device is rotating the table itself, the designation for the rotary axis is *always* the B axis (no machine-tool builder we know of strays from this standard). But on a vertical machining center, if the rotary axis is parallel to the X axis, as it normally is, the rotary axis is called the A axis. However, with vertical machining centers, we have seen some machine tool builders that stray from this standard. Some call the rotary axis on the vertical machining center the B axis. Still others call it the A axis. The axis name (A, B, C, etc.) is not nearly as important as your knowing what the axis name is for your particular machine. It really doesn't matter if your machine tool builder sticks to the standard as long as you know how your machine's rotary axis is designated.

For the purpose of this text, we will call the rotary axis the B axis for horizontal machining centers and the A axis for vertical machining centers. If you understand our presentations, you will be able to easily apply what you have learned to your particular machine, no matter what the rotary axis is named. Also note that some machine-tool builders designate the rotary axis (on the machine panel) as the fourth axis and do not put the letter address on the machine panel at all.

How to program a rotary departure. The rotary axis command allows a decimal point to be programmed (if the control model allows decimal point programming). Also, all rotary commands are given in degrees of rotation. For example, if you want to designate a rotary axis position of 45° and the axis letter address is B, here is the way it will be specified:

N050 B45.

However, it is important to know that the rotary axis can be programmed only to three places after the decimal point (*not* four, like X,

Y, and *Z*). Also, if you are designating an angular departure that includes a value less than 1°, you *must* make the designation in *decimal portions of a degree.* Many workpiece drawings designate portions of a degree in minutes ' and seconds " of a degree. This *must* be converted to decimal format. Here is the formula to convert minutes and seconds to decimal format:

Degrees (in decimal format) = degrees + (minutes divided by 60)
+ (seconds divided by 3600)

For example, you have to specify the value of 13°27'37" to the CNC control. If you want to input this value into the rotary axis command of a program, it will have to be converted to decimal format.

To make the conversion, you will first divide 27 by 60. The result is 0.45°. Then you will divide 37 by 3600. The result is 0.010°. Finally, you would add 13 + 0.45 + 0.010, and the result would be 13.46°. If your rotary axis is designated with the B letter address, you will specify this command as

N045 B13.46

It can be a little frustrating if your drawings are dimensioned with minutes and seconds for the angular dimensions. Unfortunately, decimal format is the only way most CNC controls will understand angular values.

Comparison to other axes.   Just about everything you know about *X, Y,* and *Z* will still apply to the rotary axis. The methods by which you command normal motion in *X, Y,* and *Z* can still be used to control the rotary axis. Here is a list of the things that we will discuss about the rotary axis that should be somewhat familiar to you if you understand motion in *X, Y,* and *Z:*

1. Reference position
2. Designation of program zero
3. Absolute mode (G90) and incremental mode (G91)
4. Rapid (G00) and straight-line cutting (G01)
5. Usage in canned cycles

Truly, you will be treating the rotary motion as an axis of motion. The next few discussions will draw on what you know about *X, Y,* and *Z* to explain the rotary axis.

Reference position.   Just like *X, Y,* and *Z,* the true rotary axis will have a reference position. Part of powering up the machine is returning the rotary axis to its reference position just as for the *X, Y,* and *Z* axes. The reference position for the rotary axis is just as accurate as it is for all other axes.

Commanding the rotary axis to go to the reference position in the program is very similar to commanding the other axes to go to the reference position. If the rotary axis is designated with a B letter address, here is the command that will send the machine to reference position in *all* axes:

```
N055 G91 G28 X0 Y0 Z0 B0
```

In this case all four axes will move to the reference position at the same time. If you want only the B axis to return, here is the command:

```
N055 G91 G28 B0
```

As with X, Y, and Z, the reference position for the rotary axis will make an excellent reference point for your program. That is, your angular program zero point in the rotary axis can be taken from the reference position.

This next point may be hard to visualize. With X, Y, and Z, the motion along these axes is linear. There are definite ends to each linear axis of motion. These axes will overtravel when the limits are reached. But with the rotary axis, the rotation can continue countless revolutions in both directions (plus or minus). That is, the rotary axis will never overtravel. This actually presents a small problem when commanding a zero return. The rotary table will often behave poorly with regard to the reference position command. For example, imagine that the rotary axis is currently resting at reference position. Now you give the commands

```
N060 G00 G91 B1.
N065 G00 G91 B-1.
```

The rotary axis would move (incrementally) 1° clockwise and then rotate back 1° counterclockwise. Since the rotary axis started out at the reference position, you would think that the control would know the rotary axis was back at reference position. However, the control is not that smart. The axis origin light for the rotary axis will *not* come on, and the control will think the rotary axis was positioned to a point away from its reference position.

There will be many times that you end your program with the rotary axis back at its starting position, and usually its starting position is the reference position. If you are including the safety command to return the rotary axis to its zero return position at the beginning of the program to be sure the rotary axis is where it should be, an undesirable rotation may occur.

Imagine the rotary axis has been rotated since the last physical return to the reference position and is now resting back at the reference position (but the rotary axis origin light is *not* on). If you give the command

    N050 G91 G28 B0

believe it or not, the control will have the rotary axis make one full revolution in the clockwise direction to reach the reference position. This is because for *all* axes, the control will always approach the zero return position going in the plus direction. To avoid this problem you can instead give the command

    N050 G91 G28 B-15.

This will force the rotary axis to move 15° minus before the machine starts searching (in the plus direction) for the reference position. In this case, the rotary axis will move only a total of 30° to reach the zero return position.

If this is somewhat unclear, don't worry about it. You almost have to be close to the machine and see the rotary axis rotate in the plus direction to the reference position before the reference position related to the rotary axis makes sense. Also, no crash will occur because of this unwarranted motion. Only the cycle time would suffer.

Designation of program zero.   If you are going to be programming the rotary axis in the absolute mode (G90), you must also designate where the program zero location is with regard to the rotary axis. As with X, Y, and Z, the control must know a point of reference (your program zero point) from which all coordinate values are to be taken. Also, as with X, Y, and Z, you will be including the B designation value in the G92 command of your program or in the fixture offsets.

The measurements will work the same way as for X, Y, and Z. For using G92, you will include the angular distance *to* the rotary table's reference position *from* the angular program zero point in the rotary axis.

If you are using fixture offsets (G54 to G59), the distance will be taken *from* the reference position in the rotary axis *to* the program zero point (just like X, Y, and Z).

Since there are typically no angular datum surfaces on the workpiece drawing, generally speaking, there is no need to be to picky or fussy about where the program zero point in the rotary axis is to be located. It is not nearly as important as for X, Y, and Z where program zero in the rotary axis is to be located. With X, Y, and Z, many program coordinates can be taken right from the print with the wise selection of the program zero point for those axes. However, for the rotary axis, this is simply not the case. For this reason, most programmers will simply designate that the side of the rotary table facing the spindle while the rotary axis is at the reference position be specified as the program zero side. That is, the G92 rotary axis value or fixture offset rotary axis value will usually be zero.

As we get deeper into the programming of the rotary axis, we will show many cases when it is wiser to program the rotary axis incrementally, especially when the rotary axis is simply used as an indexing device. If the rotary axis is programmed in the incremental mode, there will be no need to assign a program zero point for the rotary axis.

Understanding the absolute mode for the rotary table. As you know, when you work from the program zero point and input all dimensions from that point, you are working in the absolute mode. We have told you several times that this is the easiest way for the beginner to write programs.

However, when it comes to the rotary axis, working in the absolute mode may not be the easiest or best way to input values. Some funny or unexpected things can happen if you do not understand how the absolute mode affects the rotary axis. Please do not misunderstand us here. The absolute mode affects the rotary axis in the same way it affects X, Y, and Z. How the absolute mode affects the rotary axis is just harder to visualize.

Figure 6.3 is a drawing of a horizontal machining center as viewed from above (note that the same points being made here will also apply to a vertical machining center with a rotary table).

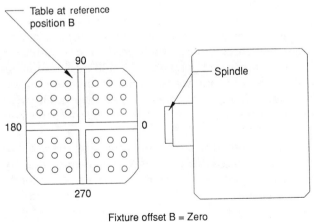

Figure 6.3 Top view of horizontal machining center with rotary table.

As you can see, we have depicted the B axis at its zero return position, and in the designation in the fixture offset word G92, B has been set to zero. This means that the table side facing the spindle while the B axis is at the reference position is the zero side.

In the absolute mode, whenever the control reads a B command of B0, the rotary table will end its rotation with the zero side facing the spindle.

Here's where it becomes a little complicated. You *must* remember that the table side commanded will end up facing the spindle at the completion of the command (in the absolute mode). The direction that the table rotates (clockwise or counterclockwise) is dependent on the position of the table prior to the command.

If rotating from a small B value to a larger one (plus), the rotation direction, as viewed from above the table, will be clockwise. If rotating from a large B value to a smaller one (minus), the rotation will be counterclockwise. This can lead to some undesirable rotation motions. Assume for example that the 270° side is facing the spindle right now. In the absolute mode, if you command G00 B0 to rotate at rapid to the 0° side, the control will rotate the table counterclockwise to make the motion. In essence, it will be taking the long way to get to the 0° side.

Figure 6.4 shows example drawings of the rotary table's motion in the absolute mode to help you get the idea. Keep in mind that these drawings apply to a rotary table in either a vertical or horizontal machining center.

We hope you are beginning to get the idea of how the absolute mode affects the rotary axis. You are probably also starting to wonder if there is an easier way to make simple indexes, and there is. We will talk about it just a little later.

Before we end our discussion of the absolute mode, we want to make one more point. The rotary axis allows you to make commands over 360°. That is, if needed, you can make the table rotate more than once in one command. For all intents and purposes, the 360° is the same as the 0° side of the table. The only time we recommend commanding rotation over 360° is when cutting movements (G01) are made with more than one rotation of the rotary table, as may be the case when cams are machined.

As a beginner, you may be tempted to continue rotating the table to higher and higher rotation values. That is, as you continue rotating the table for the various tools in your program, you may be tempted to rotate well past the 360° side. In fact, if the table has to be rotated five times during the program for operations on all four sides of the table, your *B* axis would have accumulated to 1800° (5 × 360) in the absolute mode! If at any time during the program, you make a command to go back to the zero side, the table will unscrew all the way back and waste a great deal of time.

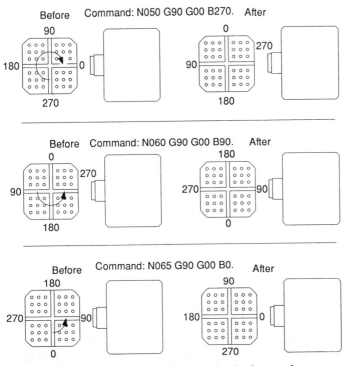

Figure 6.4 Examples of rotary motions in the absolute mode.

The incremental mode—best for simple indexing. The incremental mode allows the programmer to be more specific about how rotation is to occur. If the rotary table is being used to simply index from one position to another, with no machining to occur during the rotation, we strongly recommend that the beginner get in the habit of temporarily switching to the incremental mode (G91) to make the rotary motion. After the rotary motion, immediately switch back to the absolute mode to continue machining with X, Y, and Z.

In the incremental mode, plus is still clockwise and minus is still counterclockwise. But your point of reference will be the table's current position, not program zero. So, if you want to simply rotate 90° clockwise from the current table position, you will simply make this command:

N050 G91 G00 B90.

But don't forget to switch back to the absolute mode in the next command. Of course, if you wanted to rotate 90° counterclockwise from the table's current position, you would make this command:

N050 G91 G00 B-90.

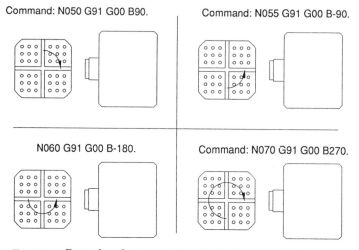

Command: N050 G91 G00 B90.

Command: N055 G91 G00 B-90.

N060 G91 G00 B-180.

Command: N070 G91 G00 B270.

Figure 6.5 Examples of rotary motions in the incremental mode.

Figure 6.5 shows some drawings which are more examples of how to make table rotations in the incremental mode.

As we think you would agree, this is easier and more logical than the previously described absolute method of commanding rotation. The only problem with making incremental table rotations has to do with X, Y, and Z. There may be times when you wish to make a table rotation at the same time you are positioning the tool in X, Y, and/or Z for its first movement on the new table side. If you intend to move in X, Y, and/or Z while rotating the rotary table, you will either have to make the rotary motion in the absolute mode or calculate the X, Y, and/or Z values incrementally. There is no way to make an absolute move in X, Y, and Z while moving incrementally in the rotary axis.

Here is one more point about using the rotary axis as a simple indexer. If heavy machining is to be done after the rotation, while the rotary axis is stationary, the drive system of the rotary axis will be under a great deal of stress. For this reason most machine-tool builders incorporate two M codes to allow the rotary device to be clamped and unclamped. After the rotation and prior to heavy machining, the programmer includes the M code to clamp the rotary device. After machining and prior to the next rotation, the programmer includes the M code to unclamp the rotary device.

Rapid and straight-line motion. The examples given so far have shown rapid commands for the rotary axis. In this mode (rapid), the rotary axis turns as quickly as it can. This mode is used for indexes only. That is, when rotated in the rapid mode, the rotary axis is being used as a simple indexer. It is being used to rotate the workpiece to allow machining on another side.

In the G01 mode, the rotary axis can be used to machine the workpiece *and* rotate at the same time. This means the workpiece can be milled while the rotation occurs. One popular use for this kind of machining is milling cams.

You must know that something strange happens to the feed rate when a rotary axis motion is commanded in a G01 mode. As you know, when X, Y, and/or Z is involved in a G01 command, the feed rate is input as the inches per minute rate. But when *any* rotary motion is included within a G01 command, the feed rate *must* be input in degrees per minute (DPM). Even when an X, Y, and/or Z motion is in the command, if the straight-line cutting command includes a rotary-axis motion, the feed rate must be in degrees per minute.

It can be somewhat difficult to calculate the degrees per minute feed rate when you know only the feed rate you want in inches per minute. The formula we will give involves calculating the length of cut and the machining time based on the inches per minute feed rate. Only then can the DPM feed rate be calculated.

Since degrees-per-minute feed rate is more difficult to calculate, many programmers cheat. They approximate the DPM feed rate and wait until the part is being machined. Then they fine-tune the feed rate by adjusting the feed rate override switch. After overriding the feed rate to the desired level, they change the programmed feed rate accordingly.

To precisely calculate the desired feed rate in degrees per minute prior to running the program requires that you apply the following formulas:

Time (in minutes) = length of cut ÷ IPM

RPM = 3.82 × SFM ÷ tool diameter

IPM = inches per revolution × RPM

DPM = incremental rotation amount ÷ time (in minutes)

As you can see, these formulas require that you first calculate the actual length of cut during machining. Unfortunately, this can sometimes be difficult to determine for rotary motions, but try to be as accurate as you can.

With the length of cut approximated, you calculate the length of time required to make the cut at the desired IPM feed rate. Finally, you can calculate the degrees-per-minute feed rate.

While we would agree that this is a cumbersome set of formulas just to calculate the proper DPM feed rate, it is important that you understand how to do this, because you will need to specify the DPM for any feed rate movement (G01) that requires a rotary movement.

Special note about G02 and G03.   It is impossible to form a true circular command that incorporates the rotary axis with simple manual programming techniques on most controls. Unfortunately, there are times when it may be necessary to form this kind of motion. In this case, usually a computer-aided manufacturing (CAM) system is required to calculate the movements.

Canned cycle usage.   Though canned cycles have not been introduced yet, we include notes related to how they affect the rotary axis. Key concept no. 6 discusses them thoroughly.

When used with canned cycles, the rotary axis can be used (just like X and Y) as hole centerline coordinates. Every X- and Y-related point made in key concept no. 6 will apply to the rotary axis. This allows you to easily specify a series of holes around the outside of the round part to be machined after an axis rotation.

How to approach rotary device programming.   You have now been exposed to the raw tools of rotary device programming. For an indexer, this simply involves specifying the proper command to make the indexer turn (usually an M or B word). For the rotary table, you now know the various ways to activate the rotary axis. Additionally, we point out once again that everything discussed to this point applies equally to vertical and horizontal machining center applications. Now we intend to present the way you should be approaching rotary device programs. For the most part, the basic method is the same for rotary tables or indexers and horizontal machining centers or vertical machining centers.

Program zero selection.   The selection of a good program zero point will make working with a rotary device much easier. To begin, Fig. 6.6 is a drawing that shows the front view of a vertical machining center that incorporates a rotary device.

As you can see, the rotary device is holding a fixture. This particular fixture is being used to hold two different workpieces. However, many fixtures used in this application hold only one part. The fixture incorporates a turned diameter on the left side to accommodate the chuck of the indexing device. On the right side of the fixture is a center-drilled hole to accommodate a tailstock for support.

Now let's take a look at the assembly drawing for the fixture itself. Figure 6.7 shows the fixture. This drawing uses standard orthographic projection techniques, so you should be able to make sense of it.

This drawing more accurately describes the fact that two different

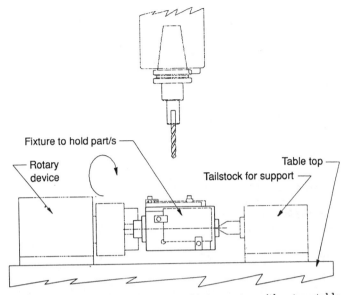

Figure 6.6 Front view of vertical machining center with rotary table.

parts are to be machined. The top view shows one of the parts to be machined, and the front view shows the other. The end views clarify the situation even more.

Though it may be difficult to visualize, notice that it would be next to impossible to assign one program zero point on the fixture that could be used for both parts. Figure 6.7 shows that the two parts do not share the same location points. The drawing in Fig. 6.8 superimposes the two parts on the same side of the fixture to demonstrate this.

Figure 6.8 lets you look at each side of the fixture as the spindle will see it. This should make it clear that after an index, the program zero

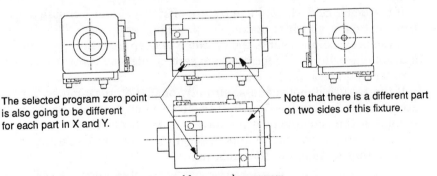

Figure 6.7 Fixture used for rotary table example program.

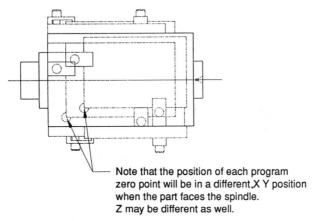

Note that the position of each program
zero point will be in a different.X Y position
when the part faces the spindle.
Z may be different as well.

Figure 6.8   Two parts superimposed in rotary device.

point for the old part facing the spindle will not apply to the new part
now facing the spindle.

This is a *very common* situation when using rotary devices. How you
handle this problem is based on three conditions:

1. Whether the machine has fixture offsets

2. Whether the programmer wants to make things as easy as possible
for the operator

3. Whether the fixture has been made accurately

There are two differing schools of thought as to where program zero
should be placed for this kind of problem. One solution is to use a different program zero point for each side of the workpiece being machined (this requires fixture offsets), and the other is to use one cen-
trally located program zero point for the entire program. We will
address both methods at length and show the merits and limitations of
both.

If the machine does *not* have fixture offsets, you must choose a central location at which to make your program zero point.

If the programmer wants the operator to keep from having to make
multiple measurements for program zero points, the central program
zero technique should be used.

If the fixture is not made accurately, or if the programmer has no
access to the fixture drawing, fixture offsets must be used, one for each
surface. If the machine does not have fixture offsets, the fixture must
be measured to find critical dimensions to be used in calculating coordinates that will go into the program.

Setting one program zero point per side with fixture offsets.   From a programmer's standpoint, this is the easiest way to handle multiple sides on

the indexer. The programmer can simply pick the most logical point on each side of the indexer to be machined as the program zero point. Then, using fixture offsets, the operator will measure the distances from the machine's zero return position to the various program zero points. These values will be placed in the corresponding fixture offsets. If you will be using this technique often, you may want to review the discussion of fixture offsets during key concept no. 4.

Unfortunately, this technique does require more work on the operator's part. The machine will be down for a longer time while the operator is taking the required measurements.

Another limitation of using this technique has to do with the number of fixture offsets available. The standard number of fixture offsets available from most control manufacturers is six. For vertical machining center applications, this may be more than enough. But for horizontal machining centers that have pallet changers, it is possible that you will not have enough. If the machine has a pallet changer and you are working on four sides of both parts to be machined, of course, you will need eight fixture offsets to use this technique. Note that most CNC controls that have fixture offsets allow many more fixture offsets as options.

Using a central program zero point. This tends to be an "old school" technique. In the past, this was the only way of handling the program zero point for a rotary device. Many programmers still use this technique even though fixture offsets are now available for multiple program zero points.

Even though this technique will require only one program zero point for the entire program, remember that you can still use a fixture offset (G54) to assign it.

With this technique, the programmer will pick one central location to place the program zero point. While this makes it somewhat difficult to calculate coordinates for the program, at least the calculations will be made in the same way for *each side of the fixture,* making calculations consistent.

With this technique, the program zero point will be in the *center of index* for two of the axes. The third-axis program zero point can be floated to a convenient position in that axis. For vertical machining centers, the program zero point will be the center of index in the Y and Z axes. The program zero point for the X axis can be floated to a convenient location.

For horizontal machining centers, the program zero point will be the center of index for the X and Z axes. In Y, it can be floated to a convenient position. Figures 6.9 and 6.10 should drive this idea home.

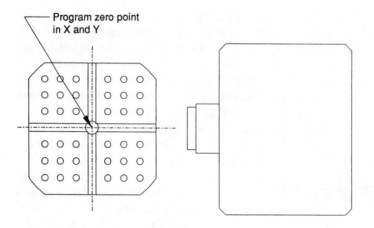

Horizontal machining center

Figure 6.9   Center of index as central program zero point on horizontal machining center.

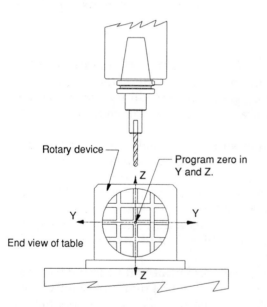

Figure 6.10   Center of index as central program zero point on vertical machining center.

Calculating coordinates.   As stated, calculating coordinates for the program becomes much more difficult. If the fixture is not made accurately, the programmer can be fooled when it comes to the values from one critical location on the fixture to another. Figure 6.11 shows how coordinates going into the program from a central program zero point must be calculated.

From Fig. 6.11, you can tell that when you are working from a cen-

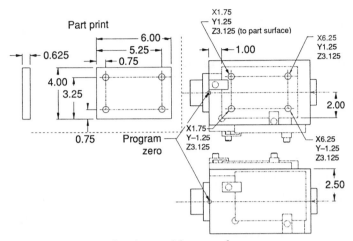

Figure 6.11 Fixture drawing used for example program.

tral program zero point with an indexer, program coordinates don't make much sense. They reflect the fixture being used to hold the part. That is, if working from a central program zero point, the programmer must take into consideration the distances from the central program zero point on the fixture to the actual datum line surface of the workpiece. This distance must be added to or subtracted from the print dimension in order to come up with the coordinate to program.

Example program using rotary device. It is difficult to develop an example program illustrating the usage of rotary devices that is comprehensive enough to show you what can be done without becoming so complicated that the beginner will not be able to follow. What follows is a relatively simple example program showing the rotary table being used as an indexer. Unfortunately, we cannot show every possible type of rotary device. This program uses a rotary table designated as the A axis (as would be the case on most vertical machining centers). However, if you have another kind of rotary device, we will depend on you to translate the rotation commands to your particular device. Note that we will be using a simple G91 incremental command to make the indexes. Also note that only three sides of the rotary device will be commanded.

The program will show the specification of four separate program zero points using fixture offsets to accomplish this. As mentioned, if you are to use a central program zero point, only the coordinates will change. As you know, using multiple program zero points is the easier way for the programmer. But the operator must measure each program zero point.

If you are a beginner to CNC and do not have to work with a rotary

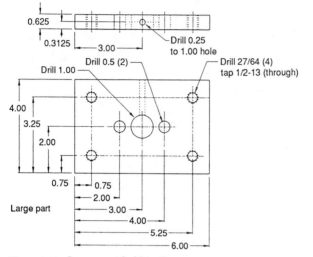

Figure 6.12 Large part held in fixture.

device for now, please just skim this program. However, if you will be working with rotary devices on a daily basis, stick with it until you understand.

Two separate workpieces will be machined. Let's begin by looking at both prints. Note this first part (Fig. 6.12) is marked *large part*. The second part (Fig. 6.13) is labeled *small part*.

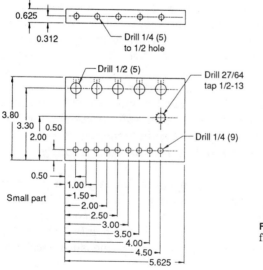

Figure 6.13 Small part held in fixture.

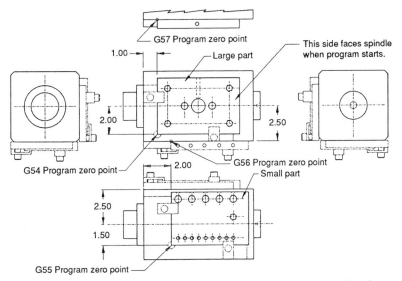

Figure 6.14 Key location points for fixture to hold two parts to be machined.

Admittedly, these parts are very simple. We intentionally kept them simple to make it easy for you to follow what is happening. Besides, our intention here is to show you rotary device programming, not to stun you with fancy fourth-axis programs.

Figure 6.14 is the fixture that will hold the parts. Though it is made for a vertical machining center, rest assured that the same techniques are used on a horizontal machining center. Only the program formatting information would change.

Note from the fixture drawing that we have marked up the print to show the locations of all program zero points. Notice that even this simple application requires *four program zero* points, one for each surface of the part being machined. Of course, this requires that the operator make 12 measurements (*X, Y,* and *Z* times 4). We have also specified on the fixture drawing the coordinate systems being used (G54 through G57).

Since this example program will be using four separate program zero points, the coordinates going into the program will be taken from the print.

Next, let's look at the sequence of operations the program will follow. Figure 6.15 shows this sequence.

There is one last point we wish to make before giving the program. The program will be using canned cycles to machine holes. While canned cycles are not discussed until key concept no. 6, we included this technique to keep the program short. At this time we are trying to

Sequence of Operations

Operation	Tool	Station	Feed	Speed
Center drill holes in plan view of large part	#3 center drill	1	3, in IPM	1200, in RPM
Center drill (5) holes in side of small part				
Index 90° CCW				
Center drill (1) hole in side of large part				
Index 180°				
Center drill all holes in plan view of small part				
Drill (5) 1/2 holes in plan view of small part	1/2 drill	2	4.5, in IPM	600, in RPM
Index 90° CCW				
Drill (2) 1/2 holes in plan view of large part				
Drill (1) 1 in hole in plan view of large part	1 dril	3	5.5, in IPM	300, in RPM
Drill (4) 27/64 holes in plan view of large part	27/64 drill	4	4.5, in IPM	600, in RPM
Index 90° CW				
Drill (1) 27/64 hole in plan view of small part				
Drill (9) 1/4 holes in plan view of small part	1/4 drill	5	2, in IPM	1000, in RPM
Index 90° CCW				
Drill (5) 1/4 holes in side view of large part				
Index 90° CCW				
Drill (1) 1/4 hole in side view of large part				
Index 90° CW				
Tap (4) 1/2-13 holes in plan view of large part	1/2-13 tap	6	17.6, in IPM	230, in RPM
Index 90° CW				
Tap (1) 1/2-13 hole in plan view of small part				
Index 90° CCW				

Figure 6.15   Process for two parts held in fixture for example on vertical machining center.

clarify rotary device programming. Once you have read about canned cycles in key concept no. 6, you may wish to review this section.

Now you should be ready to study the program.

```
O0021
N005 G91 G28 A-15. (ensure that plan view of large part is facing spindle)
N010 G54 G90 S1200 M03 T02 (center drill)
N015 G00 X.75 Y.75
N020 G43 H01 Z2.
N025 M08
N030 G81 R.1 Z-.25 F3. G99
N035 X2. Y2.
N040 X3.
N045 X4. G98 (close to clamp!)
N050 5.25 Y.75 G99
N055 Y3.25
N060 X.75 G98 (stay above clamps for move to side view holes)
N065 G56 X.5 Y-.313 R.1 Z-.25 G99 (note that the G56 is allowed in the canned
cycle command)
N070 X1.5
N075 X2.5
N080 X3.5
N085 4.5 G98
N090 G80 (cancel cycle for index)
N095 G00 Z4. (get some extra clearance in for index)
```

N095 G91 G00 A-90. (index 90° counterclockwise to side view of large part)
N100 G90 G57 G81 X3. Y-.3125 R.1 Z-.25 G98
N105 G80
N110 G00 Z4.
N115 G91 G00 A-180. (index 180° to plan view of small part)
N120 G90 G55 G81 X.5 Y.5 R.1 Z-.25 G99
N125 X1.
N125 X1.5
N130 X2.0
N135 X2.5
N140 X3.0
N145 X3.5
N150 X4.0
N155 X4.5
N160 Y2.
N165 Y3.3
N170 X3.5
N175 X2.5
N180 X1.5
N185 X.5
N190 G80 M09
N195 G91 G28 Z0 M19
N198 M01
N200 T02
N205 M06
N210 G55 G90 S600 M03 T03 (½-in drill) (still on plan view of small part)
N215 G00 X.5 Y3.3
N220 G43 H02 Z.1
N225 M08
N230 G81 R.1 Z-.85 F4.5
N235 X1.5
N240 X2.5
N245 X3.5
N250 X4.5
N255 G80
N260 G00 Z4.
N265 G91 G00 A-90. (index 90° counterclockwise to plan view of large part)
N270 G90 G54 G81 X2.0 Y2.0 R.1 Z-.85 G99
N275 X4.0
N280 G80 M09
N285 G91 G28 Z0 M19
N290 M01
N295 T03
N300 M06
N305 G54 G90 S300 M03 T04 (1-in drill)
N310 G00 X3. Y2.
N315 G43 H03 Z.1
N320 M08
N325 G73 R.1 Z-.95 Q.1 F5.5
N330 G80 M09
N335 G91 G28 Z0 M19
N340 M01
N345 T04
N350 M06
N355 G54 G90 S600 M03 T05 (27/64-in drill) (still on plan view of large part)
N360 G00 X.75 Y.75

```
N365 G43 H04 Z2.
N370 M08
N375 G73 R.1 Z-.85 F4.5 G98 (note clamp!)
N380 X5.25 G99
N385 Y3.25
N390 X.75
N395 G80
N400 G00 Z4.
N405 G91 G00 A90. (index 90° clockwise to plan view of small part)
N410 G90 G55 G73 X4.5 Y2. R.1 Z-.85 G99
N415 G80 M09
N420 G91 G28 Z0 M19
N425 M01
N430 T05
N435 M06
N440 G55 G90 S1000 M03 T06 (¼-in drill) (still on plan view of small part)
N445 G00 X.5 Y.5
N450 G43 H05 Z.1
N455 M08
N460 G73 R.1 Z-.8 Q.1 F2.
N465 X1.
N470 X1.5
N475 X2.
N480 X2.5
N485 X3.
N490 X3.5
N495 X4.
N500 X4.5
N505 G80
N510 G00 Z4.
N510 G91 G00 A-90. (index 90° counterclockwise to side view of small part)
N515 G90 G56 G81 X.5 Y-.313 R.1 Z-.5 G99
N520 X1.5
N525 X2.5
N530 X3.5
N535 X4.5
N540 G80
N545 G00 Z4.
N550 G91 G00 A-90. (index 90° counterclockwise to side view of large part)
N555 G90 G57 G83 X3. Y-.3125 R.1 Z-1.75 Q.5
N560 G80 M08
N565 G91 G28 Z0 M19
N570 G91 G00 A90. (index 90° clockwise to plan view of large part)
N575 M01
N580 T06
N585 M06
N590 G54 G90 S230 M03 T01 (½-13 tap)
N595 G00 X.75 Y.75
N600 G43 H06 Z.1
N605 M08
N610 G84 R.25 Z-.85 F17.6
N615 X5.25
N620 Y3.25
N625 X.75
N630 G80
```

N635 G00 Z4.
N640 G91 G00 A90. (index 90° clockwise to plan view of small part)
N645 G90 G55 G84 X4.5 Y2. R.1 Z-.85 G99
N650 G80 M09
N655 G91 G28 Z0 M19
N660 G91 G00 A-90. (index 90° counterclockwise back to starting point)
N665 M01
N670 G91 G28 X0 Y0
N675 T01
N680 M06
N685 M30

As stated, programs for rotary devices tend to get a little complicated. And the above program was relatively simple. In the above example, remember that we are stressing rotary device programming,
and commands related to the rotary device should be well-understood
at this point. We are machining on only three sides of the fixture and
the parts were quite simple. We hope that, if you study the above program well enough, it will make sense to you. If you will be doing this
kind of work, be sure to stick with it until it does make sense.

Conclusion to rotary device programming. While we cannot hope to prepare you for every possible condition related to rotary device programming, we hope that your knowledge has been expanded to the point
where you can approach the rotary device programming with confidence. Again, most applications requiring the use of the rotary device
require only simple indexing, so it is a simple matter of rotating the
workpiece to an attitude that allows machining.

Because it can be difficult to visualize the part being rotated, it is
wise to make sketches and come up with a clear sequence of operations
before you start writing the program (as the example has shown).
Once you can truly visualize what is going on, writing the program
becomes *much* simpler.

Turning Centers

Before we can show the basic formatting of programs for CNC turning
centers, we must point out that there are two popular G code *assignment groups* used by control manufacturers for turning centers.
There is less consistency related to what the various G codes mean
between turning center controls than there is for other types of CNC
machines.

As this problem relates to program formatting, there are only two
areas of concern. One has to do with the command used to assign program zero and the other has to do with how incremental and absolute
modes are commanded. While the format shown here is the most com-

monly used in the United States, we want you to be aware of these minor differences. Before we dig into program formatting for turning centers, we must discuss these slight discrepancies.

Command to assign program zero

The most commonly used G code to assign program zero is the G50. It is also the one we use in our given format. But you should be prepared for the possibility that the CNC turning center requires a G92 command for the same purpose. Some CNC turning center control manufacturers, especially American builders, want to maintain consistency between turning center controls and machining center controls, so they use the same command (G92) to assign program zero for both types of machines. Rest assured, no matter which G code is used to assign program zero, they all work the same way.

How incremental and absolute modes are commanded

The other difference is related to incremental and absolute mode. By now you should understand both modes. Again, the format we show is the most commonly used in the United States. With this method, no actual G code is required to assign incremental or absolute mode. The letter address used to command the motion tells the control whether the motion is intended to be in incremental or absolute mode. If X and/ or Z is used, the motion is in the absolute mode. If U and/or W is used, the motion is in the incremental mode. Note that U specifies an incremental motion in the X axis and W specifies an incremental motion in the Z axis. Also, remember that U must be a *diameter departure,* not radius, on most CNC controls.

The other method used by some manufactures also maintains consistency between CNC turning centers and machining centers. A G90 and G91 command is used to designate absolute and incremental modes. G90 commands the absolute mode and G91 commands the incremental mode. Then, only X and Z are used to command axis departure.

The four kinds of format for turning centers

Our given format assumes the machine will be at its reference point when the program is activated. Also, the reference point is the position at which tool changes will occur. While this may not be the most efficient location at which to change tools because of the distance the turret may have to travel, it is a very safe method. This method allows beginners to become familiar with the machine since indicator lights on the control panel will be illuminated when an axis is at the reference point.

Program start-up format.

O0001 (program number)
N005 G28 U0 W0 (safety command to assure the machine is at the reference point)
N010 G50 X10.3233 Z6.4335 S3000 (assign program zero, limit spindle to 3000 RPM)
N015 G00 T0101 M41 (index to first tool, select spindle range)
N020 G96 S350 M03 (select spindle mode, speed, and turn spindle on)
N025 G00 X3. Z1. M08 (first motion to workpiece, turn on coolant)
N030 G01 X ____ Z ____ F.015 (first cutting motion, don't forget feed rate)

Tool ending format.

N075 G00 X10.3233 Z6.4335 T0100 (go back to starting point, cancel tool offset)
N080 M01 (optional stop)

Tool start-up format.

N140 G50 X10.6332 Z6.0042 S2500 (set program zero, limit spindle to 2500 RPM)
N145 G00 T0202 M42 (index to station 2, select the spindle range)
N150 G97 S600 M03 (select spindle mode and speed, turn on spindle)
N155 G00 X0 Z.1 M08 (rapid up to workpiece, turn coolant on)
N160 G01 Z _____ F.010 (don't forget feed rate)

Program ending format.

N210 G00 X10.6332 Z6.0042 T0200 (rapid back to start point, cancel tool offset)
N215 M30 (end of program)

Understanding the format

While we placed messages in parentheses to help you understand what is going on in the format, this format bears further explanation. Here we will list the various words involved in the format and describe them. Note that they are in numerical order for easy reference.

G words.

G00 This is the rapid positioning command. Whenever you wish motion to occur at the machine's fastest possible rate, the G00 is the command to use. G00 is discussed at length in key concept no. 3.

G01 F_____ Though it may not be readily apparent, the feed rate in the first cutting command is also part of the program start-up and tool start-up format. To make each tool independent from the rest of the program, the feed rate must be included for each tool. Note that G01 is discussed in key concept no. 3.

G28 This command is used by most controls to send the machine to its reference point. For turning centers, the reference point makes an excellent position at which to change tools, especially for beginners. As you know, the reference point is a very accurate location along the machine's travels in each axis. For most turning centers, this position is usually quite close to the plus limit for each axis. The reference point makes an excellent position from which to start the program. It facilitates easy part loading and allows the operator to easily check and see if the machine is in the proper starting position (control panel lights will come on when the machine is at its reference point). We include a G28 as a safety command to ensure that the machine is at its proper starting position at the very beginning of the program. While not all controls require this, it is not a bad habit for the programmer to get into. The method by which the G28 is commanded will vary from control to control. Most require an intermediate point motion to be included in the G28 command. While somewhat difficult to understand, the purpose for this intermediate point is to allow the programmer to have the tool clear obstructions before it moves to the machine's reference point. Note that only the axes included in the G28 command will move. If an axis (U for X, W for Z) is left out of this command, the machine will not move in the axis left out. You can think of the G28 command as two commands in one. First the machine will move to the intermediate point, then it will move to the reference point. If you study our given technique, you will find that we pulled a trick on the control. We included a U and W (incremental movement in X and Z) in the G28 command. For each axis included in the G28 command, we also made the value of axis departure zero. This tells the machine to move incrementally nothing, then to go to the reference point. We rarely recommend that the incremental mode be used; this technique is used only in the G28 command. All other motion commands are made in the absolute mode.

G50   This command is used on most turning centers to assign the program zero point. You should note that some turning center controls use a G92 command for this purpose. The X and Z values included in the G50 command are the measured distances from program zero to the tip of the tool to be used. These values will change from tool to tool. If using our given format, these values *must* be repeated later in the program at the end of the tool. The command that rapids the tool back to the tool change position must include the same X and Z values as were in the G50 command. This will assure that the tool change position is the same for every tool in the program. Along with assigning the program zero point, the G50 command also provides another function. It allows the programmer to assign a *spindle RPM limitation*. If an S word is included in the G50 command, it tells the control not to allow the spindle to run faster than this limiting RPM. The application for this spindle limiter is related to the constant surface speed mode (see G96). When in the constant surface speed mode, the control is constantly and automatically calculating the required RPM based on the SFM spindle speed programmed and the diameter being machined. (See key concept no. 1 for more information.) The smaller the diameter, the faster the spindle will run. In the constant surface speed mode, if the tool is brought to the spindle centerline (while facing), the spindle will accelerate to its maximum. Without a spindle limiter, the control will give no consideration to the setup being made.

There are times when the programmer must not allow the spindle to exceed a given RPM. For example, say a heavy, out-of-balance casting is to be machined. During setup, the operator tests the workpiece in the setup, turning the spindle on at varying RPM. Assume, for example, when the spindle reaches an RPM of 1200, the machine starts vibrating, and the operator is worried that the part might be thrown from the setup if the spindle runs much over 1000 RPM. An S1000 could be included in each G50 command to assure the spindle could not run faster.

G96 and G97 These commands set the spindle mode. G96 sets the constant surface speed mode, and G97 sets the RPM mode. Any S word following a G96 will be taken as surface feet per minute (or meters per minute in the metric mode). Any S word following a G97 will be taken as RPM. (See key concept no. 1 for more information.)

M words.

M01 This word is called an *optional stop*. It works in conjunction with an on/off switch on the control panel labeled *optional stop*. When the control reads the M01 word, it looks to the position of the on/off switch. If the switch is on, the machine will stop executing the program. The spindle, coolant, and anything else still running will be turned off. The machine will remain in this state until the cycle is activated again by pressing the cycle start button. Then the control will continue executing the program from the point of the M01. If the toggle switch is off, the control will ignore the M01 command entirely. It will be as if the M01 was not in the program at all. The purpose for including this command in the tool and program ending format is to give the operator a way to stop the machine at the completion of each tool. Many times, especially during a program's verification, it will be necessary for the operator to stop the machine after each tool to confirm that the tool has done its job correctly. In some cases, if the tool has not machined as intended, the subsequent tools in the program may be damaged. For example, say a hole was to be drilled and tapped. Of course, the drilled hole must be deep enough for the tap to reach its desired depth. After the drill has machined the workpiece, if the optional stop switch is on, the operator can rest assured the machine will stop when the drill has finished. The operator can check to confirm that the drill has machined the hole deep enough for the tap. Note that there is another program-stopping M code called simply *program stop*. The program stop command is programmed by an M00 word and *forces* the machine to stop no matter what. There is no on/off switch involved and the operator will have no option; the machine will stop.

M03 This M word is used on CNC turning center controls to turn the spindle on in a forward or clockwise direction. When you actually see the spindle running in this direction, it may appear that this statement is wrong. That is, it may appear to you that the spindle is running counterclockwise. But you must evaluate the M03 from the perspective of the spindle nose, not from below the spindle. Generally speaking, the M03 is used with right-hand tools as long as the machining direction is toward the spindle in the Z axis. If using left-hand tools, the spindle must be started in a counterclockwise or reverse direction. The command to start the spindle in the reverse direction is M04. You may have noticed by now that we have never had to command that the spindle be stopped. When making tool changes, it is wise to leave the spindle running. There is no reason to turn it off. To do so would be a waste of cycle time and electricity. The M30 command at the end of the program will shut the spindle off.

M08   This M code is used on CNC turning centers to turn on the (flood) coolant. Note that the given format never turns the coolant off. This will happen automatically during a turret index. Also, the M30 program ending command will turn the coolant off at the completion of the program.

M30   This is a program ending word. It tells the control to turn off anything still running (spindle coolant, etc.) and rewind the memory back to the beginning. Then the control stops. While the M30 command we have shown is very popular among control builders, some builders require an M02 as the program ending word. In this case, the M02 would do everything the M30 would.

M41 and M42   Many turning centers have more than one spindle range. This allows power for heavy machining operations in the low range and high speed for finishing operations in the high range. (See key concept no. 1 for more information.) Two very common M codes to handle the spindle range changes are M41 and M42 (M41 for low range and M42 for high range). We must warn you, however, that it is quite likely the machine you must work with may use different M codes to control the spindle range changes. You must check in your programming manual to find the M codes related to spindle ranges.

Other M words related to turning centers.   While not directly related to program formatting, there are other M codes of concern to the turning center programmer. There are several machine functions that can be activated by M codes on turning centers. These things include tailstock (body and quill), chuck jaws (open and close), bar feeder (if so equipped), and possibly even chip conveyor. You must check the programming manual for the machine tool to find the list of related M codes.

Other words in the format.

End-of-block word   Also discussed during machining center formatting, this code is used to terminate each command.

O word   Most controls allow more than one program to be stored in the control's memory. Controls that allow this require a special word to designate which program is which. The O word is one of the most common ways to do this. The O word assigns the program's number. The operator can easily scan from one program to another by the O word. It is usually the very first word of the program.

S word   The S word tells the control the desired RPM or SFM (meters per minute in metric mode) for the spindle. The S word, by itself, does not actually turn on the spindle, the M03 does that. It just lets the control know what spindle speed is desired. The G96 constant surface speed mode and G97 RPM mode tell the control the usage of the S word.

T word   The T word tells the turning center control two things: the tool station number, and the offset number to be used with the tool. The T word is a four-digit word. The first two digits tell the control the turret station number to be indexed to and the second two digits tell the control the offset number to instate. For more information, see key concept no. 4.

F   The F word on turning centers tells the control the desired feed rate. There are two ways of commanding feed rate: inches per minute and inches per revolution. By far, inches per revolution is the more common way to specify feed rate on turning centers. The only time we would recommend using the inches per minute mode would be if you intend to make a feed rate movement with the spindle stopped. G98 specifies the inches per minute mode and G99 specifies the inches per revolution mode on most turning centers. G99 is initialized when the power is turned on, so it is not necessary in the program. The control will assume the inches per revolution mode unless G98 is programmed.

Other notes about the format

We do *not* claim this to be the only way to format programs for a CNC turning center. In fact, the format required for your particular turning center may differ substantially. We would be the first to admit there are *many* ways to successfully format CNC programs. But this information should truly get you on the right track, giving you the basics of program formatting. You may find certain CNC controls to require special commands we have not discussed. You may also find more M codes are required for your machine's application. However, the basic structure we have shown and the logic we have given can be used on *any* CNC turning center.

Example program showing format for turning centers

Figure 6.16 is the simple drawing we use to stress turning center format. To keep it simple, only the finish turning and drilling operations are being performed.

Program:

```
O0001 (program number)
N005 G28 U0 W0 (assures machine is at the reference point)
N010 G50 X _____ Z _____ (set program zero for turning tool)
N015 G00 T0101 M42 (select tool station and spindle range)
N020 G96 S350 M03 (turn spindle on clockwise at 350 SFM)
N025 G00 X.6 Z.1 M08 (rapid to point 2, turn coolant on)
N030 G01 Z0 F.008 (feed to point 3)
N035 X.8 Z-.1 (feed to point 4)
N040 Z-.5 (feed to point 5)
N045 X.6 Z-.6 (feed to point 6)
N050 Z-.8 (feed to point 7)
N055 X1.2 (feed to point 8)
N060 X1.4 Z-1.1 (feed to point 9)
N065 Z-1.5 (feed to point 10)
N075 G0Z X1.6 Z-1.6 R.1 (circular move to point 11)
N080 G01 X2. (feed to point 12)
N085 X2.25 Z-1.8 (feed to point 13)
N090 G00 X _____ Z _____ T0100 (rapid to tool change point,
cancel offset)
N095 M01 (optional stop)
```

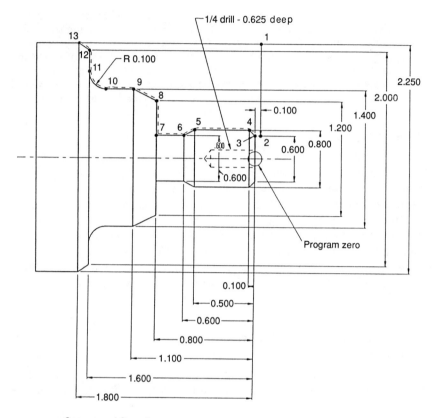

Figure 6.16   Turning center format example.

N100 G50 X _____ Z _____ (set program zero for drill)
N105 G00 T0202 M42 (index to tool 2 and select spindle range)
N110 G97 S1000 M03 (turn spindle on at 1000 RPM)
N115 G00 X0 Z.1 M08 (rapid to clear part, turn on coolant)
N120 G01 Z-.625 F.006 (drill hole)
N125 G00 Z.1 (rapid out of hole)
N130 X _____ Z _____ T0200 (rapid back to tool change position,
cancel offset)
N135 M30 (end of program)

Wire EDM Machines

Wire EDM equipment has only two forms of program format: program
start-up and program ending. However, what is just as important as

understanding these two format types is understanding the methods by which the wire EDM machine must be programmed for various applications. Most of our discussions in this section will be aimed at familiarizing you with these methods.

Before we give examples of program formatting, we want to discuss the methods by which different types of workpieces are machined on a wire EDM machine. You may also want to review the tips given during concept no. 1.

As you know, accuracy and finish are directly related to the number of trim passes being made. In some cases, one pass around the shape will be good enough when accuracy and finish are not critical. But when trim passes must be made, you need to understand how to do them correctly for the various parts you will be machining. We will discuss the different types of workpieces that are typically machined on a wire EDM machine and try to present the most important implications.

Punches

Punch machining requires that an area of the shape be left unmachined. This area is called the *glue stop area*. The purpose for the glue stop area is to allow the punch to remain stable during trim passes. The size of the glue stop area is directly related to the size of the part. Additionally, the location of the glue stop area should be planned in a way to allow easy grinding of the tab that will be left after the punch is cut off. Figure 6.17 is a drawing that shows the programmed movements for a typical punch *before the cutoff*.

After the punch is machined, the programmer will tell the control to stop (with an M00). This will allow the operator to apply magnets or glued tabs between the punch shape and the rough stock. The magnets or glued tabs will support the punch during the cutoff. Then the programmer will command the punch to be cut off with a straight cut across the glue stop area. It is wise to cut the punch off in a way that allows a minimum amount of stock for grinding (leave 0.010-in stock or so).

If trim passes are required on a punch, it makes the programming

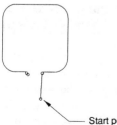

Figure 6.17 Wire EDM punch before cut off.

Start point

task much more difficult. The trim pass *must* be made in a reverse direction around the punch, meaning that the cutting path must be reversed. This requires that the program include a series of motion commands to go the other way around the shape.

One way to simplify multiple trim passes is to use subroutine techniques (discussed a little later).

Dies

Dies present a slightly different problem. On the first roughing pass, the machine must be stopped just before the point where the slug in the die is to be cut off, otherwise the slug will interfere with the movement back to the start point after cutoff and cause wire breakage. Then, if trim passes are necessary, it must be stopped again after cut off, when the wire returns back to the start point. This will allow the operator to remove the slug so that it will not interfere with the trim passes.

If trim passes are necessary, it is best (but not mandatory) to make them in a reverse direction. If the trim passes are made in the same direction as the roughing pass, the wire will have the tendency to follow the motions of the roughing pass. If any washout (washout is wire deflection) occurs during the roughing pass, the wire will try to follow the washout. But if reverse-direction trim passes are made, the wire will have a tendency to start fresh as it enters any washout areas. However, reverse-path trim passes are more difficult to program manually and most manual programmers will simply program trim passes in the same direction as the roughing pass unless tolerances are extremely tight.

Form tools

With a form tool, the cutting path is not in an enclosed shape. Generally, a form tool is cut from one end to the other. During the roughing pass, the wire should be stopped just before the wire reaches the end of the shape. This will allow the operator to apply magnets to support the slug end of the stock while the form tool is cut in half. If trim passes are necessary, after the roughing pass is completed another stop is made to allow the operator to remove the slug end.

If trim passes are necessary, it is wise to make them in a reverse direction for the same reasons as for a die. But if you want to make the trim passes in the same direction as the roughing pass (because it is easier), you can rapid the wire back to the start point and make the first trim pass. This can be repeated for as many trim passes as necessary.

Understanding how to make trim passes easily

Before we give examples of each type of shape, we want to show you how to develop subroutines in a way that simplifies making trim passes. Subroutines keep you from having to write redundant commands a second time. Basically, trim passes are redundant commands. Your goal with subroutines will be to isolate the trim passes from the main program in a way that lets you simply command that the trim pass be made when desired. At first this will seem a little complicated, but when you see an example, it should not seem so bad.

Subroutines involve four programming words:

M98   Jump to subroutine

M99   Go back (to the command *after* the calling command)

P word   Subroutine sequence number

L word   Number of times (seldom used)

The M98 command will *always* include a P word to tell the control where to go. When the control reads M98 P1000, for example, it jumps to sequence number N1000 and continues executing from there. Then, when the control reads a subsequent M99 command, it jumps back to the command *after* M98 P1000 and continues execution. The commands that are between the N word (specified by the P word) and the M99 will include the movement commands to drive the wire through a pass around the workpiece.

The subroutine (beginning with the N word specified by the P word) will *always* be included *after* the end of the main program, after the M02 command.

At this point, this can't make much sense, but now let's look at some examples. This should make it clearer.

Punch example

Say you are machining a punch that requires three total passes—one rough and two trims. For the material, material thickness, and wire diameter, you look in the builder's condition manual and find that this set of conditions and offsets is recommended:

Rough pass	Condition C207 and 0.0085-in offset
First trim pass	Condition C120 and 0.0056-in offset
Second trim pass	Condition C121 and 0.0053-in offset

Figure 6.18 is the print for the punch.

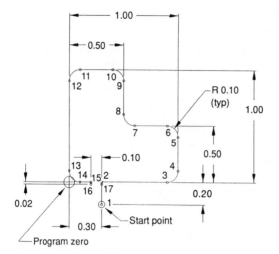

Figure 6.18   Example for punch format on wire EDM machine.

Here is the program for the punch using subroutine techniques, with full documentation:

N010 G92 X.3 Y-.2 (set program zero)
N015 G90 (select absolute mode)
N020 C207 (select condition for rough pass)
N025 G42 H0.0085 (select compensation right and 0.0085 offset, rough pass)
N028 T84 (turn on high-pressure flushing pump)
N030 M98 P1000 (jump to N1000 and make roughing pass)
N033 T85 (turn off high-pressure pump)
N035 C120 (select condition for first trim pass)
N040 G41 H0.0056 (select compensation left and 0.0056 offset, first trim pass)
N045 M98 P2000 (jump to N2000 and make first trim pass)
N050 C121 (select condition for second trim pass)
N055 G42 H0.0053 (select compensation right and 0.0053-in offset for second trim pass)
N058 M98 P1000 (jump to N1000 and make second trim pass)
N060 M00 (stop to apply magnets for cutoff)
N065 C207 (select roughing condition for cutoff)
N070 G01 X.3 (cut off to point 17)
N075 Y-.2 (move back to start point)
N080 M02 (end of main program)
(Subroutine to make counterclockwise pass)
N1000 G01 Y0 (move to point 2)
N1005 X.9 (move to point 3)
N1010 G03 X1. Y.1 J.1 (move to point 4)
N1015 G01 Y.4 (move to point 5)
N1020 G03 X.9 Y.5 I-.1 (move to point 6)
N1025 G01 X.6 (move to point 7)
N1030 G02 X.5 Y.6 J.1 (move to point 8)
N1035 G01 Y.9 (move to point 9)
N1040 G03 X.4 Y1. I-.1 (move to point 10)
N1045 G01 X.1 (move to point 11)

N1050 G03 X0 Y.9 J-.1 (move to point 12)
N1055 G01 Y.1 (move to point 13)
N1060 G03 X.1 Y0 I-.1 (move to point 14)
N1065 G01 X.2 (move to point 15)
N1070 G40 Y-.02 (cancel offset and move to point 16)
N1075 M99 (end of subroutine)
(Subroutine to make clockwise pass)
N2000 G01 Y0 (move to point 15)
N2005 X.1 (move to point 14)
N2010 G02 X0 Y.1 J.1 (move to point 13)
N2015 G01 Y.9 (move to point 12)
N2020 G02 X.1 Y1. I.1 (move to point 11)
N2025 G01 X.4 (move to point 10)
N2030 G02 X.5 Y.9 J-.1 (move to point 9)
N2035 G01 Y.6 (move to point 8)
N2040 G03 X.6 Y.5 I.1 (move to point 7)
N2045 G01 X.9 (move to point 6)
N2050 G02 X1. Y.4 J-.1 (move to point 5)
N2055 G01 Y.1 (move to point 4)
N2060 G02 X.9 Y0 I-.1 (move to point 3)
N2065 G01 X.3 (move to point 2)
N2070 G40 Y-.02 (cancel offset and move to point 17)
N2075 M99 (end of subroutine)

If you study this program, it should make sense. Notice that using the subroutine techniques saved you (in this case) from having to write one complete pass (the second trim pass) in the program.

We must point out that some controls require that subroutines be stored separately from the main program. In this case, the P word must point to a separate program number, not a sequence number.

Notes about the punch program.

1. The conditions and offsets are just examples. You *must* check with your condition manual to obtain the correct conditions and offsets for your application.

2. Note that the main program (through the M02) will be very similar from one program to the next. Only the subroutines will change dramatically from one punch program to the next.

3. Parentheses are allowed in the program at any time to allow you to include a message to the operator. Although this program takes them to extreme, they are *very* helpful and allow you to document your program. You can include a lengthy description of the program's intention at the beginning of the program to help you remember what the program is intended to do (part name, customer, number of trim passes, etc.). Also, you can include a message next to any M00 (program stop) to tell the operator what to do (apply magnets, remove slug, etc.).

Die example

This example shows how to program for a die. In this example, we make two total passes, one for the tapered area and one for the land (tapered area first). Note that this is the die opening for the previous punch example and the same set of coordinates are used. While two passes are being made, we make them in the same direction. We are also tricking the control with the optional block skip technique for the glue stop during the roughing pass. See if you can figure out what is happening! (Optional block skip techniques are shown in Key Concept No. 6)

Say you look in the condition manual and find the conditions and offsets to be as follows:

Rough pass (with taper)   Condition C207 and 0.0085-in offset
Trim pass (straight portion)   Condition C120 and 0.0053 offset

Figure 6.19 shows the print.

Die program:

N010 G92 X.5 Y.2 (set program zero)
N015 G90 (select absolute mode)

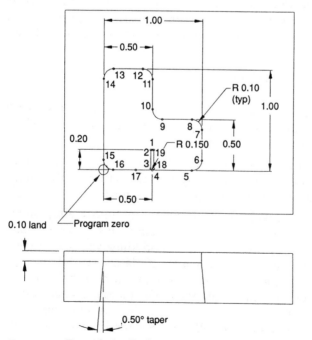

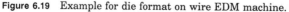

Figure 6.19 Example for die format on wire EDM machine.

N020 C207 (select condition for rough pass)
N025 G41 H0.0085 (select compensation left and 0.0085 offset for rough taper pass)
N028 G51 A.5 (set taper to left, 0.5°)
N029 G12 (turn optional block skip off to activate glue stop)
N027 T88 (select submerse mode for submersible machines only)
N028 T84 (turn on high-pressure flush pump)
N030 M98 P1000 (jump to N1000 and make roughing pass)
N033 T85 (turn off high-pressure pump)
N035 C120 (select condition for straight portion trim pass)
N040 G41 H0.0053 (select compensation left and 0.0053 offset for straight portion trim pass)
N043 G11 (turn optional block skip on to ignore glue stop)
N045 M98 P1000 (jump to N1000 and make straight portion trim pass)
N080 M02 (end of main program)
(subroutine to make counterclockwise pass)
N1000 G01 X.485 (move to point 2)
N1005 Y.015 (move to point 3)
N1010 G03 X.5 Y0 I.015 (arc in to point 4)
N1025 X.9 (move to point 5)
N1030 G03 X1. Y.1 J.1 (move to point 6)
N1035 G01 Y.4 (move to point 7)
N1040 G03 X.9 Y.5 I-.1 (move to point 8)
N1045 G01 X.6 (move to point 9)
N1050 G02 X.5 Y.6 J.1 (move to point 10)
N1055 G01 Y.9 (move to point 11)
N1060 G03 X.4 Y1. I-.1 (move to point 12)
N1065 G01 X.1 (move to point 13)
N1070 G03 X0 Y.9 J-.1 (move to point 14)
N1075 G01 Y.1 (move to point 15)
N1080 G03 X.1 Y0 I-.1 (move to point 16)
/N1085 G01 X.4 (move to point 17)
/N1090 M00 (stop to apply magnets; note "/" code)
N1095 G01 X.5 (move to point 4)
N1100 G03 X.515 Y.015 J.015 (arc out to point 18)
N1105 G01 Y.2 (move to point 19)
N1110 G40 G50 X.5 (cancel compensation and taper, move back to point 1)
/N1015 M00 (stop to remove slug)
N1120 M99 (end of subroutine)

Notes about the die program.

1. The conditions and offsets are just examples. You *must* check with your condition manual to obtain the correct conditions and offsets for your application.

2. Note that the main program (through M02) will be very similar from one program to the next; only the subroutines will change dramatically from one die program to the next.

3. Parentheses are allowed in the program at any time to allow you to include a message to the operator. Although this program takes them to extreme, they are *very* helpful for documenting your program. You can include a lengthy description of the program's intention at the beginning of the program to help you remember

what the program is intended to do (part name, customer, number of trim passes, etc.). Also, you can include a message next to any M00 (program stop) to tell the operator what to do (apply magnets, remove slug, etc.).

4. The trick with the optional block skip code (/) concerned the program commands G11 (skip on) and G12 (skip off). The first time through the subroutine, we wanted the M00 commands to be executed, so G12 was commanded prior to the command executing the subroutine. The second time through, we did not want to execute the M00 command, so the G11 was commanded prior to the second call to the subroutine. For controls that allow the optional block skip switch to be turned on and off by program command, this is a very helpful technique.

7

Key Concept No. 6: Special Features of Programming

Up to this point, you have been exposed to many of programming's raw tools for all kinds of CNC machine tools. While these raw tools are extremely important to your manual programming career, it would be quite difficult to program any kind of CNC machine tool if you were limited to using only what you have learned so far.

Most CNC control manufacturers go to great lengths to make manual programming easier and faster. They study the application for which their control was intended for the purpose of finding ways to minimize programming commands. For example, one very common type of operation for machining centers is hole machining. Drilling, tapping, boring, and reaming are among the hole machining operations commonly performed. Knowing this, machining center control manufacturers have designed a whole series of "canned" cycles aimed at simplifying hole machining operations.

Indeed this is just one example. Almost every form of CNC control has some special features to help the programmer in one way or another. It is our intention in key concept no. 6 to acquaint you with many of them. We must point out that these features will vary dramatically from control builder to builder. Competition among the builders warrants that each will strive to improve current features. When it comes to features that are common to more than one builder, each builder will do its best to apply the feature. While this makes for some incompatibility between controls, most of these features warrant the time it takes for the programmer to learn them completely.

Since there will be so much variation from control to control with regard to specifically how these features are programmed, we will only introduce the features and give you a general idea as to how each feature is utilized. We will leave it up to you to check the builder's manual to find the specific techniques to program the feature being discussed.

Some of the features discussed here are almost taken for granted in today's CNC technology. That is, they have been around for a while, and may not be considered by experienced programmers to be all that special. However, for the sake of completeness, we include them.

We will begin by discussing those features that are common to all CNC controls. Then we will go into each type of CNC control and discuss its specific features.

Note that some of the features we discuss may be considered options by certain control manufacturers. This means that you will have to check the programming manual for the particular CNC machine with which you will be working to confirm that the feature is available. We will try to point out whether the feature is usually a standard feature or an option.

Special Features Found on Most CNC Controls

We will begin our discussion of special programming features with the features you will find on almost all kinds of CNC controls. These features are somewhat generic, meaning that their usage is not dependent on the application for which the CNC control was designed.

We must point out again that the specific codes involved with these features will vary dramatically from control to control. We limit our scope in this chapter to acquainting you with what a feature is, and for what it is used. When feasible, we may give a simple example, using commands we know to exist on at least one type of CNC control. Again, you must read the control's programming manual to learn more about the specific usage and programming of the feature as it applies to your particular control.

Decimal Point Programming

We have been emphasizing the use of decimal point programming throughout this book. Almost all current CNC controls have this feature. Decimal point programming allows the programmer to input most numeric values with a decimal point, making it clear to the control where the decimal point is to be placed.

Note that certain CNC words do *not* allow a decimal point even when decimal point programming is allowed. Those words that do not allow decimal points are normally words in which a decimal point is not needed. That is, they are words that require only whole numbers or integers. G words and M words are examples of words that do not allow decimal points, even if the control allows decimal point programming.

The word types that do allow a decimal point to be used include: axis letter addresses (X, Y, Z, etc.), the radius designator for circular commands (R, I, J, and K), and the feed rate word (F).

Even when decimal point programming is allowed, it is possible to make commands without using decimal points. This allows control manufacturers to maintain compatibility between older controls that do not allow decimal point programming and more recent controls. Programming without decimal points is more difficult and error-prone than programming with decimal points, and we strongly recommend that the programmer use decimal points whenever allowed. Programming without decimal points requires the programmer to understand how the control interprets values without decimal points. The control requires a fixed format for all words without a decimal point. Even the fixed format will vary from control builder to builder. One common fixed format requires that four places be programmed to the right of where the decimal point should be. In this version of the fixed format, an X value of X3.5 would be programmed as

X35000

As you can see, programming without a decimal point requires much more thought. On older controls, when decimal point programming was not allowed, this was a constant source of error on the programmer's part. Many times, the programmer would misplace the intended location of the decimal point by not programming the correct number of places to the right of where the decimal point should be located.

If a decimal point is left out of a value in which the control allows a decimal point, the control will revert to the fixed format for the word. This is a common cause of mistakes for beginning programmers. Say, for example, you intend to make an X departure of 3 in. The correct word will be

X3.

But if the decimal point is left out, and only X3 is programmed, the control will use the fixed format, taking the X3 as

X.0003

G04 Dwell Command

The dwell command is a standard feature on almost all CNC controls. It is a feature that causes all axis motion to pause for a specified length of time. Its reason for use will vary from machine to machine, and so will its programming format. Here are some possible applications for which you may consider using the dwell command.

On machining centers, there are times when you may elect to use a dwell command to relieve tool pressure. For example, say you were using a center cutting end mill to plunge into and open up a pocket. During the plunging motion, while the end mill is feeding straight into the pocket in the Z axis, the tool is under a great deal of tool pressure. Most programmers would agree that it would be wise to relieve the tool pressure by pausing for a moment, before moving the tool in X and Y.

On turning centers, the dwell command is also often used to relieve tool pressure. For example, when plunging a groove into a large diameter, it is wise to make the grooving tool pause for a moment at the bottom of the groove. This will relieve tool pressure and guarantee that the grooving tool has machined all the way around the workpiece.

The dwell command will only make axis motion pause. That is, all other functions of the machine will continue to operate (coolant, spindle, etc.).

On almost all CNC controls, the code used to program a dwell is G04. However, how the dwell time itself is programmed will vary from control to control. Some controls will have the G04 command specify dwell time in number of spindle revolutions. This is a common method for turning centers. Others will have the G04 command specify a dwell time in seconds.

The actual letter address used in the dwell command will also vary from control to control. One popular word is the X word. In this case, the X word will have nothing whatsoever to do with the X axis. Here's an example:

G04 X5.

If the control requires dwell time in seconds, this command would specify a 5-second dwell. If the control requires the number of spindle revolutions, this command would specify a 5-revolution dwell.

Coordinate Manipulation Features

There are several special features of programming that have to do with the coordinates that actually go into the program. Some of these features manipulate those coordinates while others simply allow a more palatable and understandable way of specifying motion commands.

Mirror image

Mirror image is a commonly misunderstood feature of CNC programming and operation. While there are limitations to the use of this feature, it is quite a handy feature to know about when the need arises. Here, we discuss the limitations and possible applications for this feature.

As the name implies, *mirror image* is used to generate a series of movements that represent the mirror of a programmed path. The motions that occur under the influence of mirror image will be just the reverse image of the motion without mirror image. Figure 7.1 shows examples of how mirror image works.

The first thing to point out with mirror image is that there is nothing magical going on when you use this feature. All that is really happening is that the control is simply reversing the sign (plus to minus or vice versa) for the axes that have mirror image in the on condition. This means that an *X*-axis position of X2.0 before mirror image is turned on would be taken as X-2.0 after it is turned on in *X*. Mirror

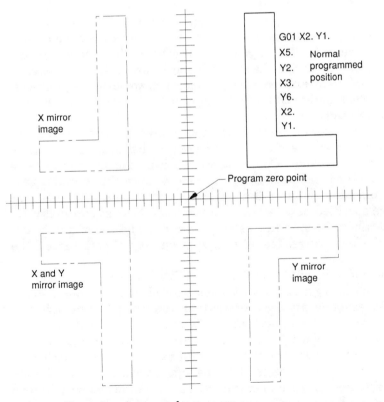

Figure 7.1 How mirror image works.

image also automatically reverses the meaning of certain motion-related commands like G02/G03 circular motion and G41/G42 radius compensation commands. That is, when mirror image is turned on, G02 becomes counterclockwise circular motion and G03 becomes clockwise circular motion. This allows the same series of commands that function well without mirror image to still properly function with mirror image in the on condition.

Applications for mirror image.   Mirror image has differing applications for different types of CNC machine tools. For machining centers, mirror image applies best to hole machining operations. Say, for example, that you have a side frame that must be machined in right-hand and left-hand versions. Mirror image allows the same program that machined the left-hand part to machine the right-hand part.

However, side milling operations present a problem for using mirror image as it relates to machining centers. While mirror image will function properly, the problem lies in basic machining practice. Conventional milling will switch to climb milling when mirror image is turned on (and vice versa). The cutter will follow the proper programmed path and truly make an opposite-hand workpiece, but the witness marks will be different from right-hand to left-hand part. Also, if tolerance is critical, you may experience differences in actual workpiece size from right-hand to left-hand part. If this is unacceptable, the only way around it is to generate two separate programs, meaning two separate cutter paths, one for the left-hand part and another for the right-hand part.

For turning centers, mirror image has limited feasible applications. There is one older type of CNC turning center that utilizes two turrets on the same cross-slide. One of the turrets is on the operator's side of the spindle, and the other is on the side opposite the operator. This kind of machine requires that all X-axis coordinates for the front turret be specified as positive values and the X-axis coordinates for the back turret be specified as minus values. The use of mirror image for the back turret makes the programming of both turrets the same.

Mirror image is quite useful on wire EDM machines. It allows the programmer to program the part in one attitude and machine it in another. For example, say the operator finds that the part cannot be run in the programmed attitude because of interference problems. Since there is none of the machining center's climb-vs.-conventional milling problem, mirror image is quite successful with wire EDM machines.

CNC turret punch presses also have application for mirror image. The turret press programmer will find many times when left-hand and right-hand parts need to be machined. In this case, the same program can be used to machine the left- and right-hand workpieces.

The two ways to activate mirror image. For almost all CNC controls, mirror image can be activated manually (through the control panel), and by program command (by a G or M code). The particular application determines whether mirror image should be activated manually or by programmed command.

Manually turning on mirror image. Imagine, for example, you had 500 left-hand parts and 500 right-hand parts to machine. If you intend to run all of the left-hand parts in one setup, then tear down and run the right-hand parts afterward, there is no need to turn mirror image on in the program. In this case, to run the hand that the program was written for, you would measure the distance from program zero in X, Y, and Z back to home position (as usual), and then input those values into your G92 command or fixture offset. So far, this is no different than what you already know.

But after you make the setup to run the opposite hand, when you measure the distances from program zero to the machine's home position, you *must* reverse the sign for the axis you are intending to use mirror image with before you enter it into the G92 command or fixture offset. Then, after you turn mirror image on manually (on the setting page of the display screen), you will be allowed to use the same program as for the first side of the part.

The key to making this work is that you *must* remember to change the sign of the axis that is being mirrored in the G92 command or fixture offset. If you intend to mirror the X axis, and the first side you run uses an X plus value in the G92 command, the second (mirrored) side will have a negative X in the G92. If you forget to reverse the G92 sign, it is easy to find during a dry run. The machine will try to go the wrong way and probably overtravel on its first attempt to move to the workpiece.

One more *very* important point. Most CNC controls will activate the mirror image *from the point where the machine is currently positioned.* With the above technique, you *must* turn mirror image on while the machine is at its reference return position (G28 or home position). If you turn on mirror image while the machine is anywhere other than the reference position, the motions of the mirrored program will not be correct.

Turning mirror image on in the program. The other application for mirror image is when you intend to run *both* the left-hand and right-hand parts *in the same setup.* Perhaps, on a vertical machining center, you are running the left-hand part on the left side of the table and the right-hand part on the right side of the table. You wish to machine both workpieces by programmed commands.

As mentioned earlier, most CNC controls have a G or M code that activates mirror image. The programming manual for the control will list the proper commands for mirror image. One popular set of mirror image commands is as follows:

G05   X-axis mirror image
G06   Y-axis mirror image
G07   Z-axis mirror image
G09   Mirror image cancel

The biggest point to make about programmable mirror image is that the machine (spindle, turret, or wire centerline) *must* be positioned at the symmetrical center of the mirror (between the two parts) whenever mirror image is turned on or off. This can be inconvenient, if not downright difficult, to program. From the programmer's standpoint, if the setup is made without concern for the position of the two parts, the position of the symmetrical center will not be known. Therefore, extra effort must be made during setup to guarantee that it is made in a way that the programmer knows where the center point is.

During setup, the G92 values (or fixture offsets) will be measured in the normal manner from the part on the table that does not require mirror image.

Knowing this, you will find it quite simple to write one program to machine both parts. In the following example, we machine one entire part and then machine the other. If you needed to machine both parts with each tool, it would be much more complicated, and using mirror image would be of minimal help. You should also notice that we are using subprogram techniques to avoid writing the machining program twice. Lastly, this example is shown in the proper format for vertical machining centers with fixture offsets.

This example program requires that you have an understanding of subprogramming technique, presented a little later in this key concept. The *symmetrical center* between the two parts is 5 in to the right of the left part and 5 in to the left of the rightmost part.

Main Program:

```
O0001
N005 G54 (½-in drill)
N010 G00 X-5. Y0 (symmetrical center point)
N015 G09 (cancels mirror image in X)
N020 M98 P1000 (run entire left part)
N025 G00 X-5. Y0 (back to symmetrical center)
N030 G05 (turns on mirror image in X)
N035 M98 P1000 (run entire right part)
N040 G00 X-5. Y0 (back to symmetrical center)
N045 G09 (cancel mirror image in X)
N050 G91 G28 X0 Y0 Z0 (go to zero return position)
N055 M30
```

Now let's look at the program that does the actual machining. Even though it is quite simple (just drilling two holes), it shows the basic idea. This technique can be incorporated no matter how complicated the actual machining program is.

```
O1000
N005 G90 S500 M03
N010 G00 X1. Y1.
N015 G43 H01 Z.1 M08
N020 G81 X1. Y1. R.1 Z-.5 F6.
N025 X2.
N030 G80 M09
N035 G91 G28 Z0
N040 M99
```

As you can see, programmable mirror image on most CNC controls has some dramatic limitations and can be complicated to work with. This, combined with the climb-vs.-conventional milling problems mentioned earlier for machining centers, makes it less helpful than you might first imagine.

X-Y–axis exchange

Axis exchange, an option on most CNC controls, allows the transformation of all X and Y values in the program. That is, any time the control reads an X, it will be taken as a Y. Any time it reads a Y, it will be taken as an X.

The result of this transformation is that the shape to be machined will be rotated 90° from its original position. Figure 7.2 shows examples of how X-Y–axis exchange can be used with mirror image to manipulate the programmed shape to any desired position.

One common command to turn on X-Y exchange is G08 (it is canceled by G09). While this command is seldom needed, one time it can be helpful is related to wire EDM equipment. The X-Y exchange command allows the programmer to easily manipulate the attitude at which the workpiece is to be machined. Many times, the wire EDM programmer will not know the exact method by which the operator will make the setup. The programmer may program the workpiece in one attitude, and the operator may decide to run the part in another. By combining mirror image and X-Y exchange, the operator can change the position of the shape to any desired attitude.

Scaling

Scaling is usually an option that must be purchased as an extra. Not all CNC controls will have this feature. Scaling allows a series of motions to be enlarged or made smaller. When scaling is activated, all

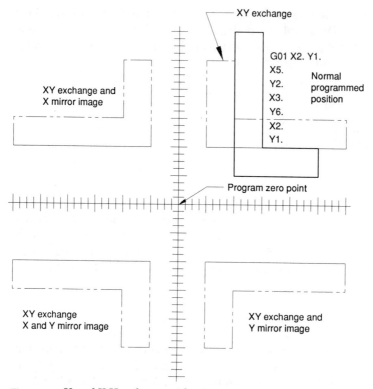

Figure 7.2 Use of *X-Y* exchange with mirror image.

the designated axis motions that the control sees will be multiplied by a scale factor. While most CNC controls with scaling will multiply all motions by the scale factor, some versions of scaling even allow the programmer to specify only one or two axes to be scaled. Also, some CNC controls will arbitrarily scale about the program zero point while others allow the center of scaling to be specified in the scaling command.

The scale factor is nothing more than the number by which the values found by the control in the program will be multiplied. If, for example, the scale factor is set to 2.0, the series of motions would be doubled in size. The scale factor of .5 would be half size. The scale factor .25 would be one-quarter size, and so on. Figure 7.3 shows what happens during scaling.

One popular method of programming a scale factor is as follows:

G68 X_____ Y_____ Z_____ R2.

where the G68 tells the control that this is a scaling command. The X, Y, and Z specify the center of scaling. The R word specifies the scale

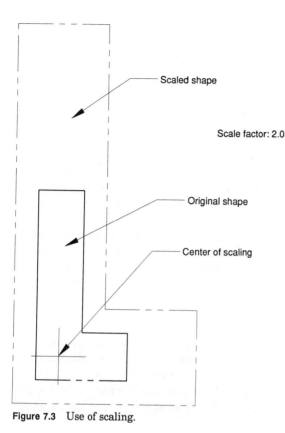

Scaled shape

Scale factor: 2.0

Original shape

Center of scaling

Figure 7.3 Use of scaling.

factor. Once scaling is programmed, it remains in effect until canceled. That is, all motions from the G68 command will be scaled until
the cancellation G code is read. One popular word for canceling scaling is G69.

Scaling is not a very useful tool for general-purpose machining, because it has limited application. In fact, about the only time you see
this feature used to any extent is in the plastic injection-molding industry. In this industry, three-dimensional shapes are machined into
a core or cavity of the injection mold. In this case, usually a CAM system is required to help produce the complicated CNC program that
forms the three-dimensional shape.

In some cases, the shape to be machined into the mold may be required in different sizes. For example, maybe a toy manufacturer intends to make a toy car in several sizes. The same program that
makes the mold for a large toy car can be used to make a smaller
model, and vice versa, if the scaling function is used.

Coordinate rotation

This feature is also an option on most CNC controls that allow rotation. It permits a series of coordinates to be rotated about a designated center point. This allows the programmer to specify the coordinates for the program from a logical attitude. Then these coordinates can be rotated to any required location. Figure 7.4 shows what happens during coordinate rotation.

Some versions of this feature even allow the rotation to be repeated a specified number of times at a given incremental angle. For example, imagine the teeth around the outside of a gear. With coordinate rotation, the programmer only needs to program the motions around one of the teeth. The programmer can even pick the tooth that is the easiest to program, possibly the one at the 3 o'clock position. Once programmed, the motions around this tooth can be rotated and repeated at given angular intervals around the gear. If, for example, there are twenty teeth, the programmed coordinates can be repeated 20 times at 18° intervals (360/20 = 18).

Polar coordinates

At this point, you know that coordinates going into the typical CNC program comply with the rectangular coordinate system, discussed in key concept no. 1. While this coordinate system allows all motions necessary for all CNC machines, there are times when it is not the most convenient coordinate system to use to specify coordinates. Polar

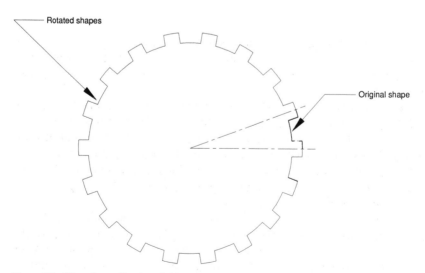

Figure 7.4   Use of coordinate rotation.

coordinates allow a different method of input to be used for coordinate values.

One common example of when polar coordinates are helpful is a bolt hole circle to be machined on a machining center. Most design engineers, while drawing and dimensioning the print, give little or no concern to the CNC programmer. They do not dimension bolt hole circles with the actual rectangular coordinate system dimensions needed in the program. Instead, they dimension the bolt hole circle with the number of holes, radius, and starting angle. They leave it up to the programmer to calculate, using trigonometry, the rectangular coordinate positions at which the holes are to be machined. Figure 7.5 shows what happens when polar coordinates are used.

The polar coordinates feature will give the programmer an easier way to input the desired coordinates. Instead of calculating the rectangular coordinate system hole positions, which is tedious and error-prone, the programmer will input the dimensions right from the print.

The polar coordinates command will include the location to be used as the center of rotation as well as the radius of the bolt hole pattern and angle at which the hole is to be machined. From there, the control will calculate the rectangular coordinate system position automati-

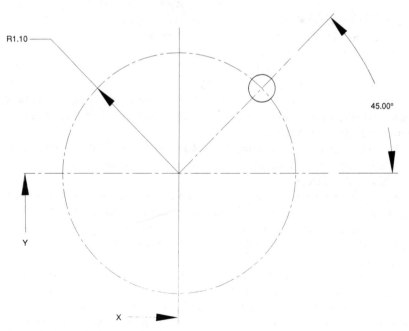

Figure 7.5 Use of polar coordinates.

cally. Here is a typical command (note that the usage for this command will vary dramatically from control builder to builder, so you will have to check the programming manual for the specific words to be used by your particular control):

G68 X _____ Y _____ R _____ A _____

This command will cause the control to calculate a coordinate position using the X and Y in the G68 command as the center of the bolt hole circle. The R value will be used as the radius of the bolt hole circle, and the A will be used as the angle on the bolt hole circle at which the hole is to be machined.

Generally speaking, on machining centers the polar coordinates command is used internal to some form of canned cycle for hole machining. That is, once the motion is made to this polar coordinate, a hole will be machined by the canned cycle.

Single-direction approach

Single-direction approach is an option for most machining center controls. It allows the programmer to assure that positioning motions in X and Y will always be coming from the same direction. This assures that any backlash that may exist in the X or Y axes does not affect the movement command. Backlash, caused by normal wear and tear on the machine, is an inaccuracy that causes problems in motion commands. When backlash exists, whenever an axis undergoes a reversal in direction, the motion will be shortened by the amount of the backlash. For example, if an axis has 0.001-in backlash, whenever the axis reverses direction, the motion will be 0.001 in shorter. While this may not sound like much, this small error can wreak havoc on the accuracy required of the workpiece.

The single-direction positioning feature will alleviate any backlash, since the control will assure that the axis movement *always* comes from the same direction.

G60 is a common G word used to turn on single-direction positioning. Once on, the control will continue to position in one direction for both the X and Y axes until single-direction positioning is turned off. A common G code used to turn off single direction positioning is G64.

The actual directions by which each axis will approach are usually set by parameters. The user will probably have control of which directions (plus or minus) are used during the single-direction approach in each axis. Figure 7.6 shows the motions that will take place when single-direction positioning is used.

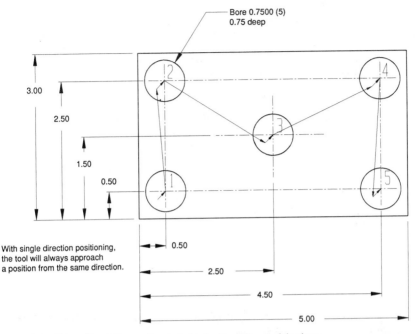

Figure 7.6   Example of the motions of single-direction positioning.

Exact-Stop Check

This feature forces the control to stop at the completion of the command. As discussed in key concept no. 4, the control is constantly scanning ahead to see what is coming up next in the command. We called this scanning ahead the *look-ahead* feature.

This scanning is important to assure there will be no physical stopping between commands. For most applications, if the control were to stop the motion between commands, the result would be detrimental to the workpiece. There would be undesirable witness marks were the motion to stop between commands.

This look-ahead feature assures that motion will flow smoothly from one command to the next. There are times, however, when you do want the control to physically stop between motion commands. For example, say you are machining a square outside shape on a machining center, and that it is important that the outside corners be sharp, meaning no radii are allowed.

Under normal operation, the CNC machining center would flow through the motions around the square and round the corners slightly. Figure 7.7 shows this. If this rounding is unacceptable, the programmer can force the motion to stop between commands. This will make the machine complete one motion before the next motion can oc-

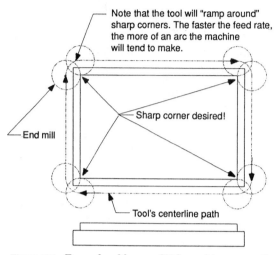

Figure 7.7 Example of how a CNC machine normally moves.

cur. One way to accomplish the pause between movements is with a feature called *exact-stop check.*

For controls that have this feature, it is usually programmed with a G09 or G61. The G09 is a one-shot G code, meaning it takes effect only in the one command in which it is included. The G61 is modal, remaining in effect for subsequent commands until canceled. A common command to cancel the exact-stop check mode (G61) is G64. In our previous square example, if a G61 were included in the first motion command for the square, the control would stop motion after each command.

In effect, this would cause the corners of the square to be sharp. After the last movement around the square, the programmer could include a G64 command to cancel the exact-stop check mode.

Slash Code (/) Optional Block Skip

The optional block skip feature can be used on almost all types of CNC equipment. If included at the beginning of a command, it tells the control to look to the position of an on/off switch on the control panel labeled *optional block skip* (sometimes called *block delete*). If the on/off switch is on, the control ignores the command in which the slash code is programmed. If the switch is off, the control executes the command with the slash code.

The basic purpose of the optional block skip feature is to allow the programmer to give the operator some kind of choice. The applications for this feature vary dramatically, but in almost all cases, when this technique is used, the operator will be asked to decide between two possible conditions.

For example, imagine that a series of castings must be machined on a machining center. A cast surface must be milled. As you may know, the casting process can sometimes be inaccurate. Some of the castings to be machined are just fine. But others have a great deal of excess stock (called *flash*) on the surface to be milled. The programmer knows that extra machining passes must be made to remove the excess stock from only those parts that have it. If extra passes are not made, the cutter could be damaged while trying to remove the excess stock. For those parts that did not have excess stock, it would be a waste of time to allow the milling cutter to make "air cutting" passes.

Knowing this problem exists, the programmer could begin each command that machines the excess stock with a slash code. Then, when the program is executed, if the optional block skip switch is left off, the control will execute the commands with the slash code (making a part with excess stock). If the optional block skip switch is turned on, the control will ignore all commands with the slash code (making a part without excess stock).

Unfortunately, when the optional block skip feature is used, it often opens the door to a possible operator error. In the above application for example, an operator who has the optional block skip switch in the wrong position may be lucky, and air cutting will occur during the unnecessary passes on a part that does not have excess stock. On the other hand, the operator may have real problems. If the optional block skip switch is left on for a part with excess stock, the milling cutter will try to machine too much stock in one pass and the cutter will be damaged.

Whenever considering whether to use the optional block skip technique, the programmer should ask this important question: "What is the worst thing that could happen if the operator makes a mistake with the position of the optional block skip switch?" In many cases, the possible damage to the machine and tool and the possibility of injury to the operator negates effective use of the optional block skip technique, and a safer alternative method of machining the part must be found.

Subprogramming (Subroutine) Techniques

Subprogramming is a feature that is found on most controls as a standard feature. There are many times when commands in a program must be repeated. Sometimes it may even be necessary to repeat commands several times in order to machine the part. These redundant commands can cause problems. It is possible that the person typing the program will make a mistake during the typing of redundant commands in the program. This means the person verifying the program must be very cautious each time the redundant commands are repeated.

With subprograms (sometimes called *subroutines*), the series of commands can be repeated, word for word, with absolutely no extra programming effort. If this technique is used, verifying the program also becomes easier because the same commands are repeated; if the commands are correct the first time, they will remain correct each time they are executed.

Here are some examples of when the use of subprogramming techniques can dramatically reduce the programming effort:

1. *Performing multiple hole machining operations on a machining center:*   For example, you have 50 holes to be machined on a machining center. Each hole has to be center-drilled, drilled, tapped, and counterbored for a shoulder bolt. You can see that the commands for the 50 holes would have to be written four times, adding up to at least 200 commands, without subprogramming techniques. With subprogramming techniques, however, the holes will have to be programmed only once. Then, the programmer can specify the coordinates to be repeated for each tool.

2. *Making trim passes on wire EDM machines:*   Subprogramming techniques are helpful in wire EDM machines (though the same techniques can be used for rough and finish milling on machining centers). Many times it is necessary to make several passes around the shape to be machined, depending on the finish and accuracy requirements of the workpiece to be machined. If wire radius compensation is used, the same coordinates can be used for the trim passes as were used for the roughing pass. This means that the use of subprogramming techniques dramatically reduces the number of commands required to machine the shape.

3. *On turning centers:*   The programming of standard shapes, such as grooves, can be made easier with subprogramming techniques. If prepared properly, the programmer can use the subprogram that machines a particular groove for any number of grooves on any number of workpieces.

These are only three possible applications for subprogramming, but there are many more. In fact, any time you find yourself writing the same series of commands a second or third time in the program, you probably should use subprogramming techniques.

Subprogramming usage. As with all special features of programming, subprogramming techniques vary from control builder to builder. The most dramatic difference has to do with where the subprogram is stored. Some controls require the subprogram to be a separate program, not part of the main program that actually uses it. Other controls require the subprogram to be included in the main program, *after* the last word of the main program (usually after the M02 or M30).

With either method, there are four words commonly related to subprogramming techniques:

M98 This command tells the control to scan to the subprogram and execute it.

P word This word is included in the M98 command and tells the control one of two things. If the subprogram is a completely separate program, the P word tells the control the program number of the subprogram to be executed. If the subprogram is part of the main program, the P word tells the control the sequence number (N word) at which the subprogram begins.

L word Though not often used, the L word can be included in the M98 command and allows the programmer to specify how many times the subprogram is to be executed. If left out of the M98 command, the control will assume the subprogram is to be executed once.

M99 This word tells the control to go back to the command *after* the M98 command that invoked the subprogram.

These four words are the programming words related to subprograms. As an example, say you have a subprogram named O1000 stored in the control's memory. In the main program, this command will call the subprogram:

    N050 M98 P1000

This command will cause the control to scan to the program named O1000 and execute it once, since there was no L word in the calling command. The program named O1000 must end with an M99 to tell the control to return to the main program after the calling M98 command.

Inch/Metric Input

Most CNC machine tools sold in the United States allow programs to be prepared in inch or metric units. While the metric mode of input is becoming more popular, most companies in the United States still allow the user to work in the inch system. A user who comes across a print that happens to be dimensioned in metric can simply convert all dimensions to inches and run the part in inches.

Unfortunately, switching to the metric mode is not quite as simple as preparing all programs in metric. Certain things unrelated to CNC also have to be considered. For example, most measuring devices are designed around and will only work in one measuring system. Measuring tools like micrometers, calipers, bore gauges, and almost every other kind of measuring tool will have to be purchased specifically in metric to allow the metric mode to be utilized.

Also, many manually operated machine tools have all designations for their handwheels in the inch mode only. This means that it can be very costly for most companies currently working in the inch system to switch to metric.

While it is not the primary intention of this book to address the pros and cons of the inch and metric systems, there is an accuracy advantage to working in the metric system.

Parametric Programming

This is a powerful programming feature that allows computer programming commands to be used within a CNC program. Note that this feature is almost always an option, meaning that it will have to be purchased in order for the control to have this feature.

Parametric programming can be compared to subprogramming techniques. Subprogramming techniques require that the commands to be repeated be totally redundant. That is, the control will simply execute the commands in the subprogram over and over, exactly as stated. If anything changes from one time to the next, subprogramming techniques cannot be used.

By comparison, parametric programming allows a kind of general-purpose subprogram to be created. Variables can be passed from the main program to the parametric program that tell the control how to behave each time the parametric program is executed.

It must be pointed out that parametric programming is the most advanced technique related to CNC programming. Also, the techniques used will vary dramatically from control to control. Generally speaking, parametric programming is like having the computer programming language BASIC or C included in the CNC control. Many of the features available in these computer programming languages are also available with parametric programming.

Applications for parametric programming

Generally speaking, most applications have to do with families of parts. That is, parametric programming techniques are most helpful when the workpieces to be programmed are very similar, having only minor differences. In this case, the programmer can prepare a general-purpose parametric program that will work for all workpieces in the family. The main advantage of parametric programming in this case is a reduction in programming and setup time. Once the parametric program is correctly prepared and verified, the operator can go from one setup to the next with relative ease. Programming may be as simple as filling in the blanks, specifying the differences from one part in the family to the next. Here are some examples:

A screw machine cam manufacturer utilizes parametric programming on a vertical machining center to handle the rises and falls of the circular cam needed in a screw machine. Based on print dimensions, the operator simply inputs the call statement. Once the parametric program is executed, the control does the thousands of calculations required to make the necessary movements around the circular cam. Programming time is under 5 minutes.

A hand-tool manufacturer uses parametric programming on a vertical machining center to machine its entire series of large sockets. The operator simply changes a few variables from one setup to the next and is ready to machine the hex shape in the next-size socket.

A piston ring manufacturer uses parametric programming on a turning center to describe and machine the various configurations for its piston ring products.

Many wire EDM users that manufacture punches and dies use parametric programming to describe and machine common shapes to be used in die buttons (like ovals, rectangles, squares, and circles).

What is a call statement?

Most CNC controls that allow parametric programming utilize a *call statement* format. The call statement is generally given in the main program and does two things. First, it tells the control the name of the parametric program to be used so that the control can find and execute it. Second, the call statement allows the user to pass the required variables to the parametric program. These variables tell the control how to behave internal to the parametric program. Here is an example of a call statement for three popular control types (however, please remember that the method used to call the parametric program will vary dramatically from control builder to builder):

G65 P1000 X3. Y3. Z0. D.75 R2.5 H8. A45.

In this case, the G65 statement tells the control that this is a call statement. The P1000 tells the control the program number to be used as the parametric program. The

X3. Y3. Z0 D.75 R2.5 H8. A45.

words are the variables to be passed to the parametric program. Each application determines what variables will be needed in the call statement and the parametric programmer determines what each will represent.

In the above example, the parametric program could have been a circle of bolt holes to be machined on a machining center. The X and Y variables represent the center of the bolt hole pattern in X and Y. The Z variable could be used to represent the surface of the part in Z to be machined. The D variable could be used to represent the depth of each hole. R could be used to represent the radius of the bolt hole pattern. H could be the number of holes, and A could be the starting angle for the first hole's location.

If the parametric program were properly prepared, just about any bolt hole pattern can be machined from one CNC command, whether it has one hole or 50 holes! This is the power of parametric programming. Truly complex applications can be simplified for use by people with little experience. Unfortunately, preparing the parametric program itself is quite another story. This is the most advanced form of CNC programming and requires work to understand.

Features of parametric programming

Here we acquaint you with some of the typical features found in parametric programming. If you have experience with any form of computer programming, you will find these features quite familiar.

Variable techniques. You have already been exposed to one form of variable used in parametric programming, the variables being passed from the call statement to the parametric program. Variables used in this manner are called *arguments*.

But there are other uses for variables in the parametric program itself. Internal to the parametric program, variables are used as storage locations, much like the memory of an electronic calculator. Whenever arithmetic calculations are being done, the result of the calculations must be stored in variables.

Once a value has been stored in a variable, it can be referenced in two basic ways. It can be used in other arithmetic calculations, or it can be used as part of actual CNC statements.

Here is a simple example. The most common form of parametric programming, but not all forms, uses a pound sign (#) to represent a variable. Following the pound sign is a number to specify the variable number, distinguishing it from other variables. The word #101 is variable no. 101, #102 is variable no. 102, and so on.

Say you wish the control to calculate the RPM for a machining center operation based on the surface feet per minute and the tool diameter. The formula to do this is as follows:

$$RPM = SFM \times 3.82 \div \text{tool diameter}$$

In the parametric program, two of the variables coming from the call statement would have to represent tool diameter and the surface feet per minute speed. Say the programmer chooses variable no. 101 (#101) to represent the SFM speed and variable no. 102 (#102) to represent the tool diameter. Here is a series of commands that shows how variables can work for the parametric programmer:

1. #101 = 300 (SFM)

2. #102 = 1.25 (tool diameter)

3. #103 = 3.82 * #101/#102 (RPM calculation)

4. S#103 M03 (start spindle at calculated RPM)

Command 3 makes the RPM calculation and stores the result in variable no. 103 (#103). In command 4, the #103 variable is referenced in the S word as the speed. The result of the calculation in command 3 will be the RPM at which the spindle would start in command 4.

While this is a very basic example, and it does not actually show the method by which variables are passed from the call statement, it should allow you to see one simple use of variables. This example barely scratches the surface of what variables allow. Variables truly give the parametric program a great deal of flexibility.

Arithmetic calculations.   The above variable example uses some basic arithmetic techniques. Equality, multiplication, and division were shown. But most forms of parametric programming allow much more. Most forms allow just about anything that can be done on a scientific calculator to be done from within a parametric program. Of course, things like addition, subtraction, multiplication, and division can be done. Most versions of parametric programming also allow trigonometry functions like sine, cosine, tangent, and arc tangent. Some even allow more powerful and helpful functions like square root, absolute function, rounding, and exponential notation. If working with parametric programming, you must check your builder's programming manual to find out just what arithmetic functions are available.

Most versions of parametric programming also allow arithmetic operations to be combined, meaning more than one operation can be used in an expression. The priority of combined operations usually follows that of computers and electronic calculators.

Conditional branching (the IF statement). Most versions of parametric programming allow the programmer to make tests internal to the parametric program. This testing is an extremely valuable tool. It allows the parametric program to make decisions by comparing two variables in the parametric program.

For example, imagine the parametric programmer wants to allow for the possibility of using left-hand and right-hand tooling for a turning center application. For right-hand tooling, the spindle must run clockwise (M03) and for left-hand tooling the spindle must run counterclockwise (M04). In this case, the parametric programmer can use one of the variables in the call statement as a flag to tell the parametric program which way to run the spindle. The IF statement can easily test this variable to determine which of two directions is desired.

Looping. Looping allows the parametric programmer to repeat a series of commands a specified number of times. Each time through the loop, some variables can be changed to suit the application. For example, a loop could be set up in a parametric program for a turning center for grooving. Knowing the width of the groove to be machined and the width of the grooving tool, the parametric program could loop through the grooving passes until the groove is machined to the desired width.

Helical Motion

While this feature can be considered a simple motion type (and we did discuss this feature during key concept no. 2), we also consider it a special feature of programming, since it is used only for special applications.

The most basic application for helical motion is thread milling. Thread milling can be very useful for holes that are too large to tap and for machining male threads on an outside diameter. Here we present many of the basic considerations you must be concerned with when thread milling.

The first point to make is that there are actually two types of thread milling cutters. One resembles a Woodruff cutter and requires several passes around the part to be threaded in order to form the thread. This type of tool has the form of the thread ground on the outside diameter and is best used when there are a limited number of parts to machine.

When production grows to the point that the previously mentioned tool is not economical to use (because of the length the tool must travel during cutting), there is a second type of thread milling cutter that actually forms the entire thread in one pass around the diameter to be threaded. This tool resembles the combination of a hog milling

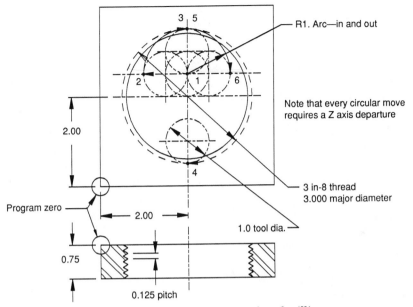

Figure 7.8 Example of helical motion used during thread milling.

cutter and a tap. With this type of tool, the total length of cut required to machine the thread is greatly reduced. Helical motion will cause a circular motion in two axes, X and Y, and a linear motion in Z. The X and Y circular motion will form the diameter to be thread-milled. The Z-axis linear motion will form the actual pitch (or lead) of the thread.

The most important point to remember is that any Z-axis motion during a circular X-Y move *must* reflect the pitch of the thread. If a circular motion is made all the way around the diameter in one command, the actual Z-axis departure will be equal to the pitch of the thread. If any arc-in and arc-out techniques are used to minimize witness marks on the thread, the Z-axis motion commanded must also reflect the pitch of the thread.

Figure 7.8 shows the motion. Notice that the arc-in circle (to point 3) forms 90° of a circle. Therefore the Z-axis departure for this command must be one-fourth of the thread pitch (since 90° is one-fourth of a circle). The same thing goes for the arc-out after the thread is machined.

Other uses

Some programmers use the helical motion commands to ramp in to a circular motion for counterboring a hole with an end mill. However, most end mills perform poorly when machining on the end of the tool at the same time they machine on the side of the tool. For this reason, the use of helical motion to machine in this way offers limited success.

Example program showing helical motion

This short program shows how a typical CNC control would be programmed for thread milling using a thread milling cutter that forms the entire thread in one pass around (the second type of tool discussed). Only the thread milling operation is shown.

```
O0001
N005 G54 G90 S400 M03 (thread milling cutter)
010 G00 X2. Y2.5 (1)
N015 G43 H01 Z-.85 (rapid through part)
N018 M08
N020 G42 D31 X1. (2)
N025 G02 X2. Y3.5 Z-.8812 R1. F5. (3)
N030 G02 X2. Y.5 Z-.9437 R1.5 (4)
N035 G02 X2. Y3.5 Z-1.0062 R1.5 (5)
N040 G02 X3. Y2.5 Z-1.0374 R1. (6)
N045 G00 G40 X2. Y2.5 (7)
N050 G91 G28 Z0 M19
N055 G28 X0 Y0
N060 M01
N065 M30
```

(handwritten annotations: "1/4 OFF (90°) OF THREAD", "FEED RATE = Pitch of thread", "F.R. 1/2 Pitch THREAD (180°)")

The only thing in this program that may not make much sense is the value of each Z-axis departure. Remember that each circular motion must take into consideration the pitch of the thread. For example, in sequence N025, since the circular motion makes a quarter circle, the amount of Z departure is one-fourth of the pitch (0.125 pitch ÷ 4 = 0.0312 and −0.85 − 0.0312 = −0.8812). In sequence number N030, the motion is half a circle, so the Z departure must be 0.0625, and so on.

As mentioned, this example shows a program for a thread milling cutter with the capability to machine the entire thread in one pass. For a single-cutting-edge-type thread milling cutter, the program would have been much longer.

Canned Cycles

For machining center applications, canned cycles are very often used. In fact, in every program you write that has holes to be machined will use this feature. While some CNC control manufacturers list this feature as an option, almost *every* machine tool builder will make canned cycles a standard feature.

Canned cycles are a way for the programmer to dramatically shorten the program. They are used for center cutting operations, like drilling, tapping, reaming, and boring. When canned cycles are used to machine holes, only one command is required to completely machine the hole.

Up to now, the only way you know to machine holes is to use the G00 and G01 commands. As you know, using G00 and G01 takes at least three commands to machine a hole (move over the hole, plunge in, rapid out). Of course, this will simply drill the hole. If the hole has to be tapped, it will require the spindle to be reversed, and the tap will have to be fed out of the hole. And what about peck drilling? Can you imagine how many commands it will take to peck-drill a deep hole with only 0.100-in pecks? And if there were many holes to peck drill, the program would get *very* lengthy.

With canned cycles, you will instate the kind of hole being machined (cycle type, depth, feed rate, first hole position, etc.) on the very first hole. Then, you will simply list the balance of the hole positions at which you need to machine the holes. When you are finished machining holes, you will cancel the canned cycle with a G80 command.

All canned cycles have many things in common and share many programming words, so once you understand how one canned cycle works, you will understand the basics of them all. Two general rules apply to all canned cycles:

1. All canned cycles are modal. Once they are instated they will remain in effect until they are canceled (by a G80 word).
2. Motions will be as follows:
   a. The X and Y motions will take place first (at rapid).
   b. Rapid motion will occur in Z to the R plane.
   c. The hole will be machined.
   d. The tool comes out of the hole.

List of canned cycles

Here is a list of all of the canned cycles and their corresponding G words. We will give a very good example showing the use of several of these canned cycles. They have so many things in common that we feel if you truly understand the canned cycles given in the example, you will easily understand the rest. Here is a list of the most common canned cycle G codes and a brief description of their meanings:

✓ **G73 peck-drill cycle for steel (chip breaking).** This canned cycle is used for drilling a material that has the tendency to produce stringy chips. That is, if the chips do not break easily and have the tendency to form a "rat's nest" around the tool, this cycle can be used to force the chips to break as the hole is drilled. The chip breaking is accomplished by slight retracts of the tool during drilling. For example, if the peck amount is set to 0.100 in, the tool will plunge into the hole to a depth of 0.100 in, then back out about 0.005 in. Then the tool will plunge another 0.100 in. It is during the back-out movement that the chip is forced to break. The back-out amount is usually adjustable and set by parameter.

✔ **G74 left-hand tapping cycle.** This cycle is used to machine with a left-hand tap, forming left-hand threads in the hole. The spindle *must* be running in a counterclockwise direction (M04) when this cycle is commanded. Once the tap enters the hole and reaches the hole bottom, the spindle will reverse and the tap will feed back out of the hole. When using this cycle, a special tension/compression tap adaptor must be used to allow the tap to float in the holder. That is, the spindle reversal will *not* be perfectly synchronized with the feed rate coming out of the hole. The tap must be allowed to move parallel to the *Z* axis in the holder.

✓ **G76 fine boring cycle with no "drag line."** This is a very nice cycle for finish boring when a "drag line," or witness mark, is not allowed in the hole being bored. With this cycle, the tool will feed into the hole to the hole bottom. Then, the spindle will be stopped and sent to its orient position (as it is for a tool change). After orient, the tool will move in *X* and/or *Y* in the direction away from the tool point. Lastly, the tool will come out of the hole.

G80 canned cycle cancel command. This command *must* be given at the completion of the hole machining operations. That is, you must remember to cancel the canned cycle *before* changing tools. If you forget to cancel the cycle, the control will continue machining holes for *every* subsequent command.

✓ **G81 standard drilling cycle.** This cycle causes the tool to simply feed into the hole, then rapid out. It is used for center drilling and for drilling when there is no problem with chip breakage. Many programmers also like to use it for reaming.

✓ **G82 counterboring cycle.** This does the same basic thing as the G81 drilling cycle, except the tool will pause at the bottom of the hole (with the spindle running) for a specified length of time. The purpose of this cycle is to allow the tool pressure to be relieved when the tool reaches the bottom of the hole. It is especially helpful when you are trying to maintain a close tolerance for the depth of the hole.

✓ **G83 deep hole drilling cycle (with pullout).** This cycle, like the G73 cycle, is a peck-drilling cycle. But its usage is different than the G73. The

G73 is used to break the chip as the hole is drilled. The G83 is used when a deep hole is being machined to allow the chips to be cleared at certain intervals. If a drill is sent too deep into the hole, eventually the chips will pack up around the flutes of the drill. If the drill continues to go deeper, the packing of chips will cause the drill to break. The programmer will specify the peck depth before the drill is sent out to clear chips. After clearing chips, the tool rapids back into the hole to a point just above where it left off and continues. Most machinists would say you can peck about 3 to 4 times the drill diameter per peck without fear of chip packing.

G84 right-hand tapping cycle. Like the G74, this cycle is used for tapping. But this cycle is *much* more commonly used than the G74. You will almost always be machining right-hand threads in the hole. The only difference between the usage of the G84 and G74 is the direction the spindle will be rotating. With the G84, the spindle *must* be running clockwise (M03) prior to this command. Also, as with G74, the speed and feed rate must be synchronized. The recommended formulas are given a little later, after the example program. As with G74, a tension/compression tap holder must be used.

G85 reaming cycle (feed-out). This cycle is identical to G81, but the tool will feed out of the hole, not rapid out. That is, the tool will rapid to position, feed into the hole, then feed back out. Some people elect to use the G81 drilling cycle for reaming, since no machining is being done while the feed-out motion occurs. They consider the feed-out motion a waste of machining time.

G86 standard boring cycle (leaves drag line). This cycle can be used for boring when a drag line or witness mark is permissible. The tool will feed into the hole, then the spindle will stop. Finally, the tool will rapid out of the hole. On the rapid-out motion, a fine line will be drawn inside the hole, left by the boring bar tip.

G87 and G88 manual cycles (try not to use). G87 and G88 are manual cycles that we do not recommend using. They perform differently, based on the settings of certain parameters. Each requires manual intervention on the operator's part. For the most part, it is just as easy (and more predictable) to program manual cycles "long-hand" with G00 and G01.

G89 counterboring cycle for boring bar (with dwell). This cycle is like a combination of the G86 and G82 commands. The boring bar will feed into the hole, and then the motion will pause for a specified time. Fi-
NOT USED MUCH.

nally, the spindle will stop and the tool will rapid out of the hole. A drag line will be left in the hole. This cycle is used when the boring bar is machining to a precise depth.

Words used in canned cycles. Canned cycles share many words in common. These words will have exactly the same meaning in all canned cycles. Although some canned cycles have no use for some of these words, when you understand how the basic words work in one canned cycle, you will know how they work in all of them. Here is a list of all words used in canned cycles for one popular control:

Word	Status	Description
N	All cycles	Sequence number
G73–G89	All cycles	Canned cycle call word
X	All cycles	X coordinate of the hole center
Y	All cycles	Y coordinate of the hole center
R	All cycles	Rapid plane above workpiece to start feeding from (taken from program zero in Z)
Z	All cycles	Z position of hole bottom (taken from program zero in the Z axis)
F	All cycles	Feed rate for the machining operation
L	All cycles	Number of repeats for holes to be machined (used with incremental techniques only)
G98	All cycles	Rapid out of the hole to the initial plane (initialized)
G99	All cycles	Rapid out of the hole to the R plane
P	G82, G89	Pause time at hole bottom (P500 = 0.5 second)
Q	G73, G83	Peck-drill amount per peck
I	G76	Amount of move-over in X at hole bottom
J	G76	Amount of move-over in Y at hole bottom

(handwritten margin note: ONLY FOR INCREMENTAL)

(handwritten: BRINGS TOOL TO n cm 2 HOME POSITION)

Understanding G98 and G99 *(handwritten: BRINGS TOOL TO JUST ABOVE WORK PIECE)*

Before we look at an example, it will be helpful to look at exactly how the G98 and G99 rapid-out commands work. While they are not found on every CNC control, many control manufacturers use these codes to control motions around clamps and other obstructions between holes without breaking out of the canned cycle. As mentioned, canned cycles cause the tool to follow this sequence:

1. Rapid to X and Y position

2. Rapid down to the R plane

3. Machine the hole

4. Move out of the hole in Z

In step 4 (coming out of the hole), the G98 and G99 words inform the control as to where you wish to leave the tool in Z after the hole is machined. If the G99 is in effect, the tool will be left at the R plane close to the work surface. The R plane is the surface in the Z axis commanded by the R word of the canned cycle command. If G98 is in effect, the tool will be left at what is called the *initial plane* (usually above any obstruction) in Z. The initial plane is defined as the last position in Z prior to the canned cycle command. So, if you get in the habit of leaving the tool 2 to 3 in above the work surface, you can easily clear the obstructions between holes. If this is somewhat unclear at this point, don't worry. When we show an example, this will make more sense.

Even though the format of the canned cycle command is not important, we recommend that the beginner get in the habit of putting the G98 or G99 at the end of the command. This will help you to remember what will happen *after* the hole is machined. That is, *after* the hole is machined, either the tool will be left at the initial plane *or* the R plane. Having the G98 or G99 at the end of the canned cycle command helps you remember this.

Note that the G98 command (initial plane) is initialized when the machine power is turned on and whenever the G80 cancel command is executed. This means that if you do not include the G98 or G99 command in the canned cycle, the tool will automatically be left at the initial plane (above any obstructions). When you think about it, this is the safe way. If you forget to include either a G98 or G99, the tool will come out of the hole to the initial plane (above the clamps).

One last point about G98 and G99 before we show an example: if you rapid the tool to the rapid (R) plane prior to the canned cycle command, the R plane and the initial plane will be at exactly the same position in the Z axis. In this case the G98 or G99 command will have no effect on where the tool is left after the hole is machined. The tool will come out to the same location in either case.

Example showing canned cycle usage

Here we will show a full example of how several of the canned cycles are used (G81—drilling, G84—tapping, G82—counterboring, G73—peck drilling, and G86—boring). The format for this program is for a vertical machining center using fixture offsets. However, it would be quite easy to convert this program's format for any other type of machining center (see key concept no. 5 for the various types of machining center formats). We assume that the first tool is in the spindle at the beginning of the program.

This program is quite lengthy, containing seven tools. We want to give you a very good example to go by when you write your own

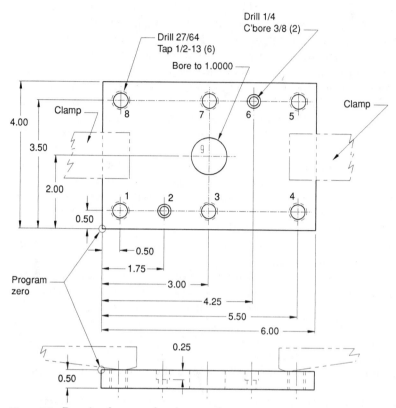

Figure 7.9 Drawing for canned cycle example program.

canned cycle programs. We recommend that you study it until you understand it completely. Also, if it seems lengthy to you, think about how long this program would have been if we had not used the canned cycles!

From the drawing in Figure 7.9, note the positions of the clamps that must be avoided during the program. Actually, only three of the tools (center drill, 27/64-in drill, and 1/2-20 tap) must avoid the clamps. We use G98 and G99 to do so. Figure 7.10 shows the sequence of operations used to machine this part. Note that we have even given the intended G code type for the canned cycle to be used for each operation. This was done to make it more clear to *you* what is going on. Of course, you would not include this information in your own sequences of operations. Also note that we are trying to show the usage of as many canned cycles as we can. While you may not agree with the necessity of some of the canned cycles we use, please keep in mind that this is simply an example.

Sequence of Operation

Seq.	Description	Code	Tool	Speed	Feed	Speed
1	Center drill all holes	(G81)	Center drill	1	3.0, in IPM	1200 RPM
2	Peck drill 27/64 holes (6)	(G73)	27/64	2	5.0, in IPM	611 RPM
3	Tap 1/2-13 holes (6)	(G84)	1/2-13 Tap	3	17.6, in IPM	230 RPM
4	Drill 1.0000 hole to 31/32	(G81)	31/32 Drill	4	5.0, in IPM	340 RPM
5	Bore to 1.0000	(G86)	1.0000 Boring bar	5	2.0, in IPM	500 RPM
6	Peck Drill 1/4 holes (deep hole)	(G83)	1/4 Drill	6	2.5, in IPM	1000 RPM
7	C'Bore 3/8 holes (2)	(G82)	3/8 C'bore	7	3.0, in IPM	600 RPM

Figure 7.10 Sequence of operations for canned cycle example program.

Program:

```
O0008
N005 G54 G90 S1200 M03 T02 (center drill)
N010 G00 X.5 Y.5
N015 G43 H01 Z2. (note 2-in initial plane)
N020 M08
N025 G81 X.5 Y.5 R.1 Z-.25 F3. G99 (center-drill hole 1)
N030 X1.75 (hole 2)
N035 X3. (hole 3)
N040 X5.5 G98 (note clamp; hole 4)
N045 Y3.5 G99 (back to R plane; hole 5)
N050 X4.25 (hole 6)
N055 X3. (hole 7)
N060 X.5 G98 (come up just in case; hole 8)
N065 X3. Y2. (hole 9)
N070 G80 M09 (cancel cycle)
N075 G49 G91 G28 Z0 M19
N080 M01
N085 T02
N090 M06
N095 G54 G90 S611 M03 T03 (27/64-in drill)
N100 G00 X.5 Y.5
N105 G43 H02 Z2. (note 2-in initial plane)
N108 M08
N110 G73 X.5 Y.5 R.1 Z-.75 Q.1 F5. G99 (hole 1)
N115 X3. (hole 3)
N120 X5.5 G98 (clear clamp; hole 4)
N125 Y3.5 G99 (hole 5)
N130 X3. (hole 7)
N135 X.5 (hole 8)
N140 G80 M09 (cancel cycle)
N145 G49 G91 G28 Z0 M19
N150 M01
N155 T03
N160 M06
N165 G54 G90 S230 M03 T04 (1/2-13 tap)
N170 G00 X.5 Y.5
N175 G43 H03 Z2. (note 2-in initial plane)
N178 M08
N180 G84 X.5 Y.5 R.25 Z-.65 F17.6 G99 (hole 1)
```

N185 X3. (hole 3)
N190 X5.5 G98 (note clamp; hole 4)
N195 Y3.5 G99 (hole 5)
N200 X3. (hole 7)
N205 X.5 (hole 8)
N210 G80 M09 (cancel cycle)
N215 G49 G91 G28 Z0 M19
N220 M01
N225 T04
N230 M06
N235 G54 G90 S340 M03 T05 (31/32-in drill)
N240 G00 X3. Y2. (hole 9)
N245 G43 H04 Z.1 (no need to clear clamps)
N248 M08
N250 G81 X3. Y2. Z-.85 R.1 F5. (since R plane and initial plane are the same, there is no need for either a G98 or G99)
N255 G80 M09 (cancel cycle)
N260 G49 G91 G28 Z0 M19
N265 M01
N270 T05
N275 M06
N280 G54 G90 S500 M03 T06 (1.0000 boring bar)
N285 G00 X3. Y2. (hole 9)
N290 G43 H05 Z.1 (no need to clear clamps)
N293 M08
⌐   N295 G86 X3. Y2. Z-.6 R.1 F2. (since R plane and initial plane are the same, there is no need for G98 or G99)
N300 G80 M09 (cancel cycle)
N305 G49 G91 G28 Z0 M19
N310 M01
N315 T06
N320 M06
N325 G54 G90 S1000 M03 T07 (1/4-in drill)
N330 G00 X1.75 Y.5 (hole 2)
N335 G43 H06 Z.1 (no need to clear clamps)
N340 M08
N345 G83 X1.75 Y.5 Z-.65 R.1 F2.5 (since the R plane and the initial plane are the same, there is no need for G98 or G99)
N350 X4.25 Y3.5 (hole 6)
N355 G80 M09 (cancel cycle)
N360 G49 G91 G28 Z0 M19
N365 M01
N370 T07
N375 M06
N380 G54 G90 S600 M03 T01 (3/8-in counterbore)
N385 G00 X1.75 Y.5 (hole 2)
N390 G43 H07 Z.5 (initial plane can be used to clear counterbore pilot)
N395 M08
N400 G82 X1.75 Y.5 Z-.25 R.1 P500 F3. G98 (comes out to clear counterbore pilot)
N405 X4.25 Y3.5 (hole 6)
N410 G80 M09 (cancel cycle)
N415 G49 G91 G28 Z0 M19
N420 G28 X0 Y0
N420 M01

N425 T01
N430 M06 (put first tool back in spindle)
N435 M30

As stated, this is a rather lengthy program. But do not let its length intimidate you. If you break it down, tool by tool, it will not seem so bad. Also, it makes an excellent example to go by when you write your first few programs that require the use of canned cycles. So study the program until it makes total sense.

Now that you have seen the entire program, there are a few points we want to make:

1. You will notice that we included the X and Y position of the hole in each first canned cycle command. But if you study the program, you will notice that the tool has already been positioned to the center of the hole in a prior G00 command. If the tool is already over the hole (as is the case for all tools in the above program), there is no need to include the X and Y hole location in the canned cycle command. That is, in each of our first canned cycle commands for each tool (the G81, G82, G84, etc.), we did *not* need the X and Y location. We only included the X and Y coordinates so that beginners could easily follow what is going on. In reality, they didn't help or hurt the program. However, they do give you the potential to make a mistake while entering the program, so we recommend that you get in the habit of *not* including them.

2. The R plane for any tap should be kept to at least 0.25 in above the part. The reason for this is a little difficult to visualize. While the tap is entering the hole and while the spindle is being reversed, there is a possibility that the tap will extend in the tension/compression holder. If it extends more than 0.100 in (the normal R plane value), it is possible that the tap will still be in the hole when finished tapping! For this reason, you should make your R plane distance quite a bit larger than for fixed tools.

3. You may be questioning the wisdom of using canned cycles to machine only one hole, especially a drilled hole. You may feel that it is just as easy to machine one drilled hole with G00 and G01 as it is to use canned cycles. But remember that there may be a necessity to change the cycle type at the machine (maybe a chip-breaking peck cycle is necessary). If you get in the habit of using canned cycles for all hole machining, the program will be in the correct format for an easy cycle-type change. If G00 and G01 are used to drill the hole, it will be more difficult to modify the program.

4. With the counterbore in the above program, we used G98 and

G99 in a somewhat unusual way. True counterbores have a pilot that extends beyond the cutting edge. The pilot guides the tool into the hole. When the tool length is measured, it is measured to the cutting edge, not the pilot. This means the pilot is sticking out farther than your programmed cutting edge. If you study the above program, you can see how G98 and G99 can be used to avoid leaving the pilot in the hole between movements. Some programmers use end mills for counterboring. An end mill has no pilot, so this problem will not exist for an end mill.

5. When calculating the depth of the hole in order to obtain the Z value, you must take into consideration the possibility of some tooling-related problems. For example, when drilling with a twist drill, you must consider the drill lead. That is, if you need a full-hole depth of 0.75 inches, the twist drill must go slightly farther into the hole because of the lead of the drill. You can easily calculate the lead for a twist drill by multiplying 0.3 times the diameter of the drill. In a similar manner, when tapping you must consider that there will always be some imperfect threads on the tap. Each tap style dictates how many imperfect threads are on the tap. In almost all cases you will have to allow for these imperfect threads.

6. The feed rate for tapping equals RPM times pitch.

7. The pitch of a tap equals 1 divided by the number of threads per inch.

8. As you can see from the above program, if there will be no need to clear clamps during the entire tool, you can rapid the tool down to the R plane in the G43 command. This means that the initial plane and the R plane will be the same. Then, no matter whether a G98 or G99 is in effect, the tool will still come out to the same place (just above the part). This allows you to leave out the G98 or G99 altogether.

Canned cycles and the Z axis

We have stated several times that the initial plane, R plane, and Z values for the program are *absolute values*. However, many times the program zero point in Z happens to be the surface of the part that is being machined (as it was in the above example). We must point out that there will be times when this will not be the case. Figure 7.11 is an example drawing that demonstrates this.

As you can see, not all of the holes are being machined into the top surface of the part. This part requires that you manipulate the R plane and Z hole bottom position cautiously. Here is the example program that would machine the part correctly.

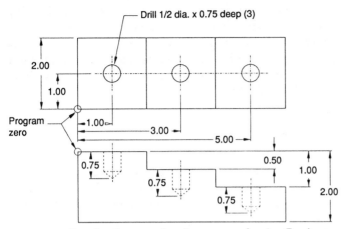

Figure 7.11   Drawing for canned cycle program showing Z-axis control.

Program:

```
O0009
N005 G54 G90 S600 M03 (1/2-in drill)
N010 G00 X1. Y1.
N015 G43 H01 Z.1 (above highest surface)
N020 M08
N025 G81 X1. Y1. Z-.75 R.1 F3. G99 (just as shown so far)
N030 X3. R-.4 Z-1.25 (note R and Z values)
N035 X5. R-.9 Z-1.75 (note R and Z values)
N040 G80 M09 (cancel cycle)
N045 G49 G91 G28 Z0 M19
N050 G28 X0 Y0
N055 M01
N060 M30
```

In block N025, we machine the hole in the top surface. There is nothing new here. But in block N030, you should notice that we included the new R and Z values. Also note that they were taken *from program zero*. This command will:

1. Move to new *X-Y* position
2. Drop down to new *R* plane
3. Machine the hole to *Z* plane
4. Come out of the hole (to *R* plane in our case)

In the same manner, the N035 command machines the third hole. Note that this program works from top surface to bottom surface. This is the way that we recommend you do it when you have multiple surfaces. If you do go from low surface to high surface, you must be very careful with the movements between holes. If, for example, we went to the lowest hole in the program first, and we let the tool come to the *R* plane between holes, a crash will result. That is, the tool will come out to the *R* plane, and then try to move to the next *X-Y* position while still at the lower surface. Surprise!

Using canned cycles in the incremental mode

You now know the function of the various canned cycles (G73–G89) as they are used in the absolute mode (G90) of programming. Most experienced programmers will agree that this mode (absolute) is the best way to input programs. For beginners, we *strongly* recommend that you keep your programs in the absolute mode. It is much easier to read and understand a program written in the absolute mode. However, there is one instance when it is helpful to know how the incremental mode affects canned cycles.

If you have a series of evenly spaced holes along a line to machine, you can machine all of the holes in one command. However, this requires that you program the holes incrementally. If you have a grid pattern of holes, using this technique will drastically reduce the number of commands necessary to machine the pattern.

Before we get in too deep, if you are a total beginner to CNC, some of this will not make much sense. This is because it is an advanced technique. You may want to simply skim this section of the text and make a mental note to come back and study it later if the application arises.

The first point is related to the L word. The L word tells the control to repeat a canned cycle command a specified number of times. In the absolute mode, this would be silly, since the same hole would be machined in the same location several times. But in the incremental mode, evenly spaced holes along a line can be easily machined in one command by using the L word. You should also note that if the L word is left out of the canned cycle command, only one hole is machined. This fact has fantastic implications. It means that you can machine hundreds of holes along a line in one command!

Before we can give an example, you must understand how the control handles the words in a canned cycle when in the incremental mode. You *must* know that each axis-related word represents the distance from the *current* position of the tool.

Here are the words in a canned cycle command as they are represented in the incremental mode:

X is the distance from the current tool position to the center of the hole in X.

Y is the distance from the current tool position to the center of the hole in Y.

R is the distance from the current position of the tool to the R plane position.

Z is the distance from the R plane to the bottom of the hole. If the R plane is 0.100 in above the part, the Z value would be the depth of the hole plus 0.100 in (in incremental canned cycles, Z is *always* minus).

The rest of the canned cycle words (G98, G99, I, K, F, P, and Q) mean exactly the same as they do in the absolute mode.

In our opinion, there is no reason to machine only one or two holes in the incremental mode. You would only consider using this incremental technique if you have several holes to machine. Now let's look at an example.

If you currently understand how to program canned cycles in the absolute mode, we think you will be impressed with how few commands are required in the incremental mode. Say you have to machine 100 holes in a grid pattern (10 holes by 10 holes). Figure 7.12 is a drawing of the part.

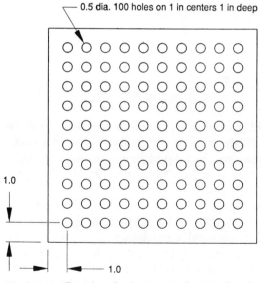

Figure 7.12 Drawing for incremental canned cycle example program.

To keep it simple, we are just drilling a 0.500-in-diameter hole (no center drilling) to a 1.0-in depth. But there are 100 holes to drill, and, as you know, if programmed in the absolute mode, this part would take at least 100 commands. But look how short the program is in the incremental mode.

Program:

```
O0001
N005 G54 G90 S600 M03 (1/2-in drill)
N010 G00 X0 Y1. (note X position)
N015 G43 H01 Z2. (sets initial plane)
N020 M08
N025 G91 G81 X1. R-1.9 Z-1.1 F5. L10 G99 (machines 10 holes; note X, R, and Z
values)
N030 Y1. (machines first hole in second row)
N035 X-1. L9 (machines nine holes in second row)
N040 Y1.
N045 X1. L9 (third row)
N050 Y1.
N045 X-1. L9 (fourth row)
N050 Y1.
N055 X1. L9 (fifth row)
N060 Y1.
N065 X-1. L9 (sixth row)
N070 Y1.
N075 X1. L9 (seventh row)
N080 Y1.
N085 X-1. L9 (eighth row)
N090 Y1.
N095 X1. L9 (ninth row)
N100 Y1.
N105 X-1. L9 (tenth row)
N110 G90 G80 M09
N120 G49 G91 G28 X0 Y0 Z0 M19
N125 M30
```

As you can see, this technique reduces the number of commands to machine the holes from 100 to about 20. If you study this, it will make sense. But it may be asking too much of the beginner to catch on right away. Remember that this shortens your program's length, but programming becomes more difficult.

We are rapiding the tool to a position 1 in to the left of the first hole (in the absolute mode, this is X0). Since the subsequent canned cycle command includes the 1-in incremental X movement (X1.), we have to stay one inch away from the first hole. The tool is then sent to a Z position 2 in above the part. The canned cycle command in block N025 sets the incremental mode (G91) and tells the control to machine a hole 1 in in the positive X direction from the current hole location (now the previous X0 position should make sense). The R word tells the control the R plane is −1.9 in down in Z from the current tool position (the current position is 2 in above the part). The hole will be drilled to a Z position −1.1 in down from the R plane. Since the R plane is 0.1 in above the part surface, the hole will be drilled 1 in deep. The whole process is then repeated ten times, since the L word is set to 10.

You should be able to follow along from there. The balance of the commands machine the rest of the holes in the grid pattern.

The major limitation with these techniques is that the holes must be evenly spaced. If they are not, this technique is almost useless, and it would be better to stay in the absolute mode.

One more point. If the holes are evenly spaced along an angular line, remember that the movement to be repeated can include an X and a Y departure.

We hope this is making sense. If you study this information, we think you'll catch on.

Multiple Repetitive Cycles for Turning Centers

Most CNC turning center controls have a series of commands that make programming many turning applications much easier. Unfortunately, their usage varies dramatically from one control type to the next. Some control builders call these features multiple repetitive cycles. Others call them lapping cycles. Yet others simply call them canned cycles.

No matter what they are called, they make the manual programmer's job much easier. While we cannot give specific examples, here are some of the things that can be accomplished with these cycles.

Making roughing passes

From what you know so far, it would be quite tedious, time-consuming, and error-prone to program rough turning, boring, and facing passes by using only G00, G01, G02, and G03. With these commands, every motion the machine makes during roughing would have to be programmed.

With multiple repetitive cycles, the control will do the entire roughing operation for you. The programmer simply commands the motions of the finish pass. Then, based on one command, the control will generate the entire roughing cycle automatically. In this command, the programmer will set up the parameters desired for roughing, like depth of cut, feed rate, and amount of stock to be left for finishing. The control does the rest.

Threading

Most turning center controls have a nice threading command that allows the programmer to specify that the thread be machined in one command. The programmer fills in the blanks about the thread, telling the control, the minor/major diameter, the thread depth, the pitch, and the tool angle, and the control does the rest.

Grooving

Some CNC turning center controls even allow grooving operations to be programmed very easily. The programmer specifies the large and small diameters to be grooved, the groove width, and the tool width. The control does the rest, machining the groove to the desired specifications.

Peck drilling

Some CNC turning center controls allow peck drilling to be done. The programmer specifies the depth of the hole and the depth per peck, and the control machines the hole in the desired fashion.

Conclusion to Special Features of Programming

As you can see, there are many helpful tools available to the manual programmer to make programming easier. As mentioned several times up to this point, the techniques required to use these features vary dramatically from one control to the next. While the examples we have shown are common to some controls, you will have to check the control builder's operation and programming manual to find out more about these features.

Conversational
Programming

Introduction to Conversational Controls

What Is a Conversational Control?

As you now know, manual programming involves a great deal of tedious work. A process must be developed, tooling must be checked, coordinate calculations must be made, feeds and speeds must be calculated, and of course, the manual program must be written. Once the manual program is prepared, it must be cautiously verified. Since the first NC controls were developed, control manufacturers have been striving to make the use of this sophisticated equipment easier. Each new control model incorporates more features to make programming easier than its predecessor. This was evidenced in Chap. 7. Many special features of manual programming were specifically designed to make manual programming easier.

Unfortunately, even with the help of these special manual programming features, manual programming is still quite tedious. These special features apply only to the type of machine being programmed. If a programmer must work with two or more different types of CNC machine tools, programming can still be difficult to master, no matter how many special features are available.

For these reasons, computer-aided manufacturing systems were developed. CAM systems take much of the tedious work out of programming CNC machines. Instead of developing a series of coordinates calculated manually, the programmer can actually draw the workpiece on the screen of a computer using a series of geometry construction elements. Generally speaking, a series of points, lines, and circles can be described to the CAM system for the desired shape to be machined. After the geometry has been described, the programmer can specify the cutter path around the geometry. Once the program is finished, the CAM system will generate the CNC coding for the particular CNC machine being programmed in much the same way a manual programmer would. This (CNC) program is loaded into the control's memory for production.

Under ideal conditions, CAM systems allow several different types of CNC machines to be programmed in one language. If the CAM system is properly developed, the programmer need only learn one system to be able to program several CNC machines. Most CAM systems have different modules for the various types of CNC machine tools they were designed to work with. If the programmer is to program a turning center, the turning module will be accessed. For programming a machining center, the machining center module will be accessed.

In the early days of NC, CAM systems were based on mainframe computers. Only very large companies could afford to purchase and support them. For smaller companies, time-sharing was a cost-effective method of attaining the benefits of computer-assisted programming at reasonable cost. One manufacturing company would purchase time on another company's computer.

As time went on, computers decreased in size and increased in capability. As minicomputers eventually gained the ability to handle the tasks related to CAM systems, more companies were able to afford to purchase them.

Today, even personal computers have the capability to be used as very powerful CAM systems, and many companies have purchased this form of CAM system.

The best way to envision a conversational control is to picture a conventional CNC control, programmed manually as discussed in key concepts nos. 1 through 6, with a single-purpose CAM system built in. Generally speaking, conversational controls have two sides, the CNC side and the programming side. While programming, the programmer will be working on the programming, CAM-like side. Once the conversational program is finished, the control will translate the conversational program into a CNC program the CNC side can understand. It is from this translated CNC program on the CNC side of the control that the machine will run production. Once the program reaches this stage, the machine is run exactly as if it had a conventional CNC control. Figure 8.1 shows a conversational turning center control.

To handle the two sides of the conversational control, some controls have two separate microprocessors, one to handle the CNC side and the other to handle the programming (CAM) side. Others use the same microprocessor to handle both sides. In most cases, conversational controls allow a program to be input on the programming side

Figure 8.1 A popular conversational turning center control. (*Courtesy Okuma Machinery, Inc.*)

while production is being run on the CNC side. This means the machine can be productive while the programmer inputs a program.

Earlier we used the term *single-purpose* in the description of conversational controls. The so-called CAM system within the conversational control is specifically designed around the one application the machine tool is intended to handle. For example, conversational turning center controls would be capable of working only with turning applications. A conversational turning center control could not be used to make programs for machining center applications.

The Controversy Regarding Conversational Controls

As with just about anything new, there has been quite a controversy in the industry related to conversational controls. Several questions have been raised as to the wisdom of using this kind of control. While we make no judgments as to these matters, we do feel the beginner should at least know what has been going on. This will give you an insight into what types of companies are using conversational controls.

One important question every company using CNC equipment must answer has to do with the programming environment. Most people in manufacturing would agree that programming at the machine tool is generally uncomfortable and not conducive to concentration. While there are exceptions, usually the area around a CNC machine is noisy and dirty. There are many distractions to the person trying to input the program. Also, the physical act of entering the program may be somewhat uncomfortable. In most cases, the control panel of the CNC machine is positioned vertical to the floor and at about shoulder height. This means the programmer's arm will fatigue if the period of time required to enter the program is lengthy.

This being the case, it is the philosophy of many shops to keep programming separate from the machine's operation. They keep the programmer in a more comfortable room or office that is free of distraction. Whether a CAM system is used or programs are prepared manually, the programmer is kept free of the machine's distracting environment.

On the other hand, smaller job shops with only a few CNC machines may not be able to afford the luxury of having an employee whose sole responsibility it is to prepare programs. Smaller shops rely heavily on the CNC machine operator to do all functions related to the CNC machine, including programming, setting up, and running production. In this case, conversational controls make an excellent choice. They allow the operator a way of quickly preparing the CNC program with a fraction of the work needed to prepare a manual program. The other distractions related to programming at the machine are endured to gain the benefits of the conversational control.

The conclusion we draw from this presentation is general in nature. While there are exceptions, larger manufacturers tend to shy away from conversational controls while smaller shops use them heavily.

Another reason for the controversy has to do with the number of machines to be programmed vs. the cost of conversational controls. If the company has several CNC machines of the same type, it can be expensive to fit them all with conversational controls. The cost comparison from conventional to conversational controls varies with style and manufacturer, but, generally speaking, conversational controls are more expensive than conventional CNC controls by a substantial margin. When multiple similar machines are involved, there will come a point when it will be less expensive to purchase the machines with conventional CNC controls *and* buy a good CAM system than it would be to equip all machines with conversational controls.

Yet another reason for the controversy has to do with the machine's productivity. Some conversational controls do not allow programming to be done while the machine is producing parts. In this case, the machine will be sitting idle while the next part is being programmed.

Even conversational controls that do allow programming to be done while production is being run do so with limited success. If the cycle time for the job currently running is short, and if the conversational program to be prepared is lengthy, the operator will be constantly breaking out of the program to load workpieces, measure the various dimensions on each finished workpiece, change tool offsets to allow for tool wear, and index tool inserts. These responsibilities will take a great deal of the operator's concentration, so for short cycle times, it may not be worth it to try to run production while programming.

Again, this is where large and small companies differ. Generally speaking, large companies that have large production quantities want to keep their CNC machines running parts for every possible minute. Any loss of time during the program's preparation that can be avoided is seen as a waste of time. On the other hand, smaller shops that run lower production quantities are more concerned with finishing the job quickly while utilizing a limited number of people. They are not nearly as concerned with the machine's dead time while programming is taking place.

Remember that most conversational controls have a CNC side that allows them to be utilized as a conventional control. Most conversational controls do *not* have to be programmed conversationally. They can also be programmed manually, or with the help of a CAM system. For this reason, more and more companies are purchasing conversational controls to gain the benefits of conversational programming while retaining compatibility with current programming methods.

Each company must decide for itself whether conversational controls fit in with its production philosophies. In any event, conversational controls are becoming *very* popular, and the number of conversational controls being used is growing. It is very likely beginners currently entering the field of CNC will be working with conversational controls in their careers at one point or another.

Understanding the Application

As with manual programming, the person inputting the conversational program must be familiar with the particular machine tool being programmed. Generally speaking, conversational controls can be applied to every kind of CNC machine discussed during key concept no. 1 of manual programming. Just as the manual programmer must be well-acquainted with the machine tool itself, so must the conversational programmer. The application for the machine, the directions of motion, and special programming functions are among the things the conversational programmer must understand. (See key concept no. 1 in Part 1 for more information.)

Also as with manual programming, the person inputting the conversational program must be well-versed with the machine's application. While programming is dramatically simplified, the programmer must still understand the basic machining practice related to the machine tool to be used. Most importantly, the programmer must still be able to come up with a process that will machine the workpiece properly. The programmer must still understand the tooling related to the processes to be applied. And, in general, the programmer must still understand the basic makeup of the machine tool itself. While conversational programming takes away much of the tedious work, the better acquainted the programmer is with the machine's application, the better off the programmer will be.

Generally speaking, conversational controls offer little guidance, if any, to machining processes. In most cases, the programmer must possess the same level of knowledge relative to machining practice as the manual programmer. This is especially true regarding the sequence of machining operations to be performed.

Advantages of conversational controls

There are several advantages that make the use of conversational controls more desirable than manual programming. Here we introduce many of them.

Interactive input.   As you know, manual programming requires a *word address* format. During key concepts 1 through 6 of manual programming, we equated manual programming to learning a foreign language. We said CNC commands in the program could be equated to sentences. The words that make up the CNC command are like the words in a sentence. With word address format, the programmer is not prompted in any way. The program is completely prepared by the programmer with no help from the control.

By comparison, conversational programming does not involve a word address language. Instead, the conversational programmer will be answering questions, inputting values, and filling in the blanks for data the conversational control requires. In most cases, the conversational control will actually draw the programmer through the programming process by prompting for required information.

This interactive input can be related to an automatic bank teller machine. If you have used an automatic teller card at an automatic bank teller, you know how to fill in the blanks in much the same way many conversational controls are programmed. The automatic teller may give you a series of multiple choice questions related to the transaction (deposit, withdrawal, money transfer, etc.). You must answer by pressing a key corresponding to an action shown on the screen. The automatic teller leads you through the transaction in much the same way a conversational control leads the programmer through a conversational program. While the conversational control usually requires *much* more input, the basic idea is the same.

Of course, the conversational programmer must understand the questions being asked and answer them correctly, just as the person using the automatic teller must understand the difference between a deposit and a withdrawal to be able to answer the automatic teller correctly. Though this is the case, conversational programming is usually much easier than the word address format required for manual programming.

Figure 8.2 shows an example of the menu-driven input a conversational control will allow. As you can see, most conversational controls will show a drawing or picture related to the kinds of inputs required.

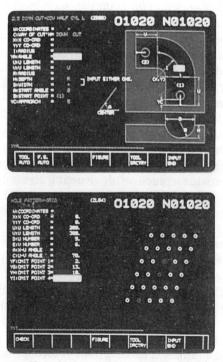

Figure 8.2 How the conversational programmer will always be prompted for information. (*Courtesy Fanuc U.S.A. Corp.*)

On the drawing, the needed information is clearly marked. While filling in the blanks, the programmer will have a clear idea of what is required.

If an incorrect answer is given during this interactive input, the conversational control will usually be smart enough to catch the mistake and alert the programmer. While the level of error trapping varies from control to control, most current conversational controls are quite sophisticated in this regard.

For example, say the programmer meant to type the value .25 as the answer to a question related to a radius to be input. By mistake, the programmer forgot the decimal point and typed the value 25. In this case, the radius would appear on the screen much larger than the intended radius. The conversational programmer, seeing this, would know something was wrong.

Graphic display.   When a manual program is being prepared, there is little to let the programmer know whether mistakes are being made. No one will know how the program will behave until it is loaded into the control's memory and executed. Unfortunately, manual programs are tedious and error-prone. It is common for complicated manual programs to take several hours to verify.

By comparison, conversational controls allow the programmer to see basic mistakes. By no means do we mean to say conversational controls are fail-safe, but they do make things more obvious than conventional CNC controls. For example, most conversational controls require that the programmer draw the part on the control screen. During this process, the programmer will see the basic configuration of the workpiece. If a mistake is made, it is likely the shape on the screen will not appear properly. If the programmer is paying attention, the mistake should be found. But as stated, conversational controls are not fail-safe. If the mistake made while drawing the part on the screen is very small, perhaps the difference between 0.100 and 0.105 in, the part may still appear to be correct on the screen, when in reality it is not.

Once the conversational program is finished, most conversational controls allow the tool path to be graphed on the control screen prior to running the program. This gives the programmer a very good idea of what the program is going to do. While there could still be problems and mistakes in the program, at least the basic cutting motions can be judged.

It is this kind of graphic capability that makes the conversational control easier to work with and makes it an excellent choice for beginners. While experienced programmers may not want to give up the intimacy of manual programming, even most experienced programmers would agree that conversational controls have a great deal to offer once they have been exposed to what the conversational control can do. Figure 8.3 shows an example of the kind of graphic display we mean. In almost all cases, the conversational programmer will know what is expected.

Reduced calculations. One of the strong points of conversational CNC controls is their arithmetic capability. As you know, the manual programmer must do a great deal of arithmetic in preparation for the programming task. Addition, subtraction, multiplication, and division are done quite regularly, and sometimes trigonometry is involved. The more complicated the part, the more difficult the math involved.

This is because conventional CNC controls can accept only coordinate values for the axes involved with the motions. Some may have helpful features like polar coordinates and coordinate rotation, but for the most part, the manual programmer must calculate all coordinates going into the program. Remember key concept no. 2 in manual programming, "Preparation for Programming." In that key concept, we said there are many times the design engineer will not dimension the part print in the way conventional CNC controls require coordinates.

On the other hand, conversational CNC controls remove the burden of doing math. While they vary in their level of math capabilities, most conversational controls have been designed to do almost all math related to the control's application. This is because they can accept dimensions right from the print and reconstruct the shape of the part from the dimensions given. The print must describe the workpiece fully in order for the conversational control to know enough about the part.

Here's an example. Figure 8.4 is a turning center part that would require the manual programmer to do quite a bit of math. The 10° angle combined with the 0.25-in radii would make the manual programmer come up with four points in X and Z, all rather difficult to calculate.

Current conversational turning center controls would handle this problem with ease. The conversational programmer would be allowed to simply enter dimensions right from the print, telling the control the desired angle and radii. The conversational control would do the rest, coming up with the starting and ending points automatically. Also, the part would appear on the screen to scale, meaning the programmer could check the shape of the workpiece. If it did not look right, the entries could be checked.

Automatic cutting conditions. Most conversational controls can automatically develop cutting conditions for the machining operations the control has been designed to handle. Usually some kind of material file is incorporated. The material file allows the user to create cutting

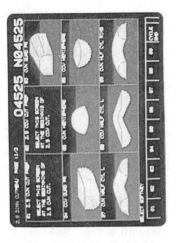

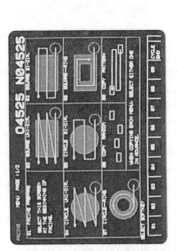

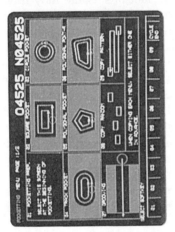

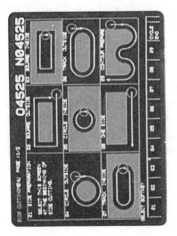

Figure 8.3 How the conversational control will always make it clear what it expects. (*Courtesy Fanuc U.S.A. Corp.*)

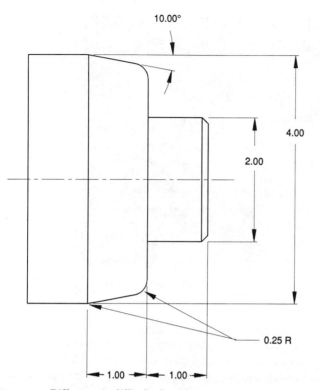

10.00°

4.00

2.00

0.25 R

1.00   1.00

Figure 8.4 Difference in difficulty level between conversational and manual programming.

conditions for several materials. The user usually has the ability to modify these materials to meet the company's requirements. The entries in the material file include the speeds and feeds for the cutting tools the machine tool is designed to use.

Once a material is in the material file and set up correctly, whenever the material must be machined, the control will generate cutting conditions for all operations automatically. If the programmer wishes to modify the automatically generated cutting conditions while programming, most conversational controls allow this to be done.

Figure 8.5 shows an example of how the conversational control displays the cutting conditions once they have been generated. This should demonstrate how most conversational controls allow the programmer to modify any of this data, if desired.

Consistent difficulty level for programming. There is a misconception in the industry related to how simple it is to work with conversational CNC controls. Some people believe no experience is required to begin

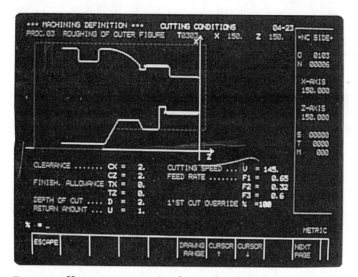

Figure 8.5 How one conversational control will display cutting conditions. (*Courtesy Fanuc U.S.A. Corp.*)

working with a conversational control. They say anyone can begin working with them with no prior training. While conversational controls are usually easier to work with than conventional CNC controls, they *do* require training and they *do* require some previous machining practice experience.

However, a programmer who masters the conversational control can usually outperform the manual programmer by a healthy margin. This is because the conversational control smooths out the difficulty related to program preparation. With a manual program, the level of difficulty is directly related to the part print. The more difficult the print, the harder it will be to write the program, especially relative to math difficulty. But with conversational programming, difficult workpieces are just as easy to program as easy ones.

Limitations of conversational controls

While conversational controls are usually easier to work with and program, *all* conversational controls have limitations with regard to their capabilities. Depending on the manufacturer, these limitations can sometimes be quite severe, requiring the conversational programmer to be well-versed in manual programming techniques in order to overcome the shortcomings of the conversational control.

Many conversational controls, however, are quite powerful, allowing the programmer to handle a wide variety of programming applications. With this kind of conversational control, the programmer need possess only a cursory knowledge of manual programming techniques.

No matter what level of capability the conversational control possesses, there will always be limitations as to what the control can handle. The limitations we are talking about stem from two possible sources, workpiece geometry limitations and machining technique limitations.

Workpiece geometry limitations.   The source for workpiece geometry is the part print. If the print is made correctly, it describes the workpiece completely. Dimensions are given to all critical surfaces, and any machinist would understand how the part is to be produced. Unfortunately, sometimes design engineers do not describe the workpiece in an easy-to-understand way. In some cases, mistakes may even be made on the part print that make it impossible for the part to be produced. Missing and conflicting dimensions are among the problems encountered by any programmer. The first point to make with *any* form of conversational control is very simple. It also applies to manual programming: "It *must* be possible to make the part from the part print!" Though conversational controls can be very smart, they cannot work miracles. The part to be programmed must be described and dimensioned in a way that allows it to be produced.

However, many conversational controls have limitations, sometimes severe limitations, as to the method of inputting the geometry for the workpiece to be machined. With some conversational controls, there will be times when the part print is correct and the part can be made from the print, yet the conversational control cannot handle the values being input.

Some conversational controls are more powerful than others in this regard. There may be times when the programmer may have to manually calculate coordinates required for the program. When you think about it, there are numerous ways for a design engineer to dimension workpieces, some more obvious than others. For example, when dimensioning a tapered surface to be machined on a turning center, the design engineer could make life easier for everyone by dimensioning both diameters as well as the length of the tapered surface. But most design engineers would not make it this simple. They would give an angular dimension related to the tapered surface and make everyone do some trigonometry to calculate coordinate values. A conversational control may be able to accept the end-point technique for dimensioning, but not the angular technique.

While control manufacturers are improving their conversational controls in this regard, the limitation still exists. Considering the limitless dimensioning techniques available to design engineers, it is unlikely this limitation will ever be totally overcome.

Machining technique limitations.   Manual programmers have always had difficulty adapting to conversational controls because of their machining technique limitations. The manual programmer is completely intimate with a CNC program. The manual programmer can be very explicit as to what the program is going to do. On the other hand, conversational programs leave a lot to the imagination as to exactly what the CNC program generated will do. The intimacy of manual programming is lost with conversational controls if programs are generated conversationally.

On the other hand, beginners in CNC who do not want to know manual programming techniques do not care about this lost intimacy. They would rather take things at face value and trust the conversational control to do what it must to generate the desired machining operations correctly.

There are limitations, however, to what conversational controls can do in machining operations. Generally speaking, the main applications for the control should be well-covered and easy to handle conversationally. But more obscure machining operations may not be possible conversationally. For example, one conversational turning center control handles the most commonly applied turning operations very easily. Rough turning, finish turning, threading, grooving, rough boring, and finish boring can be done with ease. But when it comes to other, less often used operations, like tailstock and bar feeder control, the conversational control is incapable of commanding these basic functions. If the tailstock or bar feeder must be commanded from within the program, the programmer *must* use manual programming techniques.

Generally speaking, you will find that those things a conversational control has been designed to do, the control handles quickly and easily. However, those things the conversational control has not been designed to do are quite difficult to handle, and must be done with manual programming techniques.

These limitations also explain part of the controversy related to conversational controls. One company may have dimensioning and machining techniques that fall within the capabilities of the conversational control. This company can conversationally program every workpiece with absolutely no problems. Another company may require techniques the conversational control was not designed to handle. In this case, every program would be difficult to conversationally program. With the above situation, one company loves the conversational control while the other hates it! You can see how arguments get started.

Program storage

With conversational controls, there are usually two programs involved with the programming process. With most conversational controls, the conversational program generates a CNC program. After the programmer inputs the conversational program telling the control about the workpiece to be machined, the conversational control creates a CNC program similar to the one a manual programmer would write. It is from this generated CNC program that workpieces are run.

It is important to know that this program generation is a one-way street. It is *not* possible to generate a conversational program from a CNC program. Only the reverse is possible.

This doubling in the number of programs required for a workpiece requires slightly more organization on the programmer's part. The two different programs are stored in separate places within the conversational control. Two distinct methods of accessing programs is necessary, one for the CNC program and one for the conversational program. Also, the CNC program can usually be transmitted to some other common storage device, like a paper-tape punch or computer. Generally speaking, the conversational program cannot. This fact limits the number of conversational programs that can be stored within the control's memory, therefore the conversational programmer must make wise decisions as to which conversational programs will be stored for future use.

Most conversational programmers will keep only active conversational programs (programs being used currently) in the control's memory. Once the job is finished, the conversational program is dumped. If the programmer expects to run the job again some day, the CNC program is stored for future use on some outside device, like a tape punch or computer.

The exception to this general rule is when the user machines a family of parts. If the company manufactures a series of similar parts, they will keep a master conversational program in the control's memory on a permanent basis. Whenever a new part in the family must be programmed, the master conversational program is simply modified to create a program for the new part in the family.

Fine-tuning a conversational control

We have already mentioned differences of opinion from user to user regarding the value of using conversational controls. Some companies feel conversational controls are very useful and use them on a daily basis. Others feel conversational controls do not help very much and rarely, if ever, use them.

In an attempt to try to please more potential users, conversational control manufacturers have increased the flexibility of their conversational controls by adding ways to fine-tune them to the user's specific needs. Most conversational controls made today allow users to tailor the control to their liking. To do this, a whole series of parameters can be manipulated to change the way the conversational control functions.

For example, say the conversational control is currently outputting the CNC program with sequence numbers (N words). But the user does not want N words in the CNC program to conserve memory space on the CNC side of the control. On most conversational controls, a parameter could be changed to tell the control not to output sequence numbers from this point on.

This is just one example of how a conversational control can be fine-tuned. Most conversational controls allow hundreds of functions to be manipulated in this manner. This manipulation allows users to change the control to their liking.

Tooling organization

Most conversational controls designed for metal cutting operations allow a method for tooling organization. They incorporate a tooling file in which tooling information can be stored. This information includes general information like tool name and station number. But most conversational controls also allow more specific information about the tool related to the geometry and capabilities of the tool. While conversational controls vary with regard to how this information is used, most have rather complex checking capabilities related to how the tool will be used in the conversational program. This information is designed to make things easier for the programmer. Once a tool is described properly in the tooling file, the conversational control will make certain assumptions based on how the tool has been described.

For example, many turning center conversational controls require a comprehensive tool description including the tool type, lead angle, nose angle, and nose radius. This information tells the turning center control what the tool looks like. With this information, the conversational turning center control can make correct decisions as to what machining operations the tool is capable of performing.

Most conversational controls require that tools be entered into the tooling file *before* a conversational program can be written. During the conversational program, when the programmer wants to reference a tool in the tooling file, the tool's identification number is entered. This causes the control to search the tooling file to find the corresponding number. Once found, the control can judge what the tool is capable of.

Most conversational controls require information for the tooling file to be organized in two ways, one for the tool's configuration, and another for the tool's setting position.

The tool's configuration tells the conversational control what the tool is capable of and what kind of machining operation the tool is to perform. For machining center conversational controls, the information required for an end mill may include the number of flutes, the flute length, and the tool material (among other things).

The tool's setting position tells the conversational control the position of the tool relative to the setup. This information is necessary for the control to calculate the tool's current position. You can think of these two ways of organizing tooling in a simplified manner. The control has to know what the tool is as well as where the tool is. Information in the tooling file related to the tool's configuration tells the control what the tool is. Information in the tooling file related to setting position tells the control where the tool is.

Chapter

9

Flow of Conversational Programming

Conversational programming can be compared to learning any computer software application program. If you have experience with a word processor, data base, spreadsheet, desktop publishing, games, or any other computer application, you know some of what goes into learning conversational programming for a CNC control.

It takes time to become familiar and comfortable with any kind of computer program. While software is getting easier to learn and use, the person sitting down with a new application program for the first time must take some time to study the program. By using the manual that came with the software and the help screens that can be commanded from within the program, the beginner can become acquainted with the new software with relative ease.

In a similar way, the person stepping up to a conversational control for the first time will have to learn the various techniques required by the control in order to input programs. Like computer software, conversational controls vary in their ease of use. Some make it easy to learn how to input programs with an intuitive human interface, while others are rather cryptic and more difficult to learn. In any case, there will be a learning curve related to understanding all that is possible with any one conversational control.

Steps to Conversational Programming

As stated, conversational controls vary widely with regard to how a program is entered. Most follow a logical series of steps. Once learned, these steps will be repeated for every conversational program to be entered. This means the various screens of conversational input will soon become familiar.

What follows is a general sequence for conversational programming input. By no means is this the only way conversational programs are entered. Quite the contrary, conversational controls made by two manufacturers will have little in common. Though the order by which conversational programs are input varies, most conversational controls require each of the steps described here in one fashion or another.

General Information

Most conversational controls begin by asking for some general information related to the program being entered. Of course the questions vary depending on application, but here are some of the things that the conversational control will want to know. (Figure 9.1 shows a display screen from a popular conversational control related to general information. Note the menu-driven technique to enter data.)

Workpiece material

This information helps the control to set up the proper feeds and speeds to be used for the machining operations within the program.

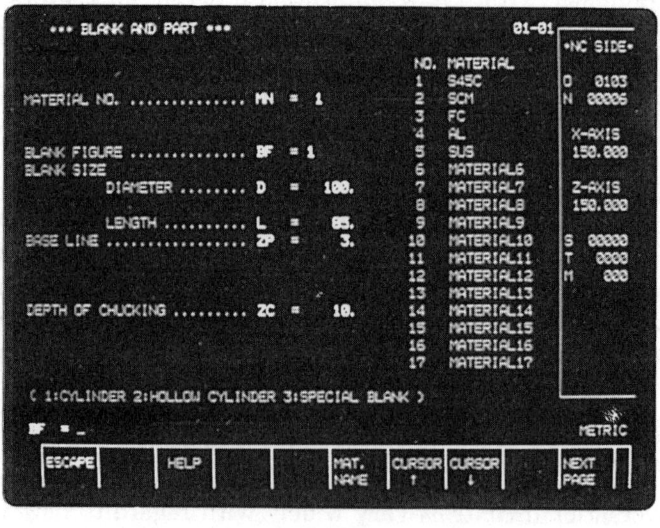

Figure 9.1 How a popular conversational control collects general information about the program. (*Courtesy Fanuc U.S.A. Corp.*)

Most conversational controls allow these feeds and speeds to be changed according to the operation being performed, but this early entry in the program sets the general speeds and feeds to be used.

Rough stock size and shape

Early in the conversational program, most conversational controls will ask for the stock size of the workpiece to be machined. This helps the control to know how large the workpiece is. This information is also used to set up the boundaries for the screen size during the program. Most conversational controls allow round and square-shaped workpieces to be easily entered in this step. If the shape is irregular, such as a casting or forging, it can be somewhat more difficult to describe the rough stock.

Other miscellaneous information

Other things the control may ask during this step include the program name (for the conversational program), the position of program zero, how much stock is to be removed from the part, the part name and part number for the workpiece to be machined, and any other general information pertinent to the job. Again, conversational controls vary dramatically in the kind of general information required.

Workpiece Geometry

Another step required of the conversational programmer also varies from control to control. At some point, the conversational programmer will be required to input the shape of the part to be machined. This step is quite dependent on the control's application. For machining center controls, this step is required in order to drive the cutting tools around the workpiece being machined. For turning center controls, it is required to let the conversational control know what the finished workpiece looks like. For turret punch presses, it is required to let the control know where holes are to be pierced. For wire EDM machines, it is required to let the control know the shape to be machined.

In all cases, this step will keep the programmer from doing tedious calculations. That is, the conversational control is designed to allow the programmer to input dimensions right from the print, with little or no regard for the math involved. As mentioned in Chap. 8, conversational controls vary with regard to how powerful this step is. Some controls can handle almost anything that comes along while others get bogged down with relatively simple tasks.

Figure 9.2 shows an example of how one popular conversational con-

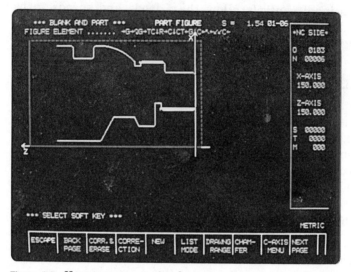

Figure 9.2 How one conversational turning center control displays the finished part drawing. (*Courtesy Fanuc U.S.A. Corp.*)

trol displays the finished drawing of the geometry. For this turning center control, only the top half of the workpiece is shown. The bottom horizontal line represents the spindle center.

The most powerful conversational controls (those that allow geometry to be described the best) use a construction method for geometry compromising a series of *point, line,* and *circle* definitions. This method of construction comes from a programming language called APT (for Automatic Programmed Tool). In its original format, APT was a very cumbersome language to work with, and required a sophisticated mainframe computer. Despite this, APT has always been considered the most powerful programming language for CNC application. With older versions of APT, the programmer was forced to learn a series of cryptic commands to describe geometry and to machine the workpiece.

The best and most powerful conversational controls take their geometry construction methods from APT. With APT-like geometry construction, the programmer can describe just about anything that can be made from the print. With the interactive graphic capability of the conversational control, it is also quite simple to do so.

With this form of geometry input, the programmer describes a series of points, lines, and circles in a way that makes up the shape of the part. Figure 9.3 shows an example. As you can see, the workpiece drawing is on top. On the left is a marked-up print showing a series of circle and line numbers representing the various geometry elements of the drawing. The drawing on the right shows each element con-

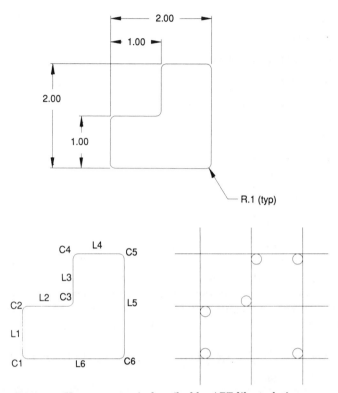

Figure 9.3   How geometry is described by APT-like techniques.

structed in its entirety. This series of drawings illustrates the fact that most workpiece geometry can be described with a simple series of points, lines, and circles.

The method and order of geometry construction is quite important. With APT techniques, the programmer will have a number of *defini-tion types* with which to describe the points, lines, and circles.

Here are some examples of possible point definitions:

- *X-Y* coordinates
- Intersection of two lines
- Intersection of line and circle
- Center of circle
- Polar coordinate

Here are some examples of possible line definitions:

- Parallel to an axis

- Passing through two points
- Tangent to a circle at a given angle
- Passing through a point at a given angle
- Tangent to two circles

Here are some examples of possible circle definitions:

- Center and radius known
- Passing through three points
- Tangent to two lines with a known radius
- Tangent to three lines
- Known center point tangent to a line

Remember that the above examples represent but a few of the possibilities as to how geometry can be constructed with APT techniques.

The decision as to which geometry definition type to use will be based on how the part print is dimensioned. In most cases, the programmer will be allowed to construct the geometry in a conversational program as it would be drawn by the design engineer on a piece of paper.

Figure 9.4 shows an example. In this drawing, the circle labeled C1

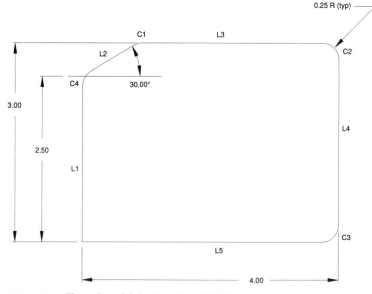

Figure 9.4 The order of definition for APT-like techniques.

would be quite difficult to describe without lines L2 and L3 being described first. With these two lines previously described, it would be easy to construct circle C1 with the definition type for tangent to two lines with a given radius.

Here is the order and construction method that could be used to describe this part:

L1	Parallel to axis
L2	Passing through a point at a given angle
L3	Parallel to axis
L4	Parallel to axis
L5	Parallel to axis
C1	Tangent to two lines with a known radius
C2	Tangent to two lines with a known radius
C3	Tangent to two lines with a known radius
C4	Tangent to two lines with a known radius

Element symbols to construct geometry

While the geometry definition techniques we have shown are based on APT techniques, there are other methods used to construct workpiece geometry. One alternative technique that is quite popular utilizes a series of *element symbols*. Figure 9.5 is a list of element symbols as they are utilized on a popular turning center control.

The arrows pointing in various directions are used to describe straight surfaces. The C is used to describe chamfers of equal proportion between two elements. The R is used to describe radii tangent to adjacent elements. The clockwise and counterclockwise arcs are used to describe arcs that are not tangent to the adjacent elements.

With this construction method, the programmer begins at one location on the contour to be described and inputs the various element symbols to go around the shape. Figure 9.6 shows an example.

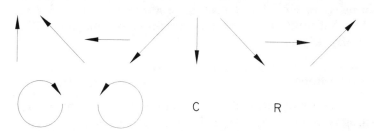

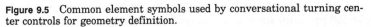

Figure 9.5 Common element symbols used by conversational turning center controls for geometry definition.

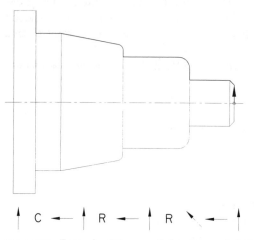

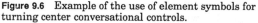

Figure 9.6 Example of the use of element symbols for turning center conversational controls.

While successfully used by several different turning center control manufacturers, this method of describing geometry is not nearly as powerful as the APT-like techniques shown earlier. This is because the programmer is locked into keeping geometry definition on the contour being described. The programmer is not free to construct any geometry that is not part of the actual workpiece being machined. For complex geometry shapes, this can be a major limitation.

Constructing geometry during machining

Both of the previous geometry definition methods given to this point involve describing geometry *prior* to the definition of machining operations. That is, the programmer was allowed to concentrate only on geometry, not considering how the workpiece is to be machined. While we consider this to be the best form of geometry definition, some conversational controls have the programmer defining geometry during the machining operations. This forces the programmer to consider machining while defining geometry, making programming more difficult and usually less flexible.

Machining Definitions

The final stage of conversational programming involves defining how the machining is to take place. As with all steps to conversational programming, this step varies with the application for the control. Most conversational controls have been specifically designed to accept information in a simple and logical way for the types of machining operations to be performed.

Most provide menu selections for the operations that can be performed. The programmer would simply choose the operation to be performed first from the set of menu choices. From there, the control will ask for pertinent information regarding the operation, leading the programmer through the critical information required.

When finished with an operation, the programmer will proceed to the next operation. The process continues until the machining definitions are completed.

Here are the kinds of things the control will require during this step. Though the order in which this information is entered varies from control to control, most conversational controls require this information.

Operation type

The first choice the programmer must make for each operation is the type of operation to be performed. From some kind of menu selection, the programmer must enter the desired operation type. The choice will match the most common operations performed by the type of conversational control. For example, with conversational turning centers, the choices will include rough turning, finish turning, rough boring, finish boring, grooving, threading, drilling, tapping, and so on. For conversational machining centers, the choices will include rough milling, finish milling, rough boring, finish boring, drilling, tapping, reaming, counterboring, and so on.

Cutting conditions

At some point during the machining definition for each operation, the control will require the cutting conditions for the operation. Cutting conditions include things like feed and speed, depth of cut, and, for roughing operations, stock to be left for finishing. Most conversational controls will have this data preset in one way or another. This means the control will make its best guess at what the cutting conditions should be. Usually this preset information is contained in the material file related to the material being machined and the tooling file related to the kind of tool being used. The programmer must view this information cautiously. There will be times when the generated cutting conditions will not be correct for the operation to be performed.

Tooling information

The tool to be used in performing each machining operation must be entered. The conversational control must know two basic things about each tool. It must know what the tool is and where it is. By "what the tool is," we mean the control must know the style of tool being used (drill, tap, reamer, etc.). Also, special information about each tool is required to let the control know the tool's capabilities. For example, typical information about an end mill would include things like number of flutes, flute length, whether it is center cutting, and its material. Most conversational controls use this kind of information to double-check the programmer, assuring that some tooling limitation is not being exceeded.

By "where the tool is," we mean the control must know the setting position for each tool so it can command motions accordingly.

Most conversational controls allow data about each tool to be permanently stored in a tooling file. During programming, the programmer is required only to enter an identification number for each tool, causing the control to search the tooling file to attain all information about the tool. Especially helpful for often-used tools, this keeps the programmer from having to enter redundant tooling information from program to program.

Information related to the surface to be machined

The conversational control will also require the programmer to specify what surface is to be machined during each operation. For example, if drilling holes on a machining center, the control will want to know which holes are to be drilled. If rough turning on a turning center, the control will want to know which surfaces are to be rough-turned. The actual techniques to accomplish this vary widely from control to control.

Other Functions of Conversational Programming

Along with the prompts during the basic flow of program entry, most conversational controls provide special features aimed at making it easier for the programmer to tell if the program is correct.

Tool path display

This feature allows the programmer to see how each tool used in the program will make its motions. Most conversational controls have this feature and even let the programmer view tool motion from several different attitudes. For example, for vertical machining centers, the programmer can view any two axes at the same time (X-Y, X-Z, or

Y-Z). Also, an isometric (three-dimensional) view is possible on most conversational machining center controls.

Besides being able to view the tool path from a variety of attitudes, in most cases the programmer has control of several other things related to how the tool path will be displayed. Most controls allow zooming, or enlarging, techniques to allow the programmer to easily specify a viewing window. This technique is helpful when an intricate area of the part is too small to be seen in its normal size.

Figure 9.7 shows the tool path display from a popular machining center conversational control, while Figure 9.8 does the same for a popular turning center control. While these are completed tool paths and you can see the completed shape, note that the conversational programmer will also be able to watch these shapes being generated. This means the programmer will also be able to see the order in which the workpiece will be machined, making it even easier to detect problems.

While presenting the tool path, the control will show only the motions of the tool. The programmer will *not* see other things related to the program (workpiece, fixture, tool, etc.). This means the program-

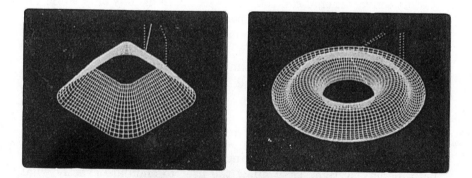

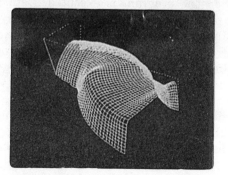

Figure 9.7 How a typical machining center conversational control will display the tool path. (*Courtesy Fanuc U.S.A. Corp.*)

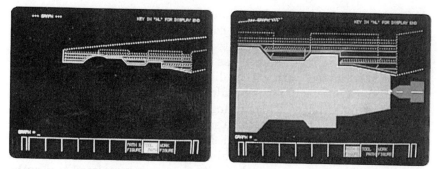

Figure 9.8 Methods by which conversational turning center controls display tool path. (*Courtesy Fanuc U.S.A. Corp.*)

mer cannot check for possible interference problems until the program is executed. During the tool path display, rapid motions will be represented by dotted lines, while cutting motions will be represented by solid lines. If the control has a color screen, most controls will change the color of the tool path display between tools, making it easier to distinguish the tool paths when the screen becomes cluttered.

Most conversational controls allow tool path display only while the conversational program is being converted to the CNC program. However, there are some conversational controls that allow the CNC program to generate the tool path display on the CNC side of the control. This is a helpful feature, since any CNC program can be displayed, including manually written and CAM-system-generated CNC programs.

Animation

Recent conversational controls incorporate more than a simple tool path display system. They allow the user to simulate exactly what will happen when the program is executed. Data about the work-holding device and tooling is first entered in a way that allows the control to actually animate the program right on the screen of the conversational control. By using this technique, the programmer will be able to check almost everything about how the program will behave, including tool interference with the workpiece and work-holding setup. During the animated execution of the program, as each tool machines its intended surface, the control will show the tool, workpiece, and setup as if the machining operation were being viewed on a television. Figure 9.9 shows the screen of an animated turning center workpiece.

If the animated program appears to be correct, just about the only things that could still be wrong would be the cutting conditions (feeds

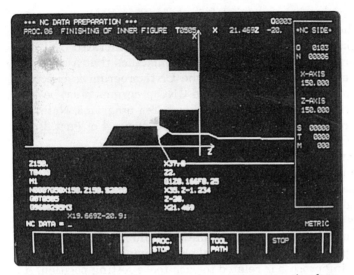

Figure 9.9   Use of animation on a turning center conversational control. (*Courtesy Fanuc U.S.A. Corp.*)

and speeds) related to each machining operation and each tool's first approach movement to the workpiece. The basic motions of the program should be correct.

Statistical data

Most conversational controls keep track of certain things a manual programmer would give little regard to. Once the conversational program is finished, most conversational controls will display information related to the cycle time of each tool used in the process. Along with this, the total accumulated cycle time is also given. While this information can be helpful, most conversational controls calculate only *motion* (rapid and cutting) time. They do not consider cycle time that is not related to motion. Tool changing time, spindle starting time, and coolant activation time are among the things most conversational controls do not take into consideration. This means the cycle time calculated by most conversational controls will not be accurate.

Program Verification

As you might expect, a conversationally generated CNC program is much easier to verify than a manually written CNC program. During the programming process, the conversational programmer is allowed to see many things related to the potential success of the program, while the manual programmer is left in the dark until the program is executed at the machine tool.

However, this does not mean conversationally generated CNC programs are perfect. There are two possible mistakes that a conversational programmer cannot see until the CNC program is executed. This means conversationally generated CNC programs must be verified in much the same way as manually written programs. Note that verification procedures are given in key concept no. 4 of Part 3, "operation" (see Chap. 13).

The first mistake a conversational programmer cannot see is related to cutting conditions. The feeds and speeds generated for each machining operation may be too aggressive. This means the operator must be cautious with the program as the first part is being run. Even if the motions generated by the program are perfect, if the speeds and feeds are not correct, a dangerous situation could result.

The second problem a conversational programmer cannot see during the programming process is related to each tool's setting position. If a mistake is made about the setting position for any one tool, or if a tool is loaded into the setup incorrectly, there could be real problem. The conversational control would believe the tool setting information and would display tool paths based on it even if mistakes were made with regard to the setting position of the tool.

For this reason, the programmer must be very cautious with each tool's first approach to the workpiece. After confirming that the tool has moved into its cutting position correctly, normally the operator can rest assured that the machining will take place properly.

Operation

10

Key Concept No. 1: Know Your Machine

The first key concept of machine operation parallels the first key concept of manual programming. You must understand the basic makeup of the CNC machine tool being used. However, during our discussion of programming, we looked at the machine from only the programmer's viewpoint. We addressed each machine tool's basic components and discussed their use from the programmer's standpoint.

Now we will look at the machine from the operator's viewpoint. The operator must have a much better understanding of the machine tool itself.

For example, the operator must understand the basic components of the machine. While the programmer must also know about these basic components, the operator will be called on to make physical adjustments to the machine on a regular basis.

The operator must know the function of *all* buttons and switches. While the programmer must also know about the various buttons and switches, the operator will be activating them regularly and must possess a much better understanding of their function.

The operator must know how to adjust tool offsets. While the programmer must also know their function, the operator will be making physical adjustments to tool offsets with every new setup made on the CNC machine.

For these and many other reasons, the operator must be more intimate with the machine tool itself than the programmer. The operator must possess a higher level of understanding about the usage of the machine tool in order to perform required tasks. Unfortunately, there are people in the machine-tool industry who believe that the operator of a CNC machine is nothing more than a button pusher. They believe that anyone who can push a button can operate a CNC machine. They perceive the operation of a CNC machine as nothing more than loading a part and pushing a button to activate the machining cycle.

While there are instances when everything is perfect and running a CNC machine is relatively simple, the informed CNC professional would agree that the operator of a CNC machine should be well-trained and possess a high level of knowledge related to all functions of CNC, including programming, setup, and operation. The better this knowledge level, the better the operator.

Just what the operator of a CNC machine will be expected to do will vary from company to company. While there are exceptions to this general statement, larger manufacturing companies tend to segment the responsibilities related to their manufacturing. One person will set up tooling. Another will make the work holding setup on the CNC machine tool. Yet another will verify the program. Once the program is deemed correct and safe, it is turned over to the CNC operator who runs the workpieces to be produced. In this case, the operator's responsibility may be limited to simply loading, checking workpieces, and adjusting offsets.

On the other hand, smaller companies cannot afford the luxury of having so many people involved with any one job. Smaller companies tend to expect more of each person. These companies may expect the operator of a CNC machine to do all functions related to the machining of a workpiece, including writing the program, setting up the tooling, making the work-holding setup, verifying the program, and running the production.

The more the operator is expected to do, the better the knowledge level should be. In any event, knowing more than may be required can never hurt. The more the operator knows, the better the CNC environment.

In this first key concept for CNC machine operation, we examine the CNC operation panels from the operator's standpoint, assuming the operator must possess a complete knowledge of all machine functions. While we cannot be specific about any one type of control, you will find the buttons and switches we discuss to be quite common among CNC controls.

As you will see, many of the functions we describe in this key concept will require an understanding of what was presented during manual programming. For example, to understand a switch labeled *optional block skip* (sometimes called *block delete*), you must first understand the function of optional block skip (the slash code) as it was presented during key concept no. 6 of programming. Throughout the presentations we make in this part on operation, we assume you have read Part 1.

The Two Most Basic Operation Panels

When the beginning CNC operator views a CNC machine control panel for the first time, it can be somewhat intimidating. There will appear to be so many buttons and switches to learn, the beginner can be tempted to give up before getting started.

Do not let the control panel intimidate you. As we go on, you will find that the CNC machine's buttons and switches are quite logical and aptly named. While it may take some memorizing, once you are exposed to a CNC machine on a regular basis, we think you will agree that operation is not difficult to master.

As with any complex problem, it is helpful to break up the seemingly difficult task of learning the CNC control panel into smaller pieces. Here we discuss the two most basic panels of buttons and switches.

Most CNC controls are designed to be used on a variety of CNC machine tools. While there are exceptions, many times the control is made by one manufacturer and the machine tool is made by another manufacturer. In all cases, the CNC control must be interfaced with the machine tool. Generally speaking, the control manufacturer is responsible for the buttons and switches required to operate the CNC control. However, the machine-tool builder is responsible for the buttons and switches related to the machine tool. This means there are two distinct control panels for almost all CNC machines: the *control* panel, designed and built by the control manufacturer, and the *machine* panel, designed and built by the machine-tool manufacturer.

While there are some exceptions, the control panel is used to manipulate data through the display screen and can be thought of as the keyboard of a personal computer. The machine panel is used to make physical adjustments as to how the machine tool behaves. It should help the beginner to segregate the various buttons and switches into two distinct categories, the machine panel and the control panel.

Figure 10.1 shows a photograph of the two panels of a popular turning center control. Note the distinct separation of buttons and switches. The keypad and display screen on the top half make up the control panel. Below, the buttons and switches, including the handwheel, make up the machine panel.

Here are some examples of when the control panel is used:

The control panel can be used to enter programs into the control's memory. While there are faster ways to do this, through the communications port, many CNC users enter their programs in this manner. Once entered, the control panel can be used to edit, or modify, a program stored in the control's memory. The display screen will constantly show the program being entered or modified, much

Figure 10.1   Two distinct operation panels: the control panel (top) and the machine panel (bottom). (*Courtesy Okuma Machinery, Inc.*)

the same as the display screen of a personal computer when used as a word processor.

Tool offsets are also entered and modified through the control panel. Once the operator selects the offset mode, the display screen will show the table of offsets. Through the control panel, the operator can manipulate tool offsets.

Yet another use of the control panel is to access all functions of the display screen. Axis position, control diagnostics and parameters, and other setting data are manipulated through the control panel.

In contrast, buttons and switches on the machine panel have little or nothing to do with the display screen. For the most part, they are used to activate machine functions. Here are some examples of machine panel functions:

One time the machine panel is used is when manually jogging the axes of the machine. Some form of joystick or a combination of push buttons can be used to make the desired axis motion. Most CNC machines even have a handwheel on the machine panel that can be used like the handwheels on a manually operated machine tool.

For machines that have a rotating spindle, the machine panel can be used to manually turn the spindle on and off. Some machines even have a rheostat that allows the operator to manipulate the speed of the spindle.

The machine panel also includes many *conditional switches* that control how the machine behaves during automatic operation. Features like *single block, dry run,* and *optional stop* are among those controlled by these conditional switches (more on these features a little later).

While there may be other operation panels on the machine to allow convenient control of other special machine features, the machine panel and the control panel are the two most basic operation panels.

Common Buttons and Switches Found on the Control Panel

As mentioned, the buttons, keys, and switches on the control panel are directly related to the control's display screen and can be thought of as the keyboard of a personal computer. Many of the keys found on a typical personal computer are also found on the CNC machine's control panel. Here we itemize them as best we can and give a brief description of their use.

Power buttons

Most CNC machines separate the power-up procedure into stages. To avoid having a surge of power, most have at least two different power-on buttons, one for the control power and another for the power to the machine tool itself. The control panel power-on button must be pressed first, and activates the control screen and control panel. Once the control panel power is on, the machine panel power-on button, usually labeled *hydraulic on* or *machine ready,* can be pressed to complete the power-up procedure.

To turn off power to the machine, the steps are reversed. First the power is turned off to the machine tool, then to the control.

Display screen control keys

These keys manipulate the control screen, allowing the operator to specify what the display screen is to show. These keys set the basic modes of the display screen. You can think of these keys as the channel selector on a television set. They allow the operator to select the function of the display screen to be viewed.

Position button. This display screen selector button allows the operator to look at the machine's current position display. In this mode, the display screen shows pertinent information about where the machine is currently positioned.

Most CNC controls allow several types of position display pages. The *absolute* position display shows the current machine position relative to program zero. The *relative* position display allows the user to set an origin at any location and take measurements. The *machine* position display shows the distance from the machine's reference position to the current position. The *distance to go* position display allows the operator to monitor the amount of movement still needed to complete an executed motion command. All of these pages can be found by pressing the position button and page-up and page-down buttons.

Program button. This display screen selector button allows the operator to monitor the active program in the control's memory. This key is pressed when editing CNC programs and when monitoring programs in automatic operation.

Offset button. This display screen selector button allows the user to display and manipulate the tool offsets. Along with the cursor control buttons, the operator can use this button to find and change offsets in memory.

Setting or parameter button. This display screen selector button allows the user to display current settings of the CNC control. Note that most of the displayed information in this selection will be difficult for beginning operators to understand. Normally this function is required only when the machine is not behaving properly and is used at the discretion of a service engineer.

Alarm button. This control screen selector button can be pressed to show the alarm condition of the control. Most CNC machines will automatically show the alarm on the control screen the moment the alarm condition exists, meaning there is usually no need to actually press this button.

Letter keys

This keypad allows alphabetical character entry. Some CNC control panels provide only those alpha keys (N, G, X, etc.) needed for CNC programming on the keyboard. On others, the full character set (A through Z) is available. Some CNC controls have these keys positioned in the same fashion as on a computer keyboard or typewriter.

Others have them positioned relative to their order of common usage within a CNC program (first N, then G, then X, Y, Z, etc.).

The shift key

As on a typewriter or computer keyboard, a shift key allows multiple usage of other keys on the control panel on many CNC controls. This allows the control manufacturer to double the number of functions for the keyboard without increasing the number of keys required. For example, the main function for a key may be the number 9, but if the shift key is used, the shifted function of the number 9 key may be the left parenthesis key [(]. When the control utilizes the shift key, normally the nonshifted keys are the ones most often used. That is, the shift key will be needed only for obscure operations.

The slash key (/)

This key serves two functions. If the control allows arithmetic functions, this is the *divide by* key. The second use for this key is as the optional block skip code. If this character is included as the first character of a CNC command, the control will look to the position of the optional block skip switch on the machine panel. If the switch is on, the control will ignore the command in which the slash code was included. If the switch is off, the control will execute the command with the slash code. This technique was discussed at length during key concept no. 6 of manual programming.

Number keys

These keys allow numeric entry. Normally located close to the letter keypad, most CNC controls have these keys positioned in much the same way as on the keypad of an electronic calculator.

Decimal point key

This key allows numeric entry with a decimal point. Setting offsets and entering CNC programs are examples of when it would be needed.

Arithmetic operator keys

All CNC controls will have *at least* the minus sign (–) available to specify negative axis position. In this case, the minus sign is not actually performing an arithmetic operation, though it is still considered an arithmetic operation key.

Some CNC controls do allow actual arithmetic operations to be done from within a program and/or through keyboard entry. For those controls that do, the control panel will include the necessary arithmetic operators. The addition sign (+), subtraction sign (–), multiplication sign (*), division sign (/), and equal sign (=) are the most basic arithmetic operators available on some CNC controls.

The input key

This key is pressed to actually enter data. Examples of when this key is pressed include entering offsets and parameter setting.

Cursor control keys

The display screen of the CNC control will often show a prompt cursor that shows the current entry position. The prompt cursor usually appears as a blinking square or underline character in the entry data and is highly visible because of its blinking. It is at the current position of the cursor that data will be entered. For example, the typical CNC turning center has a table of tool offsets. Suppose a turning center has 99 tool offsets, each with at least two required pieces of information (X and Z position offsets). The operator must position the cursor to the desired offset number in order to enter data in the correct location. So that this can be done, the control panel will have at least two keys. These keys allow the operator to move the cursor forward and back throughout the entry positions until the correct position is achieved.

Note that offset is but one example of when it is necessary to position the cursor. Other examples include editing a CNC program, executing a CNC program from the beginning of a specific tool, and entering setting data and parameters.

The way these cursor control keys are marked depends on the control manufacturer. One common technique is to label one key with an arrow down and the other with an arrow up. When pressed once, the arrow-down key moves the cursor one position forward within the entry data. The arrow-up key moves the cursor backward one position within the entry data. Most CNC controls give the operator a fast way of moving the cursor over longer distances, instead of only one position at time. One common way is to incorporate a page-up and page-down function. When pressed, the page-down key causes the cursor to jump forward one full page of the display. The page-up key causes the cursor to jump backward one full page of the display.

Program editing keys

There are many times when a program stored in the control's memory must be altered. Especially during a program's verification, the operator will be required to make alterations to the program being executed. These keys allow program entry and modification.

The actual techniques required to make modifications vary from builder to builder. Most allow at least the basic functions we show in this section.

Insert key. *Not to be confused with input,* this program editing key allows new information to be entered into a program. Most CNC controls will insert the entered program data *after* the current position of the cursor.

Alter key. This program editing key allows data in a program to be altered. After positioning the cursor to the incorrect word in the CNC program, the operator can enter the new word and press the alter key. This will cause the current data in the program to be changed to the entered data.

Delete key. This program editing key allows program data to be deleted. Most CNC controls allow a word, a command, a series of commands, or even an entire program to be deleted with this key.

Reset button

This *very important button* serves three basic functions:

- While you are editing CNC programs, this key will return the program to the beginning.
- While the control is executing programs, this key will clear the look-ahead buffer and stop execution of the program. This is required when it is determined that there is something wrong in the program and you wish to stop execution of the program. However, it can be dangerous to press this key during a program's execution in other cases. If the program is executed immediately after the reset key is pressed, the control will have forgotten the commands in the look-ahead buffer. In effect, this will cause several commands to be skipped. The control will pick up and continue running, but severe problems could arise because of the missing commands. When in automatic operation, be careful with this key.
- When the machine is in alarm state, this key will cancel the alarm if the problem causing the alarm has been solved.

Other keys possibly on the control panel

The above keys are found on most CNC controls and named as stated. There may be other keys, depending on the control manufacturer and style of control. Also, these keys are not named consistently from one manufacturer to the next.

Input/output (I/O) keys. Almost all current CNC controls allow communication of programs, offsets, and parameters to outside devices such as computers and tape reader/punches. This means the control panel will include keys needed to send and retrieve data to and from these devices. Usually two keys are involved—one to send data and the other to receive data. The send key may be labeled as *output* or *punch*. The receive key may be labeled as *input* or *read*.

Graphic keys. On controls with graphic capabilities, there will be a series of control panel keys related to the control's graphic functions (scaling, plane display, rotation, etc.). Unfortunately, graphic functions vary dramatically from CNC control to control. Therefore, we cannot be specific about their usage.

Soft keys. More and more CNC controls are adopting soft keys in order to minimize the number of buttons and switches required on the control panel. Soft keys are mounted close to the display screen, usually directly below and directly to the right of the control screen. The function of these keys will change depending on the modes of the display screen. For example, in the offset mode of the display screen, the soft keys will have one meaning. In the program mode, the soft keys will have a totally different meaning. In *all* cases, there will be boxes close to each soft key telling the operator the *current* function of the soft key. The F1 through F8 keys directly under the display screen shown in Fig. 10.1 are soft keys.

When the control uses soft keys, many of the previously described keys (program, position, offset, etc.) will be replaced with soft keys on the control's display screen.

Common Buttons and Switches Found on the Machine Panel

Now let's look at common buttons and switches found on the machine panel of a typical CNC machine tool.

Mode switch

The mode switch is the heart of a CNC machine. It should be the very first switch an operator checks before performing any function on the machine. In many cases, the mode switch must be positioned properly and according to the operation to be performed. If it is not, the control will not respond to the operator's command.

You can think of the mode switch on a CNC machine as like the function selector for a high-fidelity stereo sound system. Most current stereo systems allow you to select from tuner, CD player, phonograph, and cassette tape player. Before you can activate any of the sound generating devices, the function selector of the stereo system must be positioned accordingly. If it is not, the desired device cannot be activated.

In a similar way, the mode switch of a CNC machine must be positioned correctly before any function can be activated. For example, if the operator wishes to make a manual movement by jogging the machine with a handwheel, the mode switch must be positioned in the manual mode. If the mode switch is not positioned properly, no manual movement will be allowed.

It is nice to know that there is no chance of damage to the machine or control if the mode switch is not positioned correctly. If the mode switch is not in the correct position, the worst that will happen is that the machine will not respond to the operator's command.

Edit. The edit mode allows an operator to enter and modify CNC programs through the keyboard and display screen in much the same way a word processor is used on a personal computer. The edit mode is also used to scan within the active program to a position at which the cycle is to be started. For example, the operator may wish to skip to the beginning of the third tool and execute the program from that point. The edit mode is used to scan to the beginning of the third tool.

Memory or auto. This is one the modes from which a program can be executed. When in this mode the operator is allowed to begin the automatic cycle, executing the active program from within the control's memory.

Tape. This mode is found on CNC machines that allow a program to be executed from the tape reader. This mode is similar to auto or memory, except the program is executed from a tape on the tape reader, not from within the control's memory.

With the advent of computers, tapes are becoming less and less common for CNC application. For all intents and purposes, the tape reader on today's CNC machines is used only to load a program into the control's memory. Then the operator will execute the program from the auto or memory mode.

There are only two times when the tape mode is used to any extent on today's CNC machines. Both have to do with programs that are too long to fit into the control's memory. With lengthy programs that cannot be loaded into the control's memory, one alternative is to run the program from the tape reader in the tape mode.

However, tape readers tend to be quite cumbersome to work with and also have limitations related to program length. For this reason, many CNC control manufacturers allow the user to trick the control into thinking a program is running from the tape reader, when in reality the program is being executed from some outside device. For example, a personal computer can be connected through the communications port which allows a program to be transferred from the computer to the control. In this case, the operator would execute the program while in the tape mode, but the program would be actually run from the computer. With this arrangement, the control can run programs much longer than would fit into the control's memory. The term for this technique is *direct numerical control* (DNC).

Manual data input (MDI).   This mode switch position allows the operator to make programlike commands manually, through the keyboard and display screen. CNC machine and control manufacturers do their best to place sufficient buttons and switches on the machine to allow easy operation. However, given the almost limitless possibilities with a CNC machine, it is next to impossible to give a button or switch for every machine function. For this reason, control manufacturers provide a way for the operator to make CNC commands manually, in the manual data input mode.

For example, most machining centers do not have any one single button or switch to activate the automatic tool changer. Instead they require that the operator give an MDI command in order to make manual tool changes, involving the T word and M06.

Almost anything that can be commanded in a CNC program is possible in MDI. After becoming comfortable with the MDI mode, an operator can make MDI commands quickly and easily, and may prefer to do so, even if the machine has a manual button or switch to activate the desired function.

Note that in the MDI mode, commands are one-shot. Unlike a CNC program in the control's memory, once an MDI command has been executed, it is forgotten. If the command must be repeated, it must be entered again.

Manual or jog.   In this mode-switch position, the CNC machine behaves most like a manual machine tool. This mode activates many of the machine panel's buttons and switches related to machine functions. For example, most CNC machines incorporate some kind of handwheel used to manually move each axis. Most CNC machines also have a jog function, allowing axis motion to be caused by a joystick or push button. Machines with a rotating spindle usually have push buttons to turn the spindle on and off, as well as a rheostat to control spindle speed.

All of these functions are activated manually, through the manual or jog mode.

Reference return. Most CNC machines have a manual method of returning each axis to its reference point. To do this, the *reference return* mode must be selected (also called *grid zero, home position,* or *zero return*). With some controls, this procedure is as simple as selecting the reference return mode and pressing a button. Other CNC machines require that the operator jog each axis to its reference point independently.

Cycle start

This button has two functions. First, it is used to activate the active program in the control's memory, causing the machine to go into automatic cycle. Second, most CNC controls use the cycle start button to activate manual data input commands.

Feed hold

While a program or MDI command is being executed, this button allows the operator to halt axis motion temporarily. The cycle start button can be used to reactivate the cycle. Note that all other functions of the machine (coolant, spindle, etc.) will continue to operate.

You should think of this button as your panic button. If you are verifying a program, you should *always* have a finger on this button, ready to push it at any time. If you suspect any mishap, press the feed hold button to pause the motion. Then check for mistakes. If a mistake is found, you will take the program out of cycle (by pressing the reset key), fix the problem, and start over. If a mistake is not found, you can continue running the program by pressing cycle start.

Some people feel the emergency stop button should be considered the panic button. However, the emergency stop button will actually turn the power off to the machine tool, which sometimes causes more problems than it solves. For example, when emergency stop is pressed and the machine power is turned off, the axes of the machine will drift until a mechanical brake can lock them in position. Axes bearing a great deal of weight are the most prone to drift. While the amount of drift is usually quite small (under 0.010 in in most cases), if a cutting tool is actually machining a workpiece when the emergency stop button is pressed, the drift could cause damage to the tool and workpiece.

Here is how you determine which button to use as your panic button. If you are cautiously verifying a program, standing close to the machine with both hands on the control panel, feed hold is your panic button, and should be pressed in case of mishap (as discussed above).

However, once a program is verified and production is being run, there is little need for a CNC operator to monitor every movement the machine is making. Usually a CNC operator will let the machine run by itself and perform some other task for the company. If this is the case, and you are away from the machine when you hear the machine making a terrible sound, you know something is wrong with the cycle (maybe a carbide tool insert is damaged on the cutting tool). In this case, time is of the essence since the machine tool is in danger. Since you have no time to diagnose the problem and need to stop the machine cycle as quickly as possible, in this case emergency stop is your panic button.

Feed rate override

This multiposition switch allows the operator to change the programmed feed rate during cutting commands (G01, G02, G03, etc.). Notice we said *feed rate.* Under normal conditions, this switch has no control over *rapid* motion. The feed rate override switch is usually segmented in 10 percent increments and will usually range from 0 to 200 percent. This means the operator can slow down programmed feed rates to nothing, stopping motion, and increase the feed rate to double its programmed value.

This switch is most helpful during the verification of a new program. While the first workpiece is being machined, this switch can be used to manipulate the programmed feed rate. For example, say a cutting tool has been cautiously brought into its first cutting position under the influence of single block and maybe dry run modes. Now the cutting tool is ready to carry out its first machining command. Having no idea as to whether the cutting conditions are correct (feeds and speeds), we recommend that the operator turn the feed rate override switch down to its lowest value. With single block still on, but with dry run *off,* the operator can press cycle start to begin the first cutting command. With the feed rate override switch set at its lowest value, the machine will not move. With the command activated, the operator can *slowly* crank up the feed rate override switch, watching and listening to what the cutting tool is doing. If everything appears to be all right, the operator brings the feed rate override switch to 100 percent and continues to monitor each command. However, if something is wrong with the cutting conditions and the operator cannot run at 100 percent, the programmed feed rate must be changed. This procedure must be repeated for every tool in the program.

A feed rate override setting of 100 percent is the target for every tool. Once a program is verified, *all* tools *must* run properly at 100 percent to allow automatic operation. If problems are found during program verification with the programmed feed rates, the operator *must* note this and eventually change the program before it can be considered verified.

Rapid traverse override

Rapid traverse override is used to slow the rapid motion rate. It comes in two different forms. In one form, it is a simple on-off switch. When on, all rapid motion is slowed to 25 percent of the normal rapid rate. While 25 percent of rapid is still quite fast, this switch can be left on during program verification (and whenever the operator is a little nervous about the cycle) to assure that the machine will not be allowed to move at its normal rapid rate.

In its second and more useful form, rapid override is a four-position switch and can be adjusted to 5, 25, 50, and 100 percent of the normal rapid rate. This form of rapid override can be used during program verification to assure that rapid movements toward the workpiece are correct.

Emergency stop

This button will turn power off to the machine tool. Usually, power remains on to the control. See the description of feed hold for more information about how emergency stop is used.

Conditional switches

There are several on-off switches on the machine panel that control how the machine behaves during automatic and manual operation. They could be toggle switches, locking push buttons, or even switches set through the display screen and keyboard.

Though the location and style of these switches vary, their meaning and usage stays amazingly similar from one type of CNC machine to the next. These switches are *very* important. If one or another is improperly set, the machine may not perform as expected. The operator should get in the habit of checking each of these switches before the CNC program is executed.

Dry run. This conditional switch is most used with new programs during program verification. When this switch is on, it gives the operator control of the motion rate at which the machine will traverse. This is extremely helpful during rapid motions. The rapid rate of current CNC machines is very fast, ranging from 100 to 800 IPM. At these extremely fast rates, the operator will not be able to stop the machine in time in case of a mishap.

With a new program, the operator will have no idea as to whether some motion mistake has been made in the program. So, by turning on the dry run switch, the operator can take control of the machine's motion rate. Dry run *always* works in conjunction with some other multiposition switch (usually, either feed rate override or jog feed rate). This multiposition switch acts as a rheostat, allowing the operator to manipulate the rate at which the machine axes move.

By turning down the multiposition switch, the motion rate can be slowed. By turning up the multiposition switch, the motion rate can be increased. At the lowest settings of the multiposition switch, most CNC machines will barely creep along, allowing even rapid motions to be cautiously checked.

Dry run is most commonly used when no actual workpiece is in the work-holding device. That is, the operator will be allowing the machine to make the motions commanded in the program without actually machining a workpiece.

Note that dry run should never be turned on during machining. With most CNC machines, dry run will also manipulate the rate of cutting commands (G01, G02, and G03). This means the programmed feed rate will be incorrect if dry run is turned on (more on dry run during key concept no. 4 of operation).

Single block. While executing a program, this conditional switch can be used to force the control to execute one command of the program at a time. When turned on, the control will stop at the completion of each command. To reactivate the cycle (execute the next command), the operator must push the cycle start button.

This switch is most helpful during program verification. With a new program, the operator will cautiously check each motion the machine makes, one at a time. With single block in the on condition, the operator can rest assured that the machine will stop at the end of each motion, giving a chance to check the motion just made.

Machine lock. When turned on, this conditional switch keeps *all* axes of the machine from moving. Many other functions of the machine will continue to operate. For example, the automatic tool changer will still change tools, the spindle will still run, coolant will still come on, etc. But axis motion will *not* occur.

Machine lock is effective at all times, even during automatic operation and manual operation. If a motion of any kind is commanded, the position displays on the screen will act as if the machine is moving, but in reality, no motion actually occurs.

The most common application for this conditional switch is during program verification. The very first time a new program is executed, machine lock and dry run are turned on. The control will quickly scan the program for errors in basic program format. During this verification procedure, the operator can be sure the axes of the machine will *not* move. The control will run through the program as if running a workpiece, except that no axis motion occurs.

If the control finds a format mistake in the program, it will generate an alarm. The kind of mistakes that can be found during a machine lock dry run include syntax mistakes. For example, say the programmer intended to program the command

N005 G01 X5. Y5. F4.

but made a mistake while typing the G01 word. Suppose it was written as

N005 G10 X5. Y5. F4.

On most CNC controls, there is no such thing as a G10. And even if there were, its format would probably not be that of G01. In this case, when the control came across G10, it would generate an alarm. After correcting the mistake, the operator would run the program again (with machine lock on) and continue to do so until the entire program can be executed without the control generating an alarm.

When the control finishes executing the program without generating an alarm, the operator will know that at least the control can interpret all commands in the program. There could still be motion mistakes, so the operator must still be cautious even after doing a machine lock dry run. But at least the basic format of the program is correct.

Optional block skip. This conditional switch works in conjunction with slash codes (/) in the program. If the control reads a slash code at the beginning of any CNC command in the program, it will look to the position of the optional block skip switch (also called *block delete*). If the switch is on, the control will ignore the command in which the slash code is included. If the optional block skip switch is off, the control will execute the command. (See key concept no. 6 of manual programming to learn more about optional block skip applications.)

Optional stop. This conditional switch works in conjunction with an M01 code in the program. When the control reads an M01, it looks to the position of the optional stop switch. If the switch is on, the control will halt the execution of the program. The operator must press the cycle start button to reactivate the program. If the optional stop switch is off, the control will ignore the M01 and continue executing the program.

The M01 is commonly used in the program at the completion of each tool. During the program's verification, these M01 commands give the operator the ability to stop after each tool and check what the tool has done.

Manual controls

The machine panel for all CNC machines will include several buttons and switches related to manual control of the machine tool's functions. Depending on application, these buttons and switches vary dramatically from one builder to the next. While we describe the most common ones, you must check the machine tool builder's manual for more specific information.

Manual axis motion. Most CNC machines have two ways to cause manual axis motion. One incorporates a handwheel that is very similar to the handwheels on a manual machine tool. The other incorporates a joystick or a series of push buttons used to jog the axis at a consistent rate.

Handwheel controls. The handwheel of a CNC machine can be used to move any of the machine's axes, one at a time. Even though the CNC machine may have up to five axes, only one handwheel is used. This is accomplished by a switch, usually right on or close to the handwheel, that is used to select the axis to be moved (X, Y, Z, etc.).

There is another switch related to the handwheel used to select the rate at which the motion will occur. This switch usually has three positions. In the × 1 (times 1) position, each increment of the handwheel is 0.0001 in (or 0.001 mm). In the × 10 (times 10) mode, each increment is 0.001 in (or 0.010 mm). In the × 100 (times 100) position, each increment is 0.010 in (or 0.100 mm). These three multipliers for the handwheel allow the operator to select how fast the motion will be.

The handwheel can be used any time the operator wishes to make manual axis motion. This includes making measurements on the machine during setup and actually machining a part manually. For example, say the operator wishes to "touch off" a surface of the workpiece with an edge finder. First, the operator would select the × 100 mode and quickly move the edge finder close to the surface to be touched in each axis. Then the operator would select the × 10 mode to bring the edge finder closer to the workpiece. But either of the two previous modes is too crude for the actual touch-off. Last, the operator would select the × 1 mode and cautiously touch the edge finder to the workpiece.

One last note about the handwheel. If you intend to use the handwheel to actually machine a workpiece, it can be difficult to rotate the handwheel at a constant rate; the feed rate for machining will not be consistent. For this reason, it may be better to use the jog technique for machining a workpiece manually.

Jog controls. *Jogging* an axis is making a manual axis movement at a consistent rate. There are two common techniques used by CNC machine tool builders that allow axes to be jogged. One involves a joystick that is very similar to the joystick used with computer games. The other involves an axis selector switch and a series of push buttons.

With both methods, there is also another multiposition switch called *jog feed rate* that controls the rate at which the jog motion will occur. Normally this switch is marked with inches per minute values as well as millimeters per minute values. The range of motion rate usually runs from 0.25 IPM all the way up to the machine's rapid rate, so that the operator has complete control of the desired rate. Before making the jog motion, the operator must set the jog feed rate switch to the desired position.

The joystick technique is usually applied to two-axis machines, such as turning centers. It will be marked with the axis directions the joystick will cause. These directions match the basic layout of the machine's axis motion. For example, for turning center applications *up* on the joystick is X plus. *Left* on the joystick is Z minus. *Down* on the joystick is X minus. *Right* on the joystick is Z plus. When you think about it, each direction of the joystick matches the actual direction of motion the machine will make.

For machines with more than two axes, most machine tool builders will use two push buttons to allow the direction of motion (plus or minus) along with an *axis selector switch* to designate in which axis the manual motion is to occur. For example, an operator who wishes manual motion along the X axis in the minus direction at 40 IPM will first set the jog feed rate selector switch to the 40-IPM position. The operator then will set the axis selector switch to the X position. Last, the operator will press the minus push button.

Since the motion rate is consistent in the jog mode, this makes a better choice for actually machining a workpiece manually than the handwheel.

One last note about jog techniques. Most CNC machines that utilize a reference position (grid zero, home, or zero return) require the operator to use jogging techniques to manually send each axis to its reference position.

Tool changing controls. Some CNC machines with an automatic tool changing device allow manual changing of tools. Usually this involves a multiposition switch with which the operator can select the desired tool station and a push button that activates tool change.

Spindle control. For CNC machines that have a rotating spindle, most allow a manual way of turning the spindle on and off. Usually push

buttons are used, one for spindle on and another for spindle off. If the
machine allows manual control of the spindle, there will also be a
rheostat that can be used to adjust the spindle speed in revolutions per
minute. If the rheostat is not available, the spindle RPM must be commanded by the manual data input mode.

Indicator lights and meters

Most CNC machines have a series of lights and meters that allow the
operator to quickly check the condition of important functions of the
machine. Consider these indicators as very similar to the gauges and
lights on the dashboard of your car.

Spindle horsepower meter. For CNC machines that have spindles,
most have two meters that show the operator key information about
the spindle. The first meter is the RPM meter, showing the operator
how fast the spindle is currently rotating. Note that some machines
show this information through the display screen instead of by an actual meter. The second spindle-related meter is a load meter. If the
machine tool has this meter, the operator can monitor how much
stress the spindle is under. Usually this meter shows a percentage of
load, ranging from 0 to 150 percent. This means the operator can easily tell to what extent a machining operation being performed is tax-
ing the spindle of the machine.

Axis drive horsepower meter. This meter allows the operator to check
how much horsepower is being drawn by any of the machine axis drive
motors. Usually there is only one meter and the operator must select
the axis (*X, Y, Z,* etc.) to be monitored by a multiposition switch.
Like the spindle horsepower meter, this meter allows the operator
to see how much stress a drive motor is under during a machining
operation.

Cycle indicator lights. Most CNC machines have two indicator lights
to show whether the machine is in cycle. One is above the cycle start
button and comes on if the machine is currently running a program.
The other is above the feed hold button and comes on when the machine is in the feed hold state.

Reference position indicator lights. Many CNC machines have a set of
indicator lights that come on if an axis is currently at its reference
point. There are as many reference position indicator lights as there
are axes on the machine. Many CNC machines are designed to be activated from their reference point. If this is the case, the operator can

easily see if the machine is in its proper starting position by checking these lights.

Optional stop indicator light.   Some CNC machines have an indicator light close to the optional stop switch. If the machine is halted by an optional stop (M01), this indicator light comes on to help tell the operator why the machine has halted operation.

Handwheel indicator light.   Most CNC machines with handwheels have an indicator light close to the handwheel that comes on when the handwheel is activated.

Other buttons and switches on the machine panel

By no means are these all the buttons and switches found on the typical CNC machine. CNC machines vary dramatically in this regard. Each machine-tool builder will decide which buttons and switches are necessary to operate its particular CNC machine. Some make operation more convenient than others. While the buttons and switches presented so far can be found on most CNC machines, you will probably find some on any one CNC machine that have not been mentioned. You *must* check the operator's manual for the machine tool to find the function of unknown buttons and switches. Or ask someone who has experience with the machine tool to explain them to you. Though some buttons and switches are seldom used, the proficient operator must know the function of every button and switch on the machine tool.

11

Key Concept No. 2: The Three Modes of Machine Operation

In key concept no. 1 of operation, we discussed the mode switch found on the machine panel. We said this switch is the heart of the CNC machine and that the operator *must* first look at this switch before any function can be performed. The mode switch must be in the proper position before the operator will be allowed to perform any desired operation. The most common mistake on the operator's part is having the mode switch in the wrong position. Fortunately, this mistake will usually not cause problems. The machine will simply not perform the desired function if the mode switch is in the wrong position.

In key concept no. 1 of operation, you found that the mode switch has several positions. However, in key concept no. 2, you will find that there are only three basic modes of operation. It should be refreshing for beginners to know that all buttons and switches on the machine and control panel can be divided into three basic modes.

Here we will introduce the three modes and give examples of when they are used. We will also discuss more about the mode switch, categorizing each position of the mode switch into one of the three basic modes.

The Manual Mode

In the manual mode, the CNC machine behaves most like the manual machine tool the CNC machine tool is replacing. The positions of the mode switch that are included in manual modes include manual or jog, handwheel, and reference return.

With the manual mode, the operator of a CNC machine is allowed to press buttons, turn handwheels, and activate switches in order to attain the desired machine function. The activation of each button or switch in the manual mode has an immediate response. For example, if the correct button is pressed, the spindle will start. If a switch is turned on, the coolant will come on. If a joystick is held in one direction or another, the corresponding machine axis will move.

CNC machines vary dramatically with regard to what they allow an operator to do manually. Also, the machine tool builder will supply buttons and switches for what it considers the most important functions of the machine. Note that those things that cannot be done manually in the manual mode must be done in the manual data input mode, the second mode of operation. Here is a detailed discussion of a common machine function that involves heavy use of the manual mode.

Measuring the distance from the program zero point to the machine's reference position

The setup stage of most CNC machines involves making a measurement in each axis to determine the location of program zero relative to the machine's reference point. As discussed during key concept no. 1 of programming, the control must know the location of the program zero point in order to make axis motions correctly during the program. Making this measurement in each axis involves using the manual mode.

In order to understand this procedure, you must also know how the axis position page of the display screen functions. On the position page of the display screen, there is a digital readout showing the current position for each axis (X, Y, Z, etc.). This position display page (example shown in Figure 11.1) works very much like a digital readout on a manual machine tool. As an axis moves, the position display for that axis follows along, indicating the current axis position. The position

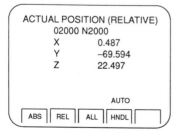

Figure 11.1 Position display screen. (*Courtesy Fanuc U.S.A. Corp.*)

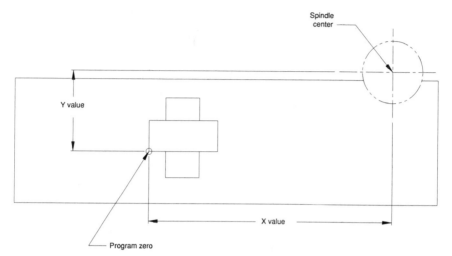

Figure 11.2   Relationship of the program zero point to the machine's reference point in X and Y.

display for each axis can be preset to any number (including zero) at any time. This presetting technique allows the operator to *calibrate* each axis position to correspond with the setup being made. Here is how. Although this example shows a vertical machining center, the technique is common to all forms of CNC equipment.

Suppose an operator had to measure the distance in X and Y from the program zero point on the workpiece to the machine's reference position on a vertical machining center. Figure 11.2 shows a drawing of the setup and the location of the program zero point relative to the machine's reference position.

Note that the program zero point is located at the lower left-hand corner of the workpiece in X and Y (X minus and Y minus corner). The first step of this procedure requires placing an edge finder in the spindle. An edge finder allows the operator to see the alignment of the spindle with the surface being touched. The spindle is then started manually. To start the spindle, the operator selects a manual mode on the mode switch (manual, jog, or handwheel), selects the desired RPM with the rheostat, and presses the button corresponding to spindle start in the forward direction. At this point the spindle comes on.

Next the operator needs to jog the X and Y axes manually, bringing the edge finder to the left (minus side in X) of the X program zero point. To do this, the operator selects the jog mode, places the axis select switch to X, selects a fast motion rate on the jog feed rate switch, and presses the button corresponding to the minus direction. The operator holds the minus button until the X axis moves far enough that the edge finder is to the left of the program zero point. This step is repeated for the Y axis, positioning the Y axis just below (to the Y minus side) the program zero point in Y. Finally, the Z axis is brought down just below program zero in the same manner.

Next, the operator selects the handwheel mode and cautiously moves the X axis until the edge finder is perfectly aligned with the surface of the workpiece in X. To do this, the operator selects the handwheel mode, places the axis select switch for the handwheel to X, and selects (at first) a fast rate of motion for the handwheel ($\times$ 100). At this fast rate, the operator brings the edge finder close to the surface to be touched in X. Once the edge finder is close, the operator selects a slower rate ($\times$ 10) for the handwheel. This technique allows the operator to "sneak up" on the surface. With this slower rate the operator brings the handwheel yet closer to the surface to be touched. When the edge finder is very close to the surface to be touched, the operator selects the slowest motion rate for the handwheel ($\times$ 1). With this very slow rate, the operator will cautiously move the edge finder until it is perfectly aligned with the surface to be touched. Now the center of the spindle is exactly the radius of the edge finder away from the program zero point in X.

At this point, the operator must calibrate the position displays, letting the control know how far it is from program zero to the center of the spindle. The operator now knows this distance. (It is the radius of the edge finder.) The operator now presets the axis display for the X axis. Note that the desired value for the axis display is the distance *from* the program zero point *to* the current centerline position of the spindle. In this case the direction is *minus,* since the edge finder is on the minus side of program zero in X. Suppose the edge finder had a 0.200-in diameter. The value to be preset in the X-axis display would be -0.100 (note minus). Once this value was preset in the X-axis display, the X-axis display would be calibrated with the actual position of the X axis relative to program zero in X.

From this point, as the X axis is moved, the position display will follow along and constantly show the distance from the program zero point to the current position of the spindle centerline in X.

This whole procedure is then repeated for the Y axis. Once the Y-axis touch-off has been made and the Y-axis position display properly calibrated, the operator manually turns the spindle off by pressing the spindle off button.

With X and Y calibrated, the operator simply moves the X and Y axes to their reference point. The X- and Y-axis displays will follow along. When the machine is resting at the reference point in X and Y, the X- and Y-axes displays will show the distances from program zero to the reference point in X and Y. These values will be needed to correctly assign the program zero point (either in the program or with fixture offsets).

While the above procedure may sound a little complicated at first, this kind of measuring is done often. A proficient operator can perform this procedure in about 10 minutes.

While measuring the program zero point is only one time when the manual mode is required, it stresses the importance of this mode, and the operator of any CNC machine will constantly have need for the manual mode.

During key concept no. 3 of operation, you will find that operating a CNC machine is little more than following a series of basic procedures. The above procedure to measure the distance from program zero to the reference point is but one such procedure. In key concept no. 3, you will be exposed to many more procedures that involve the manual mode.

The Manual Data Input Mode

This mode includes two positions on the mode switch—the edit position and the manual data input position. With both positions, the operator will be entering data through the keyboard on the control panel and display screen.

Though these two mode switch positions have substantial differences, we consider them together for two reasons. First, both provide manual capabilities that can be done in a more automatic way. With the edit mode switch position, an operator can enter CNC programs into the control's memory. This can also be accomplished by loading the program from some outside device, such as a computer or tape reader. With the MDI mode, CNC commands are entered through the keyboard and display screen manually and can be executed once. If the command must be executed a second time, the operator must enter the MDI command a second time. However, if the same command is included in a program, it can be executed automatically, over and over, without having to be retyped.

Second, both mode switch positions involve entering data through the keyboard. With the edit position of the mode switch, a program is entered or modified. With the MDI position of the mode switch, CNC commands are entered and executed.

The manual data input position of the mode switch

Manual data input is used to enter and execute CNC commands, one at a time. Once the CNC command has been executed, it is forgotten.

If the same command must be executed again, it must be entered again. Almost anything that can be done within a CNC program with CNC commands can be done in the MDI mode. An operator who is proficient with CNC commands often can perform functions more quickly with MDI than by other manual means. For example, CNC machines that utilize a reference point (home position, grid zero, or zero return) usually provide a manual means of returning each axis of the machine to the reference point (using the reference return position of the mode switch). This usually requires each axis to be returned separately. However, if the operator knows the CNC command used within programs to return all axes to the reference point, the command can be entered and executed through the MDI mode. Usually the machine will perform this function more quickly in MDI than by completely manual means.

The most important reason for using the MDI mode is to allow the operator to perform manual operations that are not possible with using buttons and switches. For example, some CNC machines that have rotating spindles have no manual buttons or switches to control the spindle (select the spindle speed, direction, and turn the spindle on and off). With this type of machine, an operator who needs to manually turn the spindle on for any reason must use MDI to do so.

For machining centers, the same applies to the automatic tool changer. Most machine-tool builders provide no way of manually changing tools on machining centers. The operator is expected to use the MDI mode to do so. Of course, this means the operator must understand the CNC commands related to the automatic tool changer (usually the T word and M06).

The manual data input mode switch position can even be used for machining a workpiece. Since almost all CNC commands are possible by MDI (including G00, G01, G02, and G03), the operator can make machining commands in the same way as in a CNC program. However, the operator must be *very careful* when using MDI to machine workpieces. The command will be executed as entered. If the operator makes a mistake while entering a CNC command in the MDI mode, disaster could result. There is no chance to verify the CNC command executed by MDI, as can be done with a CNC program.

More on the edit position of the mode switch

The edit position of the mode switch can be compared to a word processor on a personal computer. If you have ever worked with a word processor, you will find the use of the edit mode of a CNC control familiar and easy to work with.

In this mode switch position, the operator is allowed to do two basic things. They can enter CNC programs into the control's memory and the operator can modify current programs. Most CNC controls allow the operator to store multiple programs. Depending on the control manufacturer, the programs can be organized by program number or file name. By one means or another the operator will be allowed to call up the desired program from within the control's memory, making it the active program.

The actual features of the edit mode depend on the control manufacturer. All will give the user the ability to do at least three basic things. The operator will be allowed to INSERT new information into the program, ALTER the current information in the program, and DELETE information within the program. Some controls also allow the operator to do mass editing, meaning that the operator can "cut and paste" and "find and replace" data, making global changes in the entire program.

Along with these basic features, the operator will be allowed to search or scan the program for key information. For example, if the program commands for the sixth tool in the program are wrong, and the operator wants to change something in the sixth tool's commands, the operator can command that the control quickly search to the next occurrence of any programming word. If the programmer used sequence numbers (N words), the operator can use the sequence number of the incorrect command as the word to be searched for.

Entering a new program. The actual techniques to enter a new program through the keyboard and display screen will vary with the control manufacturer. Most require that the operator first name the program with a program number or a file name. Once the program is named, the operator simply types the program, ending each command with an end-of-block symbol. The end-of-block signal is usually given with a special key on the keyboard, close to the number and letter keys.

If the operator makes a mistake while typing a command, a backspace key on most controls lets the operator remove the last character entered to try again.

After each command is entered, the operator presses the INSERT key to have the command entered as part of the program.

Once the entire program is entered, some controls require that the program be saved internal to the control. Possibly the control uses a device similar to a floppy-diskette drive to which programs must be saved. You must check the control's operation manual for specifics on program saving.

Generally speaking, the CNC control makes a *very* expensive typewriter. For many CNC controls, the machine cannot be running workpieces while a new program is being entered. Even for those controls that do allow programs to be entered while production is being run, it is a cumbersome procedure. The control panel is usually mounted in a vertical attitude, and the operator's arm will become quite fatigued during the entry of a lengthy program.

For these reasons, most companies that utilize CNC equipment will prepare programs off-line, away from the machine tool. Currently, most use a personal computer into which they type their programs. The program is saved on the computer's floppy disk or hard disk. A printer listing (hard copy) of the program can also be printed on the line printer. When the CNC machine requires the program, it can be quickly transmitted from the computer to the CNC machine through a cable connected to the computer's serial port and the CNC control's communications port.

During verification, there will almost always be changes to the CNC program. After the changes have been made, the program can be sent back to the computer for permanent storage at the computer.

Modifying a previously entered program.   Once a new program has been entered into the CNC control's memory, it must be cautiously verified. It is very rare indeed for a manually prepared program to run perfectly without requiring modification.

The editing keys ALTER, INSERT, and DELETE allow the operator to modify anything about the program. To change words in the program, the operator positions the cursor to the position in the program to be modified, types in the modification, and presses the ALTER key. To add information to the program, the operator positions the cursor to the word just before the addition, types the additional words and presses the INSERT key. To delete words from the program, the operator positions the cursor to the word to be deleted and presses the DELETE key.

While these techniques vary from control to control, the basic idea remains the same. The operator of a CNC control is allowed to manipulate the program from within the control's memory. However, if changes are made at the control and if the company wishes to maintain a corrected copy of the program, after any modification, the program must be sent from the control to an outside device (tape punch or computer) for permanent storage.

Note that some controls have a key switch called *memory protect*. This switch is designed to prevent accidental modification to the program. It involves a key (similar to a house key) that can be removed and taken away from the CNC machine once memory protect has been turned on. Once a program has been verified and production is being run, there will be no need to modify the program further. At this time the memory protect switch can be turned off and the key removed, preventing the operator from modifying the program.

The Program Operation Mode

The third and last mode of operation involves actually running programs. The two positions of the mode switch are memory or auto and tape. In this operation mode, the control is actually executing programs. The operator uses this mode to verify programs and run production.

In both mode switch positions, the cycle start button is used to activate the program and the feed hold button can be used to stop axis motion temporarily at any time during the cycle.

Note that several conditional switches (discussed in key concept no. 1 of operation) manipulate how the machine behaves in the program operation mode. The dry run conditional switch gives the operator control of the motion rate. Single block forces the control to execute only one command at a time. Optional stop (when on) will cause the control to halt the program when an M01 word is read. Optional block skip (when on) will cause the control to skip commands beginning with a slash code (/). Machine lock will cause the control to execute the program, but keep all axes from moving. (See key concept no. 1 of operation for more information on these conditional switches.)

When executing a program in the program operation mode, most CNC controls will look ahead several commands into the program in order to make decisions about the current command being executed. This feature is needed to keep the control from pausing between commands and is also required for compensation-related decisions. It is important to know that this look-ahead buffer is constantly being filled during the execution of a program. This buffer is emptied when the reset key is pressed. The operator *must* know that if reset is pressed while a program is executing, the program's execution is canceled. That is, immediately pressing the cycle start button after the reset key has been pressed can cause problems, since several commands of the program will have been lost. To restart the program after reset, the operator must follow the proper procedure (given in key concept no. 3 of operation).

The memory (or auto) mode switch position

The memory (or auto) mode switch position is the position most used on current CNC machines to execute programs. As long as the control's memory is of sufficient size to hold the CNC program, the memory (or auto) position of the mode switch should be used to execute the program.

In this mode switch position, the control executes the active program from within the control's memory. While there could be several programs stored in the control's memory, only one is active. This is the program that will be run when the cycle start button is pressed. On most controls, the edit mode is used to choose the active program.

With most CNC controls, while a program is being executed from the memory (or auto) mode switch position, the programmer is allowed to see one full page of the program on the control's display screen. As the program is executed, the cursor will scroll through the program, letting the programmer see the commands that follow the command currently being executed.

The tape mode switch position

At one time, the tape mode was the only way to activate programs. But with the advent of the computer, the need for the tape mode switch position has decreased. In fact, many control manufacturers do not even equip their controls with the ability to run programs from the tape reader. While they may equip the control with tape readers, the sole purpose of the tape reader is to load programs into the control's memory.

For those controls that still allow programs to be executed from the tape mode, there are only two occasions when this is helpful. Both involve programs that are too long to fit into the control's memory. Some CNC controls have relatively small memory capacity. For those programs that are too long to fit into the control's memory, a tape can be run from the tape reader. However, the tape reader also has a limitation related to the length of the program.

For programs that are extremely long (even too long to be run from the tape reader), most CNC control manufacturers allow programs to be run from an outside device such as a personal computer. With this technique, the control is fooled into thinking it is running a program from the tape reader, but in reality it is running a program being transmitted from the computer through the control's communications port. To accomplish this, the mode switch position is set to tape while the program is executed. But the control's parameters are set in a way that tell the control not to run from the tape reader but to run from the communications port.

This allows programs of immense length to be run. The main application for this technique is three-dimensional-sculpturing CNC programs generated by CAM systems. With these programs, very tiny motions are generated to machine elaborate contoured shapes. Generally, these programs are extremely long and cannot fit into any CNC control's memory. *Direct numerical control* is the term used to describe this technique to run programs from a personal computer.

The tape position of the mode switch has two severe limitations compared to the memory (or auto) position of the mode switch. First, in the tape position, the operator will be allowed to see only one or two commands of the program being executed on the display screen. With the memory position of the mode switch, the operator will be allowed to monitor a whole page of the program. This is most helpful during the program's verification, when the operator wishes to check the commands coming up.

The second severe limitation of the tape mode is that modification of the program is *not* possible from within the CNC control. This means the program *must be perfect* before it can be run. If changes must be made, they must be made away from the CNC control. When programs are stored in the control's memory, it is easy to make changes to the program right at the control.

12

Key Concept No. 3: The Key Sequences of Machine Operation

Experienced CNC operators would agree that operating any CNC machine tool is little more than following a series of basic procedures. From powering the machine up in the morning to turning it off at night, the operator will be following a series of basic sequences in a step-by-step manner.

From the beginning operator's standpoint, half the battle of learning how to operate a CNC machine of any kind is simply knowing *when* each procedure is required. With this known, it is a simple matter to follow the basic sequence to attain the machine's desired function.

For example, if an operator wishes to load a program into the control's memory from a computer, there is a series of switches and buttons that must be activated in the proper order. If the operator wishes to change an offset, a specific procedure must be followed. While the actual buttons and switches will vary with the machine and control, once the operator has documented the most important sequences, it is relatively easy to make the machine function in the desired manner.

Figure 12.1 shows an example of a blank form that can be used to help you develop each key sequence of operation for the particular CNC machine tool you will be working with. As you can see, there is sufficient space to include all information about the desired procedure.

The Most Important Sequences

Each CNC machine tool will have those most-often-used sequences that the beginning operator must become familiar with. Procedures like powering up, powering down, loading tools, setting offsets, and

```
Machine: _____                    Page: __
   Procedure: _____

   Description: _____
              _____
              _____
              _____
              _____
              _____

Procedure:
        1. _____
        2. _____
        3. _____
        4. _____
        5. _____
        6. _____
        7. _____
        8. _____
        9. _____
       10. _____
       11. _____
       12. _____
   Notes: _____
          _____
          _____
          _____
```

Figure 12.1 Sample sequence of operations form you can use to document procedures required for your CNC equipment.

editing programs are among the things an operator will be doing on a regular basis and should strive to memorize. However, there are also procedures that are less often used that should also be documented. In this key concept, we will give you the most important procedures for three types of CNC machine tools, the machining center, the turning center, and the wire EDM machine.

Note that each series of procedures is for one popular control type related to each type of machine, but will vary dramatically from one control builder to the next. Here we are stressing the various important procedures needed by a beginning operator more so than the procedures themselves. While we do give specific techniques for each procedure, keep in mind that this information should be used only as a model for your own key sequences. That is, the actual buttons and switches, as well as the sequence of the procedure itself, will vary from one CNC control and machine tool to the next, but the needed procedures will not. As you begin to operate any CNC machine that is new to you, we strongly recommend using the blank form in Fig. 12.1 to help you develop your own series of procedures. While you may need the help of an experienced person to fill in these forms, once this is done, operating the machine and control becomes much easier.

As you will see, the series of procedures we give can be thought of as an operation handbook—quick and easy reference material about operating your CNC machine. While it would be nice if each CNC machine-tool builder incorporated this kind of handbook for every machine it sells, few builders do. For most CNC machines, you will have to develop this information on your own.

You will see that the basic sequences are divided into logical categories such as:

1. Manual sequences

2. MDI sequences

3. Program loading and saving sequences

4. Program editing and display sequences

5. Setup sequences

6. Program operation sequences

This should make it very easy to find the desired sequence.

Machining Center Sequences

This series of sequences works for a popular machining center control. While the specific techniques will vary from one machining center to the next, this should give you a good idea of the kind of operation handbook you should make for yourself. Note how specific each procedure is. A beginning operator could easily follow each procedure and make the machine do the desired function.

Manual sequences.

Sequence to start machine

Sequence to do a manual reference return

Sequence to manually start spindle

Sequence to manually jog axes

Sequence to use the handwheel to cause axis motion

Sequence to manually load tools into spindle

Sequence to manually load tools into magazine

Sequence to manually turn on coolant

Sequence to make axis displays read zero or any number
Sequence to enter tool offsets (length and radius)
Sequence to manually turn on mirror image
Sequence to manually select inch or metric mode

Manual data input sequences.

Sequence to use MDI to change tools
Sequence to use MDI to turn on spindle
Sequence to use MDI to do a reference return
Sequence to use MDI to move axes

Program loading and saving sequences.

Sequence to load programs into memory by tape
Sequence to load programs into memory by communications port
Sequence to load programs into memory through keyboard
Sequence to load punch programs from memory to tape punch or computer

Program editing and display sequences.

Sequence to display a directory of the programs in memory
Sequence to delete a whole program from memory
Sequence to search other programs in memory
Sequence to search words inside a program
Sequence to alter words in a program
Sequence to delete words and commands in a program
Sequence to insert words and commands in a program

Setup sequences.

Sequence to measure program zero positions in X, Y, and Z
Sequence to measure tool lengths

Program running sequences.

Sequence to verify programs
Sequence to run verified programs in production

Sequence to run from the beginning of any tool (requires that the program format be as taught in the programming key concepts!)

Sequences

Now we present the actual sequences for machining center operation. Keep in mind that this is just an example of the kinds of things you will need for your particular type of machine. While we give specific techniques, you will need to modify these techniques to suit the machine and control you are working with.

Sequence to start the machine.   The power-up sequence will vary from one machine to another. Here we show a common technique used by many manufacturers.

1. Turn on main breaker (usually located at rear of machine).
2. Press control power-on button.
3. Press machine ready or hydraulic on button. (Not necessary on some machines. Be sure that no emergency stop button is locked in.)
4. Follow sequence to do a manual reference return. (Many controls require that a manual reference return be done as part of machine start-up.)
5. Machine is now started.

Sequence to send the machine to its reference position.   The reference return procedure is often necessary. Many CNC machining centers require that the machine be at the reference position to begin executing a program. While there are some short cuts to this procedure, we recommend that the beginner follow the technique shown, for safety reasons.

1. Place mode switch to reference return.
2. Using the axis select switch, select the X axis.
3. Using the minus direction push button or joystick, move the machine about 2 to 3 in in the minus direction.
4. Using the plus direction push button or joystick, hold plus until the reference return origin indicator light for the X axis comes on.
5. Repeat steps 2 to 4 for all other axes. (Y, Z, and C/B axis if your machine has a fourth axis.)

The most common cause of a *crash* is pressing the cycle start button to activate a cycle when the machine is not where it should be. Almost all programs are planned to start from the reference return position, so the operator must always check to be sure that the machine is at this position before pressing cycle start.

Sequence to manually start the spindle.   Most machines have a way to manually start the spindle. However, usually you must at least select the RPM by the MDI mode, meaning that the sequence to start the spindle by MDI will usually be more useful.

1. Place the mode switch to a manual mode (jog, manual, reference return, etc.).
2. If the machine has a rheostat to adjust spindle RPM, set it to the desired position.
3. Press the spindle-on button. (Spindle comes on at the rheostat setting or last programmed RPM.)
4. To stop spindle, press the spindle-stop button.

Note: Most machines will also let you select the desired spindle direction.

Sequence to manually jog axes.   This technique will be used often. A few examples of when this technique would be used include making setups, moving the table to examine workpiece or tool, and actually manually cutting a workpiece.

1. Place the mode switch to manual or jog.
2. Place axis select switch to desired axis (X, Y, Z, or C/B).
3. Place the jog feed rate switch to the desired position to select the feed rate amount.
4. Using plus or minus push buttons or joystick, move the axis in the desired direction and amount.

Sequence to use the handwheel.   This is very helpful on machines that have a handwheel. It lets you move the machine a precise amount as you would on a manual milling machine. You can even machine workpieces using the handwheel as you would on a manual machine.

1. Place mode switch to handwheel.
2. Place axis select switch to desired position (X, Y, Z, or B/C).
3. Place handle rate switch to desired position ($\times$ 1, $\times$ 10, or $\times$ 100).

4. Using handwheel, rotate plus or minus to cause desired motion.

Note: On machines with the handwheel, × 1 = 0.0001 in per increment of the handwheel, × 10 = 0.001 in, and × 100 = 0.010 in).

Sequence to manually load tools into the spindle

1. Place mode switch to manual or jog.
2. If there is already a tool in the spindle, hold the tool with one hand and press the unclamp button with the other. The tool will drop out of the spindle and the spindle will be unclamped.
3. To load a tool into the spindle, place the tool into the spindle with one hand and press the clamp button with the other.

Note: On some machines it is important that the keys be lined up a certain way. If the tool does not seem to fit into the spindle, try rotating it 180° and try again.

Sequence to load tools into the tool changer magazine. This procedure will vary widely from one machine to another. If you are having problems here, please refer to the machine-tool builder's operation manual for more information.

1. Place the mode switch to manual or jog.
2. Go to the tool changer and rotate it to the desired position.
3. Load the tool into the desired position by unclamping a lever or simply snapping it into position.
4. Continue rotating magazine and loading tools for all tools to load.

Note: It may be *very* important to align the keys properly on the tool changer magazine. If you look at a tool holder you will notice that one key slot has a dimple and the other does not. We recommend always loading a tool into the magazine so that the key is always the same way. Also check with the builder's operation manual to find the correct alignment.

Sequence to manually turn on the coolant. Most machining centers allow the operator to activate the coolant manually. A simple toggle switch is used.

1. Place mode switch to manual or jog.
2. Place the coolant toggle switch to on (coolant comes on).
3. To turn the coolant off, turn off the toggle switch.

Note: Many machines will also have an auto position for the coolant switch. When the switch is in this position, the coolant will come on only when an M08 is commanded.

Sequence to make the axis displays read zero or any number. This technique is especially helpful for measuring tool length offsets and G92 positions. Two ways are shown—one for zeroing and the other to make the axis displays read any number.

To make axis displays read zero:

1. Mode switch can be in any position.
2. Press the extreme left soft key until *position* appears at the bottom of the screen.
3. Press the soft key under position.
4. Type the letter address of the axis you wish to make zero (X, Y, or Z).
5. Press the soft key under origin.

To make the axis displays read any number:

1. The mode switch can be in any position.
2. Press the extreme left soft key until *position* appears at the bottom of the screen.
3. Press the soft key under position.
4. Type the letter address for the axis you wish to set and type the value you wish.
5. Press the soft key under preset. (The value and axis you set will be displayed on the position page.)

Note: These techniques assume that the machine has been previously moved to the desired position.

Sequence to enter and change tool offsets. Before attempting this procedure, it is necessary to understand the function of tool offsets for tool length compensation and tool radius compensation as presented in the programming session.

1. The mode switch can be in any position.
2. Press the extreme left soft key until *offset* appears at the bottom of the screen.
3. Press the soft key under offset.

4. Using the arrow keys, position the cursor to the offset you would like to enter.

5. Type the value of the offset.

6. Press the soft key under input or the hard input key.

Sequence to manually turn on and off mirror image

1. Place the mode switch to MDI.

2. Press the extreme left soft key until *setting* appears at the bottom of the screen.

3. Press the soft key under setting.

4. Using cursor arrow key, position the cursor to the desired position for X or Y mirror image. (Usually on second page.)

5. To turn on mirror image, type 1 and press input.

6. To turn off mirror image, type 0 and press input.

Sequence to manually select inch or metric mode

1. Place the mode switch to MDI.

2. Press the extreme left soft key until *setting* appears at the bottom of the screen.

3. Press the soft key under setting.

4. Using cursor control arrow keys, bring the cursor to the desired inch/metric position.

5. Type 1 and press input to select inch mode or type 0 and press input to select the metric mode.

Sequence to use MDI to change tools. This sequence requires that you have an understanding of how the tool changer is programmed. The T word selects the tool to be placed in the waiting position and the M06 word makes the tool exchange.

1. Place the mode switch to MDI.

2. Press the extreme left soft key until *program* appears at the bottom of the screen.

3. Press the soft key under program until MDI appears at top of screen.

4. Type T and the number of the tool you wish to load into the spindle.

5. Press the end-of-block key and press the soft key under insert.

6. Press the start key or cycle start button. (Tool rotates into waiting position.)

7. Type M06.

8. Press the end-of-block key and press the soft key under insert.

9. Press the start key or cycle start button. (Tool change takes place.)

Sequence to use MDI to turn the spindle on and off

1. Place the mode switch to MDI.

2. Press the extreme left soft key until *program* appears at the bottom of the screen.

3. Press the soft key under program until *MDI* appears at top of screen.

4. Type S and the desired RPM. (Example: S500 = 500 RPM.)

5. Press the end-of-block key and press the soft key under insert.

6. Type M03 for clockwise or M04 for counterclockwise.

7. Press the end-of-block key and press the soft key under insert.

8. Press start key or cycle start button. (Spindle starts.)

9. To stop spindle, type M05, press the end-of-block key, and press the soft key under insert, then press start key or cycle start button.

Note: The last programmed RPM will also be the selected RPM for the manual mode when you turn the spindle on.

Sequence to use MDI to do a reference return

1. Place mode switch to MDI.

2. Press the extreme left soft key until *program* appears at the bottom of the screen.

3. Press the soft key under program until *MDI* appears at top of screen.

4. Check that the machine can go straight home with no interference in all axes!

5. Type

   G91 G28 X0 Y0 Z0

   then press the end-of-block key and press the soft key under insert.

6. Press start key or cycle start button.

Sequence to use MDI to move axes

1. Place mode switch to MDI.
2. Press the extreme left soft key until *program* appears at the bottom of the screen.
3. Press the soft key under program until MDI appears at top of screen.
4. Type the mode of motion (G90 = absolute, G91 = incremental).
5. Press the end-of-block key and press the soft key under insert.
6. Type the kind of motion (G00 = rapid, G01 = linear).
7. Press the end-of-block key and press the soft key under insert.
8. Type *X, Y,* or *Z* and the value to be commanded.
9. Press the end-of-block key and press the soft key under insert.
10. Press start key or cycle start button. (Motion takes place.)

This technique allows any kind of motion that is allowed in a program. However, it is quite error-prone and beginners should concentrate on moving the axes manually only!

Sequence to load programs into memory from the tape reader

1. Place mode switch to edit.
2. Press the extreme left soft key until *program* appears at the bottom of the screen.
3. Press the soft key under program until a program appears on the screen.
4. Place tape on the tape reader. (Tape goes from right to left and the leader holes are toward you.)
5. Place the toggle switch next to the tape reader to auto.
6. Type the letter O and type the program number of the program to be loaded.
7. Press the extreme right soft key until *read* appears at the bottom of the screen.
8. Press the soft key under read.

Note: Some machining centers do not have a tape reader and cannot execute this sequence.

Sequence to load a program into memory by the communications port. This sequence requires that some outside device such as a computer or portable reader punch is available and that several control parameters are properly set.

1. Place mode switch to edit.
2. Press the extreme left soft key until *program* appears at the bottom of the screen.
3. Press the soft key under program until a program appears on the screen.
4. Connect the outside device to the machine.
5. Type the letter O and the program number to be loaded.
6. Press the extreme right soft key until *read* appears at the bottom of the screen.
7. Press the soft key under read.
8. Go to the computer or tape reader and send the program.

Sequence to load programs into memory through the keyboard

1. Place the mode switch to edit.
2. Press the extreme left soft key until *program* appears at the bottom of the screen.
3. Press the soft key under program until a program appears on the screen.
4. Type the letter O and the program number of the program to be loaded.
5. Press the soft key under insert.
6. Typing one command at a time and pressing the soft key under insert after each command, enter the balance of the program.

   Note: Don't forget to end each command with an end-of-block.

Sequence to punch programs from memory to a tape punch or computer.
This sequence requires that some outside device such as a computer or tape punch is connected to the machine and that several parameters in the control be set properly.

1. Place mode switch to edit.
2. Press the extreme left soft key until *program* appears at the bottom of the screen.
3. Press the soft key under program until a program appears on the screen.
4. Connect outside device and get it ready to receive a program.

5. Type the letter O and the program number of the program to be sent.

6. Press the extreme right soft key until *punch* appears at the bottom of the screen.

7. Press the soft key under punch.

Sequence to display a directory of programs in memory

1. Place the mode switch to edit.

2. Press the extreme left soft key until *program* appears at the bottom of the screen.

3. Press the soft key under program until a program appears on the screen.

4. Press the soft key under program again.

Note: We recommend that the operator keep good records of the programs in memory. Then, when it comes time to delete programs in memory, the programs to be retained are readily identifiable.

Sequence to delete an entire program from memory

1. Place mode switch to edit.

2. Press the extreme left soft key until *program* appears at the bottom of the screen.

3. Press the soft key under program until a program appears on the screen.

4. Type letter O and the program number to be deleted.

5. Press the soft key under delete.

Note: Use caution with this sequence so as not to delete a needed program.

Sequence to search programs in memory

1. Place mode switch to edit.

2. Press the extreme left soft key until *program* appears at the bottom of the screen.

3. Press the soft key under program until a program appears on the screen.

4. Type the letter O and the program to be searched to.

5. Press the extreme right soft key until *fw srch* appears at the bottom of the screen.

6. Press the soft key under fw srch.

Sequence to search words in memory

1. Place mode switch to edit.
2. Press the extreme left soft key until *program* appears at the bottom of the screen.
3. Press the soft key under program until a program appears on the screen.
4. Press the reset key to return the program to the beginning (not always desired).
5. Type the word you wish to search to.
6. Press the extreme right soft key until *fw srch* appears at the bottom of the screen.
7. Press the soft key under fw srch.

Note: It is possible to search to the first occurrence of any letter address word by simply typing the letter address of the word and pressing the down arrow key. Example: To scan to the first occurrence of an F word, type F and press the soft key under fw srch.

Sequence to alter words in a program

1. Place mode switch to edit.
2. Press the extreme left soft key until *program* appears at the bottom of the screen.
3. Press the soft key under program until a program appears on the screen.
4. Search to the word to be altered.
5. Type the new word.
6. Press the soft key under alter.

Sequence to delete words in memory

1. Place mode switch to edit.
2. Press the extreme left soft key until *program* appears at the bottom of the screen.
3. Press the soft key under program until a program appears on the screen.
4. Search to the word to be deleted.
5. Press the soft key under dlt word.

Sequence to insert word in a program

1. Place mode switch to edit.

2. Press the extreme left soft key until *program* appears at the bottom of the screen.

3. Press the soft key under program until a program appears on the screen.

4. Search to the word just prior to the word you want to insert.

5. Type the word you wish to insert.

6. Press the soft key under insert.

Sequence to measure the program zero position.   This procedure will help you come up with your program zero positions. Remember that there are many ways to set up parts. We will be showing one way to pick up a square workpiece. First we give the technique for X or Y. Then we give the technique for Z.

For X and Y:

1. Make setup and load a workpiece.

2. Place an edge finder of known diameter in the spindle.

3. Start the spindle at a convenient RPM.

4. Using jog and the handwheel, pick up the X side of the workpiece where you want program zero to be.

5. Make the axis display read zero for the X axis.

6. Move away in Z and then move the center of the tool over program zero (knowing the radius of the edge finder and monitoring the axis display).

7. Make the X-axis display read zero again.

8. Repeat steps 4 to 7 for the Y axis.

9. Send the machine to its reference return in X and Y. (The axis displays will read the G92 dimensions to be used in the G92 command in X and Y.)

For the Z axis:

1. Remove any tool from the spindle.

2. Using jog and handwheel, cautiously position the nose of the spindle so that it touches the program zero position in Z.

3. Make the Z-axis display read zero.

4. Send the machine to its reference return position in the Z axis. (The Z-axis display will read the G92 Z dimension.)

Sequence to measure tool lengths.   Remember that there are two ways to use tool length compensation. We will be showing both methods here. The first (recommended) way will show how to measure the tool length. The second way will show how to measure the distance from the tip of the tool down to the program zero position in Z.

Measuring the tool length:

1. Place a flat square block on the table.

2. With no tool in the spindle, touch the nose of the spindle to the block.

3. Make the Z-axis display read zero.

4. Move the Z-axis away enough to load the tool to be measured.

5. Load the tool to be measured into the spindle.

6. Manually touch the tip of the tool to the block.

7. The Z-axis display will be showing the tool's length.

To measure the distance from the tip of the tool down to Z zero:

1. Send the machine to reference return position in the Z axis.

2. Make the Z-axis display read zero.

3. Touch the tip of the tool to the program zero point in Z.

4. The Z-axis display will show the distance from the tip of the tool at the reference point to program zero in Z (a big negative number).

Sequence to verify programs.   To minimize the possibility of problems, we recommend that you get into the habit of following these procedures to verify your programs.

Machine lock dry run.   This procedure will let the machine scan your programs for very basic mistakes. While the program could still have problems, at the completion of a machine lock dry run, at least the programmer knows the control can understand the program.

1. Send the machine to its starting point (usually reference return).

2. Place the mode switch to edit.

3. Press the extreme left soft key until *program* appears at the bottom of the screen.

4. Press the soft key under program until the program appears on the screen.

5. Press the reset key (check program number).

6. Turn on the machine lock toggle switch.

7. Place the mode switch to memory or auto.

8. Press cycle start button. (Machine runs cycle without moving the axes. Other functions will activate—spindle, tool changer, etc.)

Note: If the machine runs the entire program without an alarm, then continue with the next procedure. If not, then fix the problem with the program and do the machine lock procedure again.

Free-flowing dry run. This procedure allows the user to take control of all motions that the machine will make during the program with jog feed rate and feed hold. Jog feed rate will perform like a rheostat. And feed hold will be the panic button. If the machine appears to be doing something wrong, then press feed hold to stop the cycle to determine what is wrong. If you wish to continue after checking, just press cycle start again. But first:

1. Be sure there is no part in position.

2. Send the machine to its starting point (usually reference return).

3. Place the mode switch to edit.

4. Press the extreme left soft key until *program* appears at the bottom of the screen.

5. Press the soft key under program until the program appears on the screen.

6. Press the reset key (check program number).

7. Turn on the dry run toggle switch.

8. Turn off the machine lock toggle switch.

9. Place the mode switch to memory or auto.

10. Press cycle start button. (Machine runs cycle, and jog feed rate will control the rate of motion.)

Air cutting normal run. During a free-flowing dry run, it is difficult to tell the difference between a rapid command and a cutting command. Therefore, we recommend that you use the following procedure to see just where the rapid commands are. In this procedure the machine will perform as if it is actually cutting a part, but no part will be in position.

1. Be sure there is no part in position.
2. Send the machine to its starting point (usually reference return).
3. Place the mode switch to edit.
4. Press the extreme left soft key until *program* appears at the bottom of the screen.
5. Press the soft key under program until the program appears on the screen.
6. Press the reset key (check program number).
7. Turn off the dry-run toggle switch.
8. Place the mode switch to memory or auto.
9. Place the feed rate override switch to 100 percent.
10. Press cycle start button. (Machine runs cycle as if it is actually running a part, and the operator will be checking for where the rapid and cutting commands are occurring.)

Actually running first part.   When all of the above procedures have been followed, and any problems have been fixed, you are ready to run your first part. We still want to be *very* cautious while running that first workpiece. We will want to take control of the machine while approaching the workpiece on each tool's approach to the part with dry run and single block.

1. Load part into the setup.
2. Send the machine to its starting point (usually reference return).
3. Place the mode switch to edit.
4. Press the extreme left soft key until *program* appears at the bottom of the screen.
5. Press the soft key under program until the program appears on the screen.
6. Press the reset key (check program number).
7. Turn on the dry-run toggle switch.
8. Turn on the single-block toggle switch.
9. Place the mode switch to memory or auto.
10. Place the feed rate override switch to 100 percent.
11. Press cycle start button repeatedly until the spindle eventually

starts and the axis motion occurs. While approaching the part, use the jog feed rate to control the motion rate. When the tool finally makes its approach to the part and the desired clearance (0.100 in) is checked, continue to step 12.

12. Turn off dry-run toggle switch.

13. Continue pressing cycle start as many times as necessary until this tool is finished.

14. Repeat steps 7 to 12 for every tool in the program, checking after each tool to be sure that it has done what it was supposed to.

Sequence to execute verified programs from the beginning. Once a program has been verified, you can follow this procedure to run parts in production:

1. Load part into setup.
2. Send the machine to its starting point (usually reference return).
3. Place the mode switch to edit.
4. Press the extreme left soft key until *program* appears at the bottom of the screen.
5. Press the soft key under program until the program appears on the screen.
6. Press the reset key (check program number).
7. Check position of all conditional switches (dry run, etc.).
8. Place the mode switch to memory or auto.
9. Place the feed rate override switch to 100 percent.
10. Press cycle start button.

Sequence to execute programs from the beginning of a specific tool. Many times it will be necessary to rerun a tool. If the tool you wish to rerun is the first tool, you can simply follow the procedure to run programs from the beginning. However, if you want to rerun the fourth or fifth tool, you do not want to rerun the entire program.

This procedure lets you run from the beginning of any tool in your job. It requires that the format taught in key concept no. 5 of programming be followed.

1. Determine the tool station number of the tool you wish to rerun.
2. Send the machine to its starting point (usually reference return).
3. Place the mode switch to edit.

4. Press the extreme left soft key until *program* appears at the bottom of the screen.

5. Press the soft key under program until the program appears on the screen.

6. Press the reset key (check program number).

7. Type T and the tool station number of the tool you wish to rerun.

8. Press the soft key under fw srch. (The control scans to the first occurrence of the T word.)

9. Type T and the tool station number of the tool you wish to rerun (again—the first time this action found where the tool was placed in the waiting position).

10. Press the soft key under fw srch.

11. Bring the cursor to the beginning of the command that begins the tool.

12. Place the mode switch to memory or auto.

13. Check position of all conditional switches (dry run, etc.).

14. Place the feed rate override switch to 100 percent.

15. Press cycle start button.

Turning Center Sequences

This series of sequences works for a popular turning center control. While the specific techniques will vary from one turning center to the next, this should give you a good idea of the kind of operation handbook you should make for yourself. Note how specific each procedure is. A beginning operator could easily follow each procedure and make the machine do the desired function.

Manual sequences.

Sequence to start machine

Sequence to do a manual reference return

Sequence to manually start spindle

Sequence to manually jog axes

Sequence to use the handwheel to cause axis motion

Sequence to manually turn on coolant

Sequence to make axis displays read zero or any number

Sequence to enter tool offsets

Sequence to manually select inch or metric mode

Manual data input sequences

Sequence to use MDI to index turret

Sequence to use MDI to turn on spindle

Sequence to use MDI to do a reference return

Sequence to use MDI to move axes

Program loading and saving sequences

Sequence to load programs into memory by tape

Sequence to load programs into memory by communications port

Sequence to load programs into memory through keyboard

Sequence punch programs from memory to tape punch or computer

Program editing and display sequences

Sequence to display a directory of the programs in memory

Sequence to delete a whole program from memory

Sequence to search other programs in memory

Sequence to search to words inside a program

Sequence to alter words in memory

Sequence to delete words and commands in memory

Sequence to insert words and commands in memory

Setup sequences

Sequence to measure program zero positions in X and Z

Sequence to bore chuck jaws

Program running sequences

Sequence to verify programs

Sequence to run verified programs in production

Sequence to run from the beginning of any tool (requires that the program format be as taught during the programming session).

Sequence to start machine.   The power-up sequence may vary from one machine to another. Here we show a common technique used by most manufacturers.

1. Turn on main breaker (usually located at rear of machine).
2. Press control power-on button.
3. Press machine ready or hydraulic on button. (Not necessary on some machines. Be sure that no emergency stop button is locked-in.)
4. Follow sequence to do a manual reference return (if required).
5. Machine is now started.

Sequence to do a manual reference return.   The reference return procedure is often necessary. Whenever a program is run using the format shown in the programming session, most machines *must* be at the reference return position (*home position*). There are some shortcuts to this procedure, but we recommend that the beginner follow this technique for safety reasons:

1. Place mode switch to reference return.
2. Using the joystick or Z-minus push button, jog machine Z minus until the turret moves about 2 in.
3. Using the joystick or X minus push button, jog the machine X minus until the turret moves about 2 in.
4. Hold joystick plus in X, or press X-plus push button and hold until reference return origin light comes on for X.
5. Hold joystick plus in Z, or press Z-plus push button until reference return origin light comes on for Z.

The most common cause of a crash is the operator pressing the cycle start button to activate a cycle when the machine is not where it should be. Almost all programs are planned to start from the reference return position, so the operator must always check to be sure that the machine is at this position before pressing cycle start.

Sequence to manually start spindle.   Most machines have a way to manually start the spindle. However, usually you must at least select the RPM by the MDI mode, meaning that the sequence to start the spindle by MDI will usually be more useful.

1. Place the mode switch to a manual mode (jog, manual, reference return, etc.).

2. If available, adjust manual rheostat to desired RPM.

3. Press the spindle-on button. (Spindle comes on at the rheostat setting or last programmed RPM.)

4. To stop spindle, press the spindle stop button.

Note: Most machines will also let you select the desired spindle direction as well.

Sequence to manually jog axes. This technique will be used often. A few examples of when it is used are making setups, moving the turret to examine a tool, and actually manually cutting a workpiece. This procedure will vary widely, based on manufacturer.

1. Place the mode switch to manual or jog.

2. Select the desired feed rate by feed rate override or jog feed rate. (*Note:* Machines will vary with manufacturer in this respect so you may have to consult the machine-tool builder's manual.)

3. Using joystick or X-Z push buttons, move the machine in the desired direction and amount.

Be extremely careful when using this procedure if you are close to workpiece or obstructions.

Sequence to use handwheel. This is very helpful on machines that have a handwheel. It lets you move the machine a precise amount, as you would on a manual engine lathe. You can even machine workpieces using the handwheel, as on a manual machine.

1. Place mode switch to handwheel.

2. Place axis select switch to desired position (X or Z).

3. Place handle rate switch to desired position (× 1, × 10, or × 100).

4. Using handwheel, rotate plus or minus to cause desired motion.

Note: On machines with the handwheel, × 1 = 0.0001 per increment of the handwheel, × 10 = 0.001, and × 100 = 0.010 inch.

Sequence to manually turn on the coolant. Most turning centers allow the operator to activate the coolant manually. A simple toggle switch is used.

1. Place mode switch to manual or jog.

2. Place the coolant toggle switch to on (coolant comes on).

3. To turn the coolant off, turn off the toggle switch.

Note: Many machines will also have an auto position for the coolant switch. When the switch is in this position, the coolant will only come on when an M08 is commanded.

Note: Some turning centers require that the spindle be on before the coolant comes on in the manual mode.

Sequence to make the axis displays read zero or any number. This technique is especially helpful for measuring program zero positions. Two ways are shown: one for zeroing and the other to make the axis displays read any number.

To make axis displays read zero:

1. Mode switch can be in any position.
2. Press the extreme left soft key until *position* appears at the bottom of the screen.
3. Press the soft key under position.
4. Type the letter address of the axis you wish to make zero (X or Z).
5. Press the soft key under origin.

To make the axis displays read any number:

1. The mode switch can be in any position.
2. Press the extreme left soft key until *position* appears at the bottom of the screen.
3. Press the soft key under position.
4. Type the letter address for the axis you wish to set and type the value you wish.
5. Press the soft key under preset. (The value and axis you set will be displayed on the position page.)

Note: These techniques assume that the machine has been previously set to the desired position.

Sequence to enter and change tool offsets. Before attempting this procedure, it is necessary to understand the function of tool offsets for dimensional tool offsets and tool nose radius compensation as presented in the part on programming.

1. The mode switch can be in any position.
2. Press the extreme left soft key until *offset* appears at the bottom of the screen.

3. Press the soft key under offset.

4. Using the arrow keys, position the cursor to the offset you would like to enter. Remember that the cursor must be at the actual offset position (X, Z, R, or T).

5. Type the value of the desired offset.

6. Press the soft key under input or plus input.

Note: If you press the soft key under input, the actual value you typed will be taken as the new offset. If you press the soft key under plus input, the value you typed will be added to or subtracted from the current value of the offset.

Sequence to manually select the inch or metric mode

1. Place the mode switch to MDI.

2. Press the extreme left soft key until *setting* appears at the bottom of the screen.

3. Press the soft key under setting.

4. Using cursor control arrow keys, bring the cursor to the desired inch/metric position.

5. Type 1 and press input to select inch mode or type 0 and press input to select the metric mode.

Sequence to use MDI to index the turret.   This sequence requires that you have an understanding of how the turret is programmed. The T word selects the tool station number to be indexed to and the tool offset number to be used.

1. Place the mode switch to MDI.

2. Press the extreme left soft key until *program* appears at the bottom of the screen.

3. Press the soft key under program until *MDI* appears at top of screen.

4. Type T, the tool station number, and the tool offset number you wish to use. (Example: T0100 indexes the turret to station number 1 and calls up no offset. While you are using this sequence, no tool offset should be called up, or the turret will move by the amount of the offset.)

5. Press the end-of-block key and press the soft key under insert.

6. Press the start key or cycle start button. (Turret indexes to desired station.)

You must remember that, if a tool offset is called up, the turret will actually move the amount of the offset. This is a major cause of machine crashes, since the operator is causing the turret to be out of position for the next cycle.

Note: Some machines will also allow the turret to index by a push button in the manual mode.

Sequence to use MDI to turn on the spindle

1. Place the mode switch to MDI.
2. Press the extreme left soft key until *program* appears at the bottom of the screen.
3. Press the soft key under program until MDI appears at top of screen.
4. Type the desired spindle speed mode. (G96 = SFM, G97 = RPM.)
5. Press the end-of-block key and press the soft key under insert.
6. Type the desired spindle range if required. (On most machines, M41 equals low range, M42 equals high range.)
7. Press the end-of-block key and press the soft key under insert, then press the cycle start button. (Nothing happens yet.)
8. Type S and the desired spindle speed. (Example: S500 means 500 SFM or 500 RPM, depending on step 4.)
9. Press the end-of-block key and press the soft key under insert.
10. Type M03 for clockwise or M04 for counterclockwise.
11. Press the end-of-block key and press the soft key under insert.
12. Press start key or cycle start button. (Spindle starts.)
13. To stop spindle type M05, press the end-of-block key and press the soft key under insert, then press start key or cycle start button.

Note: The last programmed RPM will also be the selected RPM for the manual mode when you turn the spindle on.

Note: Be sure to check the M41 and M42 codes for spindle range prior to attempting this procedure (for step 6). Some machines may have only one spindle range, therefore steps 6 and 7 may be omitted.

Sequence to use MDI to do a reference return

1. Place mode switch to MDI.
2. Press the extreme left soft key until *program* appears at the bottom of the screen.

3. Press the soft key under program until MDI appears at top of screen.

4. Check that the machine can go straight home with no interference in both axes!

5. Type

G28 U0 W0

then press the end-of-block key, and press the soft key under insert.

6. Press start key or cycle start button.

Sequence to use MDI to move axes

1. Place mode switch to MDI.

2. Press the extreme left soft key until *program* appears at the bottom of the screen.

3. Press the soft key under program until MDI appears at top of screen.

4. Type the kind of motion (G00 = rapid, G01 = linear).

5. Press the end-of-block key and press the soft key under insert.

6. Type X or Z and the value to be commanded.

7. Press the end-of-block key and press the soft key under insert.

8. Press start key or cycle start button. (Motion takes place.)

This technique can be used for any kind of motion that is allowed in a program. However, it is quite error-prone and beginners should concentrate on moving the axes manually only!

Sequence to load programs into memory by tape

1. Place mode switch to edit.

2. Press the extreme left soft key until *program* appears at the bottom of the screen.

3. Press the soft key under program until a program appears on the screen.

4. Place tape on the tape reader. (Tape goes from right to left and the leader holes are toward you.)

5. Place the toggle switch next to the tape reader to auto.

6. Type the letter O and type the program number of the program to be loaded.

7. Press the extreme right soft key until *read* appears at the bottom of the screen.

8. Press the soft key under read.

   Note: Some turning centers do not have a tape reader, and this sequence is not possible.

Sequence to load a program into memory through the communications port. This sequence requires that some outside device like a computer or portable reader punch is available and that several control parameters are properly set.

1. Place mode switch to edit.

2. Press the extreme left soft key until *program* appears at the bottom of the screen.

3. Press the soft key under program until a program appears on the screen.

4. Connect the outside device to the machine.

5. Type letter O and the program number to be loaded.

6. Press the extreme right soft key until *read* appears at the bottom of the screen.

7. Press the soft key under read.

8. Go to the computer or tape reader and send the program.

Sequence to load programs into memory through the keyboard

1. Place the mode switch to edit.

2. Press the extreme left soft key until *program* appears at the bottom of the screen.

3. Press the soft key under program until a program appears on the screen.

4. Type the letter O and the program number of the program to be loaded.

5. Press the soft key under insert.

6. Typing one command at a time and pressing the soft key under insert after each command, enter the balance of the program.

   Don't forget to end each command with an end-of-block signal.

Sequence to punch programs from memory to a tape punch or computer. This sequence requires that some outside device like a computer or tape punch is connected to the machine and that several parameters in the control be set properly.

1. Place mode switch to edit.
2. Press the extreme left soft key until *program* appears at the bottom of the screen.
3. Press the soft key under program until a program appears on the screen.
4. Connect outside device and get it ready to receive a program.
5. Type the letter O and the program number of the program to be sent.
6. Press the extreme right soft key until punch appears at the bottom of the screen.
7. Press the soft key under punch.

Sequence to display a directory of programs in memory

1. Place the mode switch to edit.
2. Press the extreme left soft key until *program* appears at the bottom of the screen.
3. Press the soft key under program until a program appears on the screen.
4. Press the soft key under program again.

Note: We recommend that the operator keep good records of the programs in memory so that when it comes time to delete programs in memory, it is clear which should be retained.

Sequence to delete an entire program from memory

1. Place mode switch to edit.
2. Press the extreme left soft key until *program* appears at the bottom of the screen.
3. Press the soft key under program until a program appears on the screen.
4. Type letter O and the program number to be deleted.
5. Press the soft key under delete.

Use caution when using this sequence so as not to delete a needed program.

Sequence to search to other programs in memory

1. Place mode switch to edit.
2. Press the extreme left soft key until *program* appears at the bottom of the screen.
3. Press the soft key under program until a program appears on the screen.
4. Type the letter O and the program to be searched to.
5. Press the extreme right soft key until *fw srch* appears at the bottom of the screen.
6. Press the soft key under fw srch.

Sequence to search to a specific word within a program

1. Place mode switch to edit.
2. Press the extreme left soft key until *program* appears at the bottom of the screen.
3. Press the soft key under program until a program appears on the screen.
4. Press the reset key to return the program to the beginning (not always desired).
5. Type the word you wish to search to.
6. Press the extreme right soft key until *fw srch* appears at the bottom of the screen.
7. Press the soft key under fw srch.

   Note: It is possible to search to the first occurrence of any letter address word by simply typing the letter address of the word and pressing the down arrow key. Example: To scan to the first occurrence of an F word, type F and press the soft key under fw srch.

Sequence to alter a word in a program

1. Place mode switch to edit.
2. Press the extreme left soft key until *program* appears at the bottom of the screen.
3. Press the soft key under program until a program appears on the screen.
4. Search to the word to be altered.
5. Type the new word.

6. Press the soft key under alter.

Sequence to delete a word within a program

1. Place mode switch to edit.
2. Press the extreme left soft key until *program* appears at the bottom of the screen.
3. Press the soft key under program until a program appears on the screen.
4. Search to the word to be deleted.
5. Press the soft key under dlt word.

Sequence to insert words within a program

1. Place mode switch to edit.
2. Press the extreme left soft key until *program* appears at the bottom of the screen.
3. Press the soft key under program until a program appears on the screen.
4. Search to the word just prior to the word you want to insert.
5. Type the word you wish to insert.
6. Press the soft key under insert.

Sequence to measure the program zero position. This procedure will help you come up with your program zero positions manually. Note that there are many ways to set up parts. We will be showing one way to do this.

For the X axis:

1. Make setup and load a workpiece.
2. Index the turret to the desired tool.
3. Manually jog the tool close to the workpiece.
4. Start the spindle manually or through MDI at the desired RPM for this workpiece material.
5. Using handwheel, skim-cut a diameter.
6. Without moving the X axis, move the turret away from the part in Z far enough to measure the diameter that was just cut.
7. Measure the skim-cut diameter.
8. Make the X-axis display read the diameter that was just cut.

9. Send the machine to its reference return position in X.

10. The X-axis display will read the G50 X value.

For the Z axis:

1. Make the setup and load a workpiece.

2. Index the turret to the desired tool.

3. Manually jog the tool close to the workpiece.

4. Start the spindle at the desired RPM for this material.

5. Using handwheel, skim-cut the face of the workpiece to Z zero.

6. Without moving Z, move the tool away from the part in X.

7. Make the Z-axis display read zero.

8. Send the machine to its reference return position in Z.

9. The Z-axis display will read the G50 Z value.

Sequence to manually bore soft jaws. CNC turning centers with hydraulic chucks require that soft jaws be gripping something while the jaws are being bored. This means that many times, the operator must come up with a practice workpiece for the jaws to grip while they are being bored. Also, the serrations in the master jaws of the chuck are sometimes quite difficult to align. The beginning operator should always be extra cautious to confirm that the workpiece is "running true."

It is wise to obtain the diameter of the tool tip of the boring bar at the machine's reference return position. If this is done, the operator can make the X-axis display read the diameter of the tool tip at reference return position. This allows the operator to monitor the X axis display to obtain the exact diameter that the boring bar is cutting at any one time. To obtain this diameter for the boring bar, the operator can use the procedure to measure the program zero X position shown in the preceding sequence.

1. Cautiously place the soft jaws on the master jaws of the chuck, confirming that they are in the serrations in each master jaw.

2. Using a chucking ring or practice workpiece, close the jaws so that the jaw stroke has moved about halfway.

3. Using jog and handwheel, bore the jaws to the specified diameter and depth.

Note: By touching the tip of the boring bar to the face of the jaws and making the Z-axis display read zero at that point, the operator can easily monitor the depth that the jaws are being bored to by monitoring the Z-axis display.

Sequence to verify programs.   To minimize the possibility for problems, we recommend that you get into the habit of following these procedures to verify your programs.

Machine lock dry run.   This procedure will let the machine scan your programs for very basic mistakes. While the program could still have problems at the completion of a machine lock dry run, at least the programmer knows that the control can understand the program.

1. Send the machine to its starting point (usually reference return).
2. Place the mode switch to edit.
3. Press the extreme left soft key until *program* appears at the bottom of the screen.
4. Press the soft key under program until the program appears on the screen.
5. Press the reset key (check program number).
6. Turn on the machine lock toggle switch.
7. Place the mode switch to memory or auto.
8. Press cycle start button. (Machine runs cycle without moving the axes. Other functions will activate: spindle, tool changer, etc.)

Note: If the machine runs the entire program without an alarm, then continue with the next procedure. If not, then fix the problem with the program and do the machine lock procedure again.

Free-flowing dry run.   This procedure allows the user to take control of all motions that the machine will make during the program with jog feed rate or feed rate override (depending on machine builder) and feed hold. Jog feed rate or feed rate override will perform like a rheostat. And feed hold will be the panic button. If the machine appears to be doing something wrong, then press feed hold to stop the cycle to determine what is wrong. If you wish to continue after checking, just press cycle start again.

1. Be sure there is no part in the chuck.
2. Send the machine to its starting point (usually reference return).

3. Place the mode switch to edit.

4. Press the extreme left soft key until *program* appears at the bottom of the screen.

5. Press the soft key under program until the program appears on the screen.

6. Press the reset key (check program number).

7. Turn on the dry run toggle switch.

8. Turn off the machine lock toggle switch.

9. Place the mode switch to memory or auto.

10. Press cycle start button. (Machine runs cycle, and jog feed rate or feed rate override will control the rate of motion.)

Air cutting normal run.   During a free-flowing dry run, it is difficult to tell the difference between a rapid command and a cutting command. Therefore, we recommend that you use this procedure to see just where the rapid commands are. In this procedure the machine will perform as if it is actually cutting a part, but no part will be in position.

1. Be sure there is no part in the chuck.

2. Send the machine to its starting point (usually reference return).

3. Place the mode switch to edit.

4. Press the extreme left soft key until *program* appears at the bottom of the screen.

5. Press the soft key under program until the program appears on the screen.

6. Press the reset key (check program number).

7. Turn off the dry-run toggle switch.

8. Place the mode switch to memory or auto.

9. Place the feed rate override switch to 100 percent.

10. Press cycle start button. (Machine runs cycle as if it is actually running a part and the operator will be checking for where the rapid and cutting commands are occurring.)

Actually running first part.   When all of the above procedures have been followed, and any problems have been fixed, you are ready to run your first part. We still want to be *very* cautious while running that first workpiece. We will want to take control of the machine while approaching the workpiece on each tool's approach to the part with dry run and single block.

1. Load part into the setup.
2. Send the machine to its starting point (usually reference return).
3. Place the mode switch to edit.
4. Press the extreme left soft key until *program* appears at the bottom of the screen.
5. Press the soft key under program until the program appears on the screen.
6. Press the reset key (check program number).
7. Turn on the dry-run toggle switch.
8. Turn on the single-block toggle switch.
9. Turn on the optional-stop toggle switch.
10. Place the mode switch to memory or auto.
11. Place the feed rate override or jog feed rate switch to its lowest position.
12. Press cycle start button repeatedly until the spindle eventually starts and the axis motion occurs. While approaching the part, use the jog feed rate or feed rate override to control the motion rate. When the tool finally makes its approach to the part and the desired clearance (usually 0.100) is checked, then continue to step 13.
13. Turn off dry-run toggle switch.
14. Continue pressing cycle start as many times as necessary until this tool is finished.
15. Repeat steps 7 to 14 for every tool in the program, checking after each tool to be sure that the tool has done what it was supposed to.

Sequence to execute verified programs from the beginning.   Once a program has been verified, you can follow this procedure to run parts in production.

1. Load part into setup.
2. Send the machine to its starting point (usually reference return).
3. Place the mode switch to edit.
4. Press the extreme left soft key until *program* appears at the bottom of the screen.

5. Press the soft key under program until the program appears on the screen.

6. Press the reset key (check program number).

7. Check position of all conditional switches (dry run, etc.).

8. Place the mode switch to memory or auto.

9. Place the feed rate override switch to 100 percent.

10. Press cycle start button.

Sequence to execute programs from the beginning of a specific tool. Many times it will be necessary to rerun a tool. If the tool you wish to rerun is the first tool, you can simply follow the procedure to run programs from the beginning. However, if you want to rerun the fourth or fifth tool, you do not want to rerun the entire program. This procedure lets you run from the beginning of any tool in your job. It requires that the format taught in key concept no. 5 of programming be followed.

1. Determine the turret station number of the tool you wish to rerun.

2. Send the machine to its starting point (usually reference return).

3. Place the mode switch to edit.

4. Press the extreme left soft key until *program* appears at the bottom of the screen.

5. Press the soft key under program until the program appears on the screen.

6. Press the reset key (check program number).

7. Type N and the sequence number of the beginning block of the tool you wish to rerun. (Usually the G50 command for that tool.)

8. Press the soft key under fw srch. (The control scans to that sequence number.)

9. Place the mode switch to memory or auto.

10. Turn on optional stop (M01).

11. Check position of all conditional switches (dry run, etc.).

12. Place the feed rate override switch to 100 percent.

13. Press cycle start button.

Wire EDM Machine Sequences

This series of sequences works for a popular wire EDM control. While the specific techniques will vary from one wire EDM machine to the next, this should give you a good idea of the kind of operation handbook you should make for yourself. Note how specific each procedure is. A beginning operator could easily follow each procedure and make the machine do the desired function.

Manual sequences.

Sequence to start machine

Sequence to manually jog axis motions

Sequence to manually thread wire

Sequence to manually adjust flushing

Sequence to manually adjust wire tension and speed

Sequence to do a sensor touch to pick up an edge

Manual data input sequences.

Sequence to use MDI to move axis (used for manual cutoffs)

Sequence to use MDI to adjust vertical alignment (U and V)

Sequence to use MDI to pick up an edge (using G80, sensor touch)

Sequence to use MDI to pick up the center of a start hole

Sequence to use MDI to pick up the corner of a workpiece

Sequence to use MDI to make axis displays read zero or any number

Program editing sequences.

Sequence to clear active memory

Sequence to type new CNC programs into memory

Sequence to save (or resave) CNC programs onto diskette

Sequence to load programs from the diskette

Sequence to delete programs from the diskette

Sequence to edit programs in the active memory
  1. Search
  2. Change
  3. Insert
  4. Delete

Program operation sequences.

Sequence to run programs from the active memory

Sequence to run programs from the diskette

Sequence to start machine

1. Turn on main breaker (inside control front panel door at the bottom).
2. Press green-lighted button near main breaker.
3. Place system diskette in drive 1.
4. Place user diskette in drive 2.
5. Press source key and wait until initial screen comes up on screen (about 4 to 5 minutes).
6. Press power button (to activate machine).

Sequence to jog axes

1. From the initial page, press the soft key next to M'AL (for manual).
2. Press the soft key next to MDI.
3. On machine panel, place the MFR (manual feed rate) dial to the desired position. (1 is the fastest rate and 4 is the slowest rate).
4. To jog X or Y, just press the desired plus or minus button for that axis.
5. To jog U or V, you will have to let the control know you wish to do this. On some machines, there is a toggle switch (labeled U/V ON) that you must turn on first. Other machines require that you hold a button (labeled U/V ON) while you press the desired plus or minus button for the axis you wish to jog (U or V).
6. If moving the U or V axis, when you're finished, be sure to turn off the U/V ON toggle switch (if your machine has one) to be sure not to accidentally move the U or V axis.

Note: If vertically aligning the U and V axes, you must also be sure to set the various position displays for U and V to zero when you are finished to avoid having your desired taper machined incorrectly.

Sequence to manually thread wire.   This sequence is for machines *without* automatic wire threaders. Even for such machines, there are varying procedures used to thread wire, based on the machine model. Some machines have a series of belts that take the wire to the disposal bin as soon as it passes through the lower guides. Other machines require that you manually pass the wire all the way through the machine until it reaches the pinch rollers just above the wire bin. With still others, you must manually bring the wire to the pinch rollers by continuing to push the wire through the lower guide.

No matter what style of machine your company owns, there are some common notes that will apply to wire threading:

1. After the wire is wrapped around the tension rollers (near the wire spool) and ready to pass through the upper and lower guides, it is wise to check that the wire is pointed. If there is any kink in the wire, you will not be able to pass it through the upper and lower guides. One very common technique to point the wire is to pull the wire apart after heating (with a cigarette lighter). As the wire is pulled apart, it forms a perfect sharp point that will pass easily through the upper and lower guides.

2. Sometimes it is helpful to release the tension on the wire while threading to make rotation of the tension rollers easier. If you do this, make sure you readjust the tension when you are finished.

3. Be sure that the wire passes under the *wire broken* sensor switch. If it does not, you will get a false *wire-broken* alarm when you start the machining cycle.

4. Wire threading can be tedious, and the first few times you do it, it will seem quite time-consuming. However, with practice, and as time goes by, you will find that you will become proficient at threading the wire quickly.

Sequence to manually adjust flushing

1. From initial page, press the soft key next to M'AL.

2. Press the soft key next to MDI.

3. Adjust the flush valves to the completely closed (clockwise) position. This will keep you from getting wet when you turn on the flushing.

4. Turn on the flushing (low or high, depending on which you intend to set) by activating the flushing switch on the machine panel.

5. To adjust the upper guide flushing, open the upper valve to the desired position.

6. To adjust the lower guide flushing, open the lower valve to the desired position.

7. Two flow meters show the actual pressure, so you can document your settings for future cutting.

Sequence to manually adjust wire tension and speed

Wire tension. If tension is not adjusted correctly, accuracy will be affected. If the wire tension is too low, there will be excessive "washout" in the corners as well as a barrel shape or taper on the workpiece. If tension is set too high, the wire will be prone to breakage.

Generally speaking, you will want the wire tension as high as possible during roughing without causing the wire to break. On subsequent trim passes, you should increase the wire tension by 15 to 20 percent, which assures that the walls of the workpiece will be straight.

Wire Speed. Generally speaking, the thicker the workpiece, the higher the wire speed must be to avoid wire breakage. This is because of the erosion of the wire during the EDM process (especially during roughing). When trim passes are being made, you can slow down the wire speed, since the erosion of the wire is much less than during roughing. Doing this will conserve wire.

Note: Two rheostats labeled *tension* and *wire speed* control these functions. They can be adjusted at any time, but to actually see the speed of the wire, you must turn the wire on manually, by going into the M'AL mode and pressing the soft key next to MDI. Then you can activate the toggle switch labeled *wire run*.

Sequence to use sensor touch to pick up an edge. This procedure is helpful to crudely touch a workpiece. It leaves the wire flush with the edge of the part. However, this procedure is not very accurate. Remember that there is a more accurate sensor touch routine that actually touches the edge three times and takes the average of the three tries (G80).

1. From initial page, press the soft key next to M'AL.
2. Press the soft key next to MDI.
3. Using the jog X and Y buttons, position the wire close to the edge to be touched.
4. Press the (red) ST button and hold it while pressing one of the jog X or Y buttons to jog the wire into the edge.
5. When the wire touches the part, it will automatically bounce back and touch the workpiece slowly. When it has finished, the buzzer will sound.
6. Press the ACK (acknowledge) button.

Sequence to use MDI to move axes. All EDM controls allow you to move any axis (one or more) by MDI as it is done in a program. Any rules you know related to programming will still apply to MDI axis motion (G54 to G59, G00, G01, G02, G03, G90, G91, etc.).

1. From initial page, press the soft key next to M'AL.

2. Press the soft key next to MDI.
3. If you intend to actually cut something during this motion, enter C and the condition number you wish to use (example: C206).
4. Press the ENTER key.
5. Type the mode of motion you wish (G90 for absolute or G91 for incremental).
6. Press the ENTER key.
7. Type the kind of motion you wish to make (G00 for rapid or G01 for straight-line cutting). *Note:* The G01 will automatically activate electricity to the wire and turn on the wire run and flushing. Also, the current condition will be in effect for cutting.
8. Type the axis and motion you wish to make (example: X1.5).
9. Press the ENTER key (motion takes place).

Notes:
  a. Machine assumes plus (+) unless minus (−) is input for axis motion.
  b. In incremental mode (G91), movement is taken from current position.
  c. In absolute mode (G90), movement is taken from program zero. A previous G92 or shifting of the axis displays is assumed.
  d. Use G00 for noncutting movements and G01 for cutting movements.
  e. You *cannot* use offset or taper when using MDI.

Sequence to use MDI to adjust wire vertical alignment. The vertical alignment is extremely important! If the wire is not vertically aligned, the workpiece will come out with undesirable taper. We recommend that you use this procedure often to test for vertical alignment, and especially when you are about to run an expensive and critical part.

1. From the initial page, press the soft key next to M'AL.
2. Press the soft key next to MDI.
3. Place the test block (that came with the machine) on the table as square as possible.
4. Type G91 and press ENTER.
5. Type C777 and press ENTER (C777 is a very light condition that will let you see the spark of the wire easily).

To adjust the *U* direction:

6. Press the soft key next to *Cutting*.
7. Using the X and Y jog buttons, position the wire close to the test block in the *X* minus direction (to the left of the test block).
8. Type X1.5 and press the ENTER key (the wire will activate and begin moving to the right (*X* plus direction).

9. If your machine has a U/V ON toggle switch, turn it on.
10. When the wire touches the workpiece, you will see the spark generated easily. Use the U + and U − buttons to form a uniform spark from the top of the test block to the bottom. Note that if your machine has a U/V ON button, you will have to hold it while pressing the U + or U − buttons.
11. Move the test block away from the wire slightly (by hand) and confirm the spark uniformity again.
12. When satisfied, press the OFF button.
13. Press the ACK (acknowledge) button.

To adjust the V direction:

14. Using the X and Y jog buttons, position the wire to the Y minus side of the test block (the side toward you).
15. Type Y1.5 and press enter (the wire will activate and move toward the test block).
16. When the wire touches the block and begins to spark, press the V + and V − buttons to make the spark uniform (just as you did with U).
17. Move the block away from the wire and confirm the spark again.
18. When satisfied, press the OFF button.
19. Press the ACK button.
20. Press the soft key next to MDI.
21. Type G54 and press ENTER.
22. Type G92 U0 V0 and press ENTER.
23. Repeat steps 21 and 22 for all other coordinate systems (G55 to G59).
24. If your machine has a UV ON toggle switch, be sure to turn it off to keep from accidentally moving the U or V axis.

Notes:
a. The ends of the test block are not acceptable for testing vertical alignment.
b. Try to always use the same side of the test block for this procedure. This will assure that you will always have a clean surface that can be placed down on the table.
c. Be sure to completely dry off the test block, table, and upper guide before doing this procedure. If anything is wet, it will affect the way the spark is generated, and the spark may be difficult (if not impossible) to see.
d. While this procedure sounds a little time-consuming, we cannot stress enough how important it is. And, as time goes on and you gain experience with the machine, you should be able to accomplish this procedure in about 10 to 15 minutes.

Sequence to use MDI to pick up and edge. As you know, the manual sensor touch (ST) is not very accurate, since the wire will contact the edge of the part only once. With the G80 command, the wire will actually touch the workpiece three times and take the average of the three touches, making the G80 much more accurate.

1. Manually jog the wire close to the surface to touch (within about 0.3 in).
2. From the initial page, press the soft key next to M'AL.
3. Press the soft key next to MDI.
4. Type the axis name and the direction of motion you desire to contact the part (choose from X, X−, Y, or Y−) depending on how the wire is positioned in relation to the edge to be picked up.
5. Press the ENTER button.
6. After the wire stops, it will be perfectly flush with the edge and the buzzer will sound.
7. Press the ACK (acknowledge) button.

Note: Before doing this procedure, be sure that the part is clean and dry, and be sure the wire tension is increased.

Sequence to use MDI to pick up the center of a hole. This is one of the Q assist routines designed by one control manufacturer to make operation easier. When you begin from a start hole (or aligning hole), this routine will automatically find the center of the hole.

1. Thread the wire through the start hole and confirm that it is not touching the workpiece (voltage reading 10 V).
2. From the initial page, press the soft key next to M'AL.
3. Press the soft key next to MDI.
4. Type "Q0145 (0,0)" and press ENTER (note that these are all zeros, not letters).
5. When the routine is finished, the wire will be positioned perfectly in the center of the hole.
6. Press the ACK (acknowledge) button.

Notes (for before pickup):
a. Increase wire tension about 20 percent.
b. Make sure workpiece is clean and dry.
c. Make sure there are no burrs on the hole to pick up.
d. Step relief holes when possible.

Sequence to use MDI to pick up a corner of the workpiece. This is another Q assist routine designed by an EDM machine manufacturer to make setting up easier. When you are trying to work from a corner of your workpiece (as your program zero point, for example), this routine will automatically pick up both edges of the corner and make the X- and Y-axis position values of that corner be any numbers you want (usually X0 and Y0).

This procedure requires that you know a code number that corresponds to the corner of the workpiece you wish to pick up. The upper right-hand corner (X+ and Y+) is corner 1. The upper left-hand corner is corner 2, the lower left corner is corner 3, and the lower right corner is corner four (all as viewed from above).

At the completion of this routine the wire will be left slightly off the corner to be picked up, and the X- and Y-axis displays will read the position *relative* to the corner.

1. Manually bring the wire to within about 0.150 in of the corner you wish to pick up (in both directions).
2. From the initial page, press the soft key next to M'AL.
3. Press the soft key next to MDI.
4. Type one of the following commands. Be sure to input the correct wire size you are using.
    a. Q0165(1,60,0,0) for upper right corner, 0.006-in-diameter wire.
    b. Q0165(2,80,0,0) for upper left corner, 0.008-in-diameter wire.
    c. Q0165(3,100,0,0) for lower left corner, 0.010-in-diameter wire.
    d. Q0165(4,120,0,0) for lower right corner, 0.012-in-diameter wire.
5. Press the ENTER button.
6. Routine automatically takes place and buzzer sounds at the end.
7. Press the ACK (acknowledge) button.
8. The axis displays will show where the wire is (relative to the corner being picked up).

Sequence to make the axis displays read zero or any number

1. From the initial page, press the soft key next to M'AL.
2. Press the soft key next to MDI.
3. To set to zero, type G92 X0 Y0 Z0 U0 X0 and press the ENTER button.
4. To set to any number, type G92 X _____ Y _____ Z _____ U _____ V _____ and press the ENTER button.

Note that all axes (X, Y, Z, etc.) need not be included. If you are only interested in setting one axis (X for example), you do not have to include the other axes in the G92 command.

Sequence to clear active memory.   Before loading or entering any new program, you must be sure that the memory is clear.

1. From initial page, press the soft key next to EDIT.

2. Press the soft key next to EDIT (again).

3. Press the memory clear soft key.

4. Press the ENTER button.

Sequence to enter programs through the keyboard. If you wish to run your own manually written programs, you can follow this procedure to enter them into memory. When you are finished, don't forget to save the program to the diskette to avoid losing your program when the power is turned off.

1. From the initial page, press the soft key next to EDIT.

2. Press the soft key next to EDIT (again).

3. Using the arrow keys, bring the cursor to the end of the start-up information and press the soft key under INSERT.

4. Enter the balance of the program.

5. When finished, press the soft key under INSERT (again).

Sequence to save or resave programs onto diskette. Whenever the power is turned off, the program in the active memory is lost. This procedure is used to permanently register the program to the diskette.

1. From the initial page, press the soft key next to EDIT.

2. Press the soft key next to SAVE.

3. To resave a program, move the cursor to the name of the desired file you wish to save the program as and press the ENTER button. If you wish to keep the same name, just press ENTER.

4. To save a new program you manually typed in, or to change the name of a program, type in the file name and press the ENTER button.

Note: Possibly a message could come up after this that states that the same file name already exists. The control is telling you that you have chosen a name that is already taken. If resaving the same program, this is fine and you will just press the ENTER button. But if you are saving a new program, *be careful!* You do not want to write over another program. In this case, you would press the ESC key and try again with a new name.

Sequence to load programs into memory from diskette

1. From the initial page, press the soft key next to EDIT.

2. Press the soft key next to LOAD.

3. Using the arrow keys, bring the cursor to the file name you wish to load.

4. Press the ENTER button.

Sequence to delete programs from the diskette.   Note that you must use caution with this technique to avoid deleting the wrong program!

1. From the initial page, press the soft key next to EDIT.

2. Press the soft key next to DELETE.

3. Using the arrow keys, bring the cursor to the program name you wish to delete (*be careful!*).

4. Press the ENTER button twice.

Sequence to edit programs in active memory.   Once you load a program into memory, you have complete control of the program. Many editing functions can be performed.

   All editing functions require that the program be loaded into the active memory and the following steps be taken:

1. From initial page, press the soft key next to EDIT.

2. Press the soft key next to EDIT (again).

   To search words in memory:

1. Press the soft key under SEARCH.

2. Type the word you wish to search to.

3. Press ENTER.

   To change words in memory:

1. Search to the word to be changed.

2. Type right over the top of the word.

3. You may have to insert or delete to match the exact amount of data.

   To insert:

1. Search to the character *after* where you want to insert data (the inserted data will go before, where the cursor currently is).

2. Press the soft key under INSERT (the data after the cursor seems to disappear, but don't worry, it is not lost).

3. Type new data.

4. When finished, press the soft key under INSERT again (disappearing data comes back).

To delete:

1. Search to the first character to be deleted.
2. Press the soft key under DELETE.
3. Bring the cursor to the last character to be deleted (characters to be deleted will be highlighted).
4. Press the soft key under DELETE again.

Sequence to run programs from memory

1. Load program into active memory.
2. From the initial page, press the soft key next to RUN.
3. Press the soft key next to MEMO.
4. Check conditional switches (dry run, single block, mirror, etc.).
5. Press the ENTER button.

Sequence to run programs from diskette

1. From the initial page, press the soft key next to RUN.
2. Press the soft key next to FILE.
3. Using the arrow keys, bring the cursor to the program name you wish to run and press the ENTER button.
4. Check conditional switches (dry run, single block, mirror, etc.).
5. Press the ENTER button.

Find Note on Operation Procedures

As you have seen from the three sample "operation handbooks" presented in this chapter—on machining centers, turning centers, and wire EDM machines—once you have documented the most important operation procedures, operating the CNC machine is little more than deciding what must be done and following a step-by-step set of instructions to accomplish it.

The most commonly used sequences will soon be memorized. For those sequences seldom used, at least you will have a crutch to rely on.

Chapter

13

Key Concept No. 4: Program Verification

The fourth and last key concept in CNC machine operation is related to verifying new programs. This is the trickiest and most dangerous part of working with CNC equipment. While some forms of CNC machine tools are relatively safe to work with, the beginner must be extremely cautious at all times, and especially when working with new programs.

For the most part, CNC controls will follow the instructions given in a program to the letter. With the exception of basic *syntax* (program formatting) mistakes, the CNC control will rarely be able to tell if a mistake has been made. While verifying any new program, the operator must be ready for anything. If the programmer has made a mistake in the program which tells the control to drive a tool at rapid into the workpiece, the control will follow the commands and do so, causing what is commonly referred to as a *crash*.

With a manually prepared program, just about anything could be wrong. The manual programmer might have meant to type X but instead typed Y. In this case, the resulting motion the machine makes will be unpredictable. Literally every command in a manually prepared program may have a mistake and should be treated as suspect.

If a CAM system was used to prepare the CNC program, generally the basic syntax of the program will be correct. The basic motion commands the control is given should also be correct. However, even CAM-generated programs will not always be perfect. While they tend to be more correct than manually generated programs, problems could still exist with cutting conditions (feeds, speeds, work holding, depth of cut, etc.). Conversationally generated CNC programs should be treated with the same respect as CAM-generated CNC programs.

We cannot overstress the need for *safe* procedures when working with CNC equipment. While we are not trying to scare the beginner, we want to instill a high level of respect for this very powerful equipment.

Safety Priorities

There are three levels of priority that you should *always* adhere to when working with any machine tool.

Operator safety

The first priority *must be operator safety*. Every step should be taken to ensure the safety of the person operating the CNC machine. When we give the actual procedures for verifying CNC programs a little later in this key concept, you will see that each procedure stresses operator safety. As time goes on and the beginner gains experience, the tendency will be to shortcut our given procedures in order to save some time. But if the recommended verification procedures are not followed, the operator is opening the door to a very dangerous situation.

Admittedly, some CNC machines are not very dangerous to operate. For example, wire EDM machines pose little danger to the operator. If a mistake is make in the wire EDM program, about the worst that can happen is wire breakage. But the operator of *any* CNC machine tool should always be on the alert for mishaps.

Machine tool safety

The second safety priority is the CNC machine tool itself. Operators must do their best to assure that no damage to the machine can occur. CNC machine operation time is very expensive. When a CNC machine goes down for any reason, the actual cost of repairing the machine is usually very little compared to that of the lost production time.

There is no excuse for a crash on a CNC machine caused by operator or programmer mistakes. If the verification procedures we give a little later are followed, we can almost assure that the operator will not be placing the machine in dangerous situations.

Running good parts

The third priority to CNC machine safety is making all workpieces to size. The cost of rough stock to be machined varies dramatically, depending on the user's application. In some cases, for very small parts being machined from completely rough stock, the cost of the material coming to the CNC machine may be as little as 10 cents per part. In this case, the company may be willing to sacrifice several workpieces in order to verify the program.

However, there are also times when the cost of the rough material coming to the CNC machine is very high. Possibly the material itself is very expensive, or previous operations might have been performed on the workpiece prior to the CNC machine operation. The user may have several hundred dollars (or more) invested in the workpiece prior to the CNC machine operation. In this case, the company cannot afford to lose even one workpiece during the program's verification.

No matter what situation exists concerning rough stock cost, there is *no excuse* to scrap the first workpiece to be machined on any CNC machine. If the verification procedures we give are followed, there will always be a way to assure the workpiece will come out with excess stock after the first time each tool machines the workpiece. Following the given techniques will allow the operator to sneak up on all critical dimensions being machined.

Typical Mistakes

Before discussing the actual techniques to verify programs, let's first examine those kinds of mistakes the programmer tends to make that will cause a program to fail. Knowing these common mistakes will help you diagnose problems as they occur. There are three levels of mistakes we will discuss.

Syntax mistakes in the program

This level of mistake is the easiest to diagnose and fix. With this kind of mistake, the control will not even be able to execute the command. For example, the programmer may have meant to type a G code as G01, but by mistake typed G10. Most CNC controls do not even have a G10 command. Even if they do, it is doubtful that the format would be the same as for a G01.

With this kind of mistake, the control will go into an alarm state the moment the erroneous command is executed. If the machine is moving through its motions, the motions will stop and the program will halt. The display screen will show a message related to the cause of the problem.

Syntax mistakes are usually silly mistakes the manual programmer makes while writing or typing the program. They are easy to diagnose because it is likely that as soon as the programmer sees the command generating the alarm, the mistake will be very obvious.

Motion mistakes in the program

This kind of mistake can be a little more difficult to diagnose. While motion mistakes may be still be silly mistakes on the programmer's part (transposing numbers or axis letter addresses, for example), generally they are caused by incorrect coordinate calculations. For example, maybe the programmer misinterpreted the dimensions on a drawing and came up with incorrect coordinate values for the tool to move through. In this case, the control would follow the program's instructions without generating an alarm. But the tool path would not be correct.

Other mistakes in this category include forgetting to instate or cancel tool offsets, reversing clockwise and counterclockwise commands, and improper mode selection for incremental vs. absolute. This category of mistake can be very serious if not found. In many cases, this kind of mistake means the program is telling the control to crash the tool into the workpiece, work-holding device, or machine.

Setup mistakes

Even a perfectly written program will behave poorly if mistakes are made during setup. In most cases, the operator will have some measurements to make related to the setup and numbers to enter into the control *before* the program can be executed.

For example, most CNC machines require that program zero be measured at the machine. Once this measurement is made, either the program must be modified, or the program zero values must be entered into a fixture offset.

If these numbers are incorrectly measured or if mistakes are made during their entry, the control will not know where the program zero point is. *All* motions the program makes will be incorrect.

Another example of a setup mistake that will cause the program to fail is related to tool offsets on a machining center. For machining centers, a tool length offset must be entered for every tool and a tool radius offset must be entered for tools using cutter radius compensation. (See key concept no. 4 of programming for more information about tool length and cutter radius compensation.) If these values are not entered, or if the values are entered incorrectly, the results could be disastrous.

Procedures for Program Verification

Now that you have an idea of the kinds of things to watch out for, let's discuss the actual procedures to verify programs. Note that we gave specific step-by-step instructions for program verification in key concept no. 3 of operation. At this time, we will discuss the logic behind those instructions.

One technique you will want to get in the habit of using has to do with the feed hold button. When you press the cycle start button to activate the program, *always* have a finger resting on the feed hold button. If an unexpected rapid motion takes place, you will not have to go looking for the feed hold button. Your finger will be ready to press it. No matter what you are doing, until you have the program verified, use this technique. We guarantee it will save a crash some day!

The machine lock dry run

This procedure will allow the control to scan the program for syntax mistakes. Once the setup is made and all information related to the program is entered into the control (tool offsets, program zero, etc.), the operator will turn on the machine lock and dry run switch. The operator will also turn up the dry run motion switch (usually feed rate override or jog feed rate) to its highest position to allow the control to move quickly through the program.

When the cycle is activated, the control will quickly scan the program for syntax mistakes. During the program's execution, the spindle will come on, the tools will change, and the control will appear to be actually running the program. However, the axes (*X, Y, Z*, etc.) will not move. This procedure gives the operator a relaxed way of assuring the control can execute the program. After confirming that no axis motion is occurring (machine lock is really on), the operator can rest easy until one of two things happens. Either the control will generate an alarm or it will complete the program without generating an alarm.

If the control finds a syntax mistake in the program and generates an alarm, the operator must diagnose the alarm, fix the problem, and execute the program again. This must be repeated until the entire program can be executed without generating alarms.

When the control completes the execution of the entire program without generating an alarm, the operator will know that the control can accept the program. While there still may be serious motion mistakes within the program, at least the program can be executed from beginning to end without generating an alarm.

Free-flowing dry run

Once the machine lock dry run procedure has been successfully completed, the operator is ready to allow the program to cause axis motion. However, there could still be very serious mistakes in the program. In fact, the main reason for doing a free-flowing dry run is to check for motion problems. So the operator must be very careful indeed. At this time, the work-holding setup has been made, but *no workpiece is in position.*

To execute this procedure, the operator will turn off the machine lock switch, turn down the dry run motion rate switch (usually either feed rate override or jog feed rate) to its lowest position, and set the rapid override switch to its slowest motion rate. When the cycle is activated, the operator must have a finger ready to press the feed hold button. Feed hold will be the panic button in case anything goes wrong and the operator wishes to stop the cycle. With the dry run motion-controlling switch set to its lowest position, the axes will barely creep along. As the operator increases the setting on this switch (by rotating it clockwise), the axes will move faster. The operator will increase the motion rate to a comfortable setting.

As each tool comes close to the work-holding device, the dry run motion rate can be turned down. If the operator is worried and wants to check something, the feed hold button can be pressed to temporarily stop the cycle. After pausing, pressing the cycle start button gets the machine moving again. If the operator wishes to cancel the cycle because of a motion mistake in the program, after pressing feed hold, the reset key can be pressed, the machine can be sent back to its reference point, and the problem can be corrected.

The free-flowing dry run allows the operator to check the basic movements generated by the program. As stated earlier, the operator is most concerned with checking for interference problems (crashes). But aside from severe problems, the operator must also check for less serious problems. For example, the operator must check to assure that, for each tool, the spindle is rotating the correct direction and that basic motions the program generates appear to be correct.

In many cases, the operator will not be quite sure after doing only one free-flowing dry run. The first time the program is executed in this way, the operator may be concerned with serious mistakes that could cause damage to the machine, and pay little attention to the details of each movement. For this reason, many times the free-flowing dry run must be repeated several times until the operator is satisfied that the motions are correct. It is possible that the free-flowing dry run will cause the machine to go into an alarm state even though a machine lock dry run has been successfully completed. Axis overtravels and problems related to offsets and other forms of compensation are among the things that can generate alarms during a free-flowing dry run that are ignored during a machine lock dry run.

Normal air cutting cycle execution

Before actually trying to run the first workpiece, there is one more important procedure to follow. The operator must execute the cycle one more time with the dry run switch turned off (and *no* part in position). This will allow the operator to see one thing that could not be seen during a free-flowing dry run. The free-flowing dry run allowed the operator to take control of all motion rates with the dry run motion rate switch. This means that, when dry run is turned on, the operator will not be able to tell the difference between rapid motions and cutting motions. While watching the motions of the machine, all motions appear the same.

Here is an example of a problem the normal air cutting cycle will allow the operator to correct. Suppose the programmer rapids a drill up to the workpiece to drill a hole. But in the command to drill the hole, the programmer leaves out (by mistake) the G01 word. Since the machine is currently in the rapid mode, the hole would be drilled at rapid, breaking the drill and possibly injuring the operator. This is the kind of thing the operator *cannot see* during a free-flowing dry run with most CNC controls. For this reason, we recommend running the program one more time as if a part were being machined, but without a part in position. This way the operator can confirm that the machine is rapiding where it is supposed to and machining where it is supposed to.

For extremely long cycles, the operator can toggle the dry run switch on and off. Once the operator is sure the current command is a cutting command, rather than wait for the command to be completed, the operator can temporarily turn on the dry run switch to allow the balance of the command to be completed quickly. When it is, dry run can be turned off again. Repeating this technique allows the operator to go through a lengthy program quickly, yet assure the cutting commands are where they are supposed to be.

Running the first workpiece

At long last, the operator is ready to run the first workpiece. Yet each tool in the program includes at least one movement that poses a potential dangerous situation for the operator. Most programmers rapid each tool within 0.100 in of the workpiece surface before machining. This very small distance is impossible to check during the previous verification procedures. For this reason, the operator must be *very* careful with each tool's first approach to the workpiece. Also, if the tool machines several surfaces of the part and rapids several times to different surfaces, the operator must be cautious with these motions as well.

For this reason, we recommend turning on dry run and single block during each tool's approach to the workpiece and during rapid motions to new machining surfaces. With dry run and single block on, the operator will have control of the motion rate during each tool's approach. As the tool gets close to the surface to be machined, the rate can be turned down to a very slow rate. And, since single block is on, the operator can rest assured the motion will stop at the end of the command. When the motion stops, the operator can check the clearance approach amount (usually 0.100 inch). If everything appears to be correct, the operator can turn off the dry run switch (*never* machine with dry run on!) and leave single block on. At the end of each command, the motion will stop and the operator will press the cycle start button to continue to the next command. When the coming command is a rapid command to another clearance position, the operator can turn on dry run to take control of the motion rate again. This procedure is repeated for every tool in the program.

The above discussion had safety as the primary concern. Yet the operator must also make good parts. While machining the very first workpiece, there are things the operator can do to assure the first workpiece will come out to size.

If the operator considers what each tool is going to be doing during its machining operation, many times the tool offsets can be adjusted in a way that forces the tool to leave excess stock on the surface being machined. After machining with the trial offset, the machined surface can be measured and the operator will know *exactly* how much stock is yet to be machined. The offset can be adjusted accordingly, and the tool can be rerun. This time the tool will machine precisely to size. This sequence can be repeated for each tool in the program to assure that the first part will come out to print dimensions.

Here's the procedure again in list form:

1. Consider what the tool will be machining.
2. Adjust the tool offsets in the direction of leaving excess stock.
3. Allow the tool to machine the workpiece.
4. Measure what the tool has done.
5. Adjust the offsets to machine to size.
6. Rerun the tool.

Here is an extended example of how this technique can be used for a turning center application. Suppose you had an outside diameter finish-turning tool machining a series of diameters. You are running this tool for the first time and the part has previously been rough-turned. Since this finishing tool will be machining an outside diameter, you can adjust the X offset in the *plus* direction to assure that excess stock will be left. Say you adjust the X offset by plus 0.015 inch. Now you allow this finish-turning tool to machine the part. When the tool has finished, you measure the part (it will be oversized). Suppose each outside diameter machined by this tool came out 0.0135 in oversized. You would adjust the X offset for this tool in the minus direction by 0.0135 in to make this tool cut on size. Then you would rerun this tool and it would machine precisely to size.

While the above example is for turning centers, the same techniques can be used for machining centers as well as all forms of CNC equipment.

Always make one tool machine to size *before* going on to the next tool. It can be very difficult to figure out what to do with the tool offsets if several tools have machined the workpiece before offsets are considered.

Running production and optimizing

Once the first part has been successfully completed, the operator is ready to run the production quantity of workpieces. It may be wise to keep the machine from rapiding for the first few workpieces with rapid override until the operator is comfortable with the cycle.

If many workpieces must be run, it is worthwhile to monitor the first few parts to be run. Many times the operator will find areas of the program that can be improved to minimize cycle time. Also, cutting conditions (feeds and speeds) can be adjusted to improve cycle time or lengthen tool life. However, if major changes are made in the program during optimizing, we recommend verifying the program again.

Conclusion to Program Verification

While the verification procedures we have given may seem like a great deal of work, we urge the beginner to use them. While it does take time to verify programs, the time can easily be justified by the possible consequences. Without verification procedures, CNC machines are *very* dangerous to operate. Even if no operator injury occurs during a crash, the time and expense during machine repair easily justifies being cautious with the machine while new programs are checked out.

It is common that a beginning operator is very cautious with the unfamiliar CNC machine. But after gaining experience, the operator becomes a little overconfident. Then the operator starts shortcutting the verification procedures. This is where problems begin. We liken this to a person learning how to snow ski. At first, the beginning skier is very concerned about getting hurt. No tricks or hotdogging are attempted. But after gaining confidence, many skiers attempt to do more than they are capable of. This is how legs get broken. The same can be said for the beginning CNC operator. As you gain confidence, you may be tempted to forego a procedure you would have done as a beginner. Watch out!

Index

ABOUT THE AUTHOR

Mike Lynch is president of CNC Concepts, Inc., a supplier of training materials and computer software to users of computer numerical control equipment. He is the author of a monthly column, "CNC Tech Talk," in *Modern Machine Shop* magazine. He also presents CNC training seminars conducted by the Society of Manufacturing Engineers. Previously, Mr. Lynch was NC operations manager for K.G.K. International Corporation and, prior to that, he held various technical positions with Cincinnati Milacron, Rockwell International, and G.T.E. Automatic Electric. He resides in Hoffman Estates, Illinois.